**Springer Proceedings in Physics**

Volume 159

More information about this series at http://www.springer.com/series/361

Jean-Yves Bigot · Wolfgang Hübner
Theo Rasing · Roy Chantrell
Editors

# Ultrafast Magnetism I

## Proceedings of the International Conference UMC 2013 Strasbourg, France, October 28th - November 1st, 2013

Springer

*Editors*
Jean-Yves Bigot
Institut de Physique et Chimie des Matériaux de Strasbourg (IPCMS)
Université de Strasbourg and CNRS
Strasbourg
France

Wolfgang Hübner
Department of Physics
University of Kaiserslautern
Kaiserslautern
Germany

Theo Rasing
Institute for Molecules and Materials (IMM)
Radboud University
Nijmegen
The Netherlands

Roy Chantrell
Department of Physics
The University of York
York
UK

ISSN 0930-8989 ISSN 1867-4941 (electronic)
ISBN 978-3-319-07742-0 ISBN 978-3-319-07743-7 (eBook)
DOI 10.1007/978-3-319-07743-7

Library of Congress Control Number: 2014946373

Springer Cham Heidelberg New York Dordrecht London



Printed on acid-free paper

Springer is part of Springer Science+Business Media (www.springer.com)

# Preface

This volume on Ultrafast Magnetism is a collection of articles presented at the international "Ultrafast Magnetization Conference" held at the Congress Center in Strasbourg, France, from October 28th to November 1st, 2013. This first conference, which is intended to be held every two years, received a wonderful attendance and gathered scientists from 27 countries in the field of Femtomagnetism, encompassing many theoretical and experimental research subjects related to the spins dynamics in bulk or nanostructured materials. The participants appreciated this unique opportunity for discussing new ideas and debating on various physical interpretations of the reported phenomena. The format of a single session with many oral contributions as well as extensive time for poster presentations allowed researchers to have a detailed overview of the field.

Importantly, one could sense that, in addition to studying fundamental magnetic phenomena, ultrafast magnetism has entered in a phase where applied physics and engineering are playing an important role. Several devices are being proposed with exciting R&D perspectives in the near future, in particular for magnetic recording, time resolved magnetic imaging and spin polarized transport, therefore establishing connections between various aspects of modern magnetism. Simultaneously, the diversity of techniques and experimental configurations has flourished during the past years, employing in particular Xrays, visible, infra-red and terahertz radiations. It was also obvious that an important effort is being made for tracking the dynamics of spins and magnetic domains at the nanometer scale, opening the pathway to exciting future developments. The concerted efforts between theoretical and experimental approaches for explaining the dynamical behaviors of angular momentum and energy levels, on different classes of magnetic materials, are worth pointing out. Finally it was unanimously recognized that the quality of the scientific oral and poster presentations contributed to bring the conference to a very high international standard.

The organization of this conference has greatly benefitted from dedicated people. First we thank our sponsors for making the event possible. Let us mention institutions like the European Research Agency (Research Advanced Grant ATOMAG), the French Ministry of Education and Research (Equipex UNION and Labex NIE), the National Scientific Research Center in France (CNRS), the University of Strasbourg and the University of Kaiserslautern. We acknowledge our friendly industrial partners who also contributed significantly to the success of the conference: Coherent, Amplitude Technologies, Newport, Fastlite, Zhinst, Femtolasers, Acalbfi. The local and international committee had to review 115 contributions which appear to be an excellent up-to-date condensate of the actual research and development in ultrafast magnetism. Our special thanks go in particular to Drs. Nabila Kadiri and Mircea Vomir (CNRS, University of Strasbourg) the organization, the web site as well as the preparation of this volume.

**Key words:** Coherent magnetic processes • Femtosecond • Magnetic recording • Magnetism • Magneto-Optics • Optics • Spin photo-emission • Spins • Ultrafast Dynamics

# Contents

## Part III Spin Waves Dynamics

## Part IV Theory of Spin Dynamics

## Part VII Spin Photo-Emission Dynamics



## Part VIII X-Ray and Far UV-Spin Dynamics

## Part IX Terahertz Spin Dynamics

# Part I
# Ultrafast Magnetism Dynamics in Semiconductors

# Femtosecond Laser Pulses Switch Magnetic States via Strongly-Correlated Spin–Charge Quantum Excitations

**Ilias E. Perakis**

Department of Physics, University of Crete, 71003 Heraklion, Crete, Greece, and Institute of Electronic Structure & Laser, Foundation for Research and Technology-Hellas, 71110 Heraklion, Crete, Greece

ilias@physics.uoc.gr

**Abstract** The technological demand to bump the Gigahertz switching speed-limit of today's magnetic memory and logic devices into the Terahertz regime underlies the entire field of spin-electronics and integrated multi-functional nano-devices. In this talk, I show how all-optical switching based on the ultrafast quantum-mechanical manipulation of spins could meet this challenge.

## Introduction

The technological demand to push the gigahertz switching speed limit of magnetic devices into the terahertz regime is met by all–optical magnetic switching based on coherent spin manipulation in strongly correlated systems, where the strength of local spin-spin, Coulomb, and electron-lattice interactions exceeds the kinetic energy and width of the electronic energy bands. Strong correlations between electrons in neighboring lattice sites determine the optically-induced coherent nonlinear dynamics and lead to time-dependence absent in weakly-correlated magnetic systems[1]. By analogy to femto–chemistry and photosynthetic dynamics, where photo-products of chemical/biochemical reactions can be influenced by creating suitable superpositions of molecular states, we show that femtosecond (fs) laser–excited coherence between spin/orbital/charge quantum states in neighboring sites can switch magnetic orders[2] by "suddenly" breaking the delicate balance between competing phases of correlated materials such as the colossal magneto–resistive (CMR) manganites[3]. We predict theoretically and observe in the experiment[2] fs photoinduced switching from antiferromagnetic to ferromagnetic ordering. Ferromagnetic spin correlation is driven by the establishment, within the ~100 fs time interval of the pulse duration, of laser-induced coherent superpositions of electronic quantum states in neighboring lattice sites. The development of ferromagnetic correlations *during a single fs laser pulse* reveals an initial quantum coherent regime of magnetism, clearly distinguished from the pre-



J.-Y. Bigot et al. (eds), *Ultrafast Magnetism I*, Springer Proceedings in Physics 159
DOI 10.1007/978-3-319-07743-7_1

viously studied picosecond lattice-heating regime characterized by phase separation without laser intensity threshold behavior. Note that, in the studied material, a transition from antiferromagnetic to ferromagnetic ordering *cannot be induced by increasing the temperature*, which excludes laser-induced heating as a possible mechanism during the 100fs pulse duration, where the observed magnetic phase transition occurs[2]. Our theory is based on equations of motion for *composite fermion density matrices*, which treat the strong Jahn-Teller, Coulomb, and Hund's rule magnetic interactions, as well as the nonlinear coherent fs optical photoexcitation. Our results, summarized in Fig. 1, show fs local spin nonlinear dynamics and underpin fast quantum spin–flip fluctuations correlated with coherent laser-induced superpositions of electronic states to initiate local ferromagnetic correlations via quantum kinetic processes beyond the statistical approach[2].

In the present work, we show fs photoinduced switching from antiferromagnetic to ferromagnetic ordering in $Pr_{0.7}Ca_{0.3}MnO_3$, by observing the establishment, within 120 fs, of a huge temperature–dependent magnetization with *photoexcitation threshold* behavior absent in the optical reflectivity (see Fig. 2). Our theory is based on equations of motion for the density matrix of composite fermion operators that describe the electronic excitations of the strongly correlated system, including local self-energy time-dependent effects and quantum spin fluctuations. Our simulations predict, in particular, fs *quantum spin–flip fluctuations* correlated with photoexcited coherent superpositions of electronic states to initiate local ferromagnetic correlations between neighboring chains that are antiferromagnetically coupled in the ground state, as shown schematically in Fig.1.

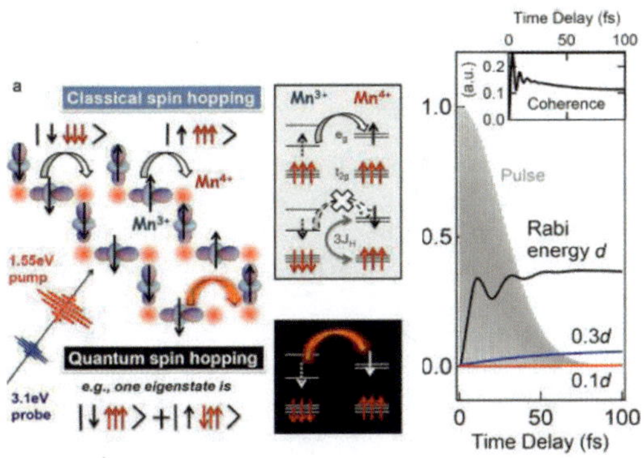


Figure 1. (a) Schematics of ultrafast excitations of a CE–type AFM/CO/OO order. Mn4+ configurations, with empty doubly–degenerate eg orbitals and t2g orbitals filled by three spin–aligned electrons, mostly populate the one–dimensional (1D) zig–zag chain corner sites. The 1D chain bridge sites are mostly populated by Mn3+ configurations, with an additional eg–orbital electron populating alternating orbitals (OO). For classical spins, electron conduction and optical transitions are restricted within the same 1D ferromagnetic chain (white arrow). Quantum spin–flip fluctuations, however, allow eg electrons to hop on sites with opposite local t2g spin, by forming non–equilibrium total spin eigenstates (red arrow). (b) Calculated photoinduced total spin and inter–atomic coherence (inset) for three Rabi energy values. The laser pulse time–dependence is superimposed, demonstrating that FM local correlations transiently build–up from the AFM ground state (total spin zero) during the nonlinear photoexcitation.

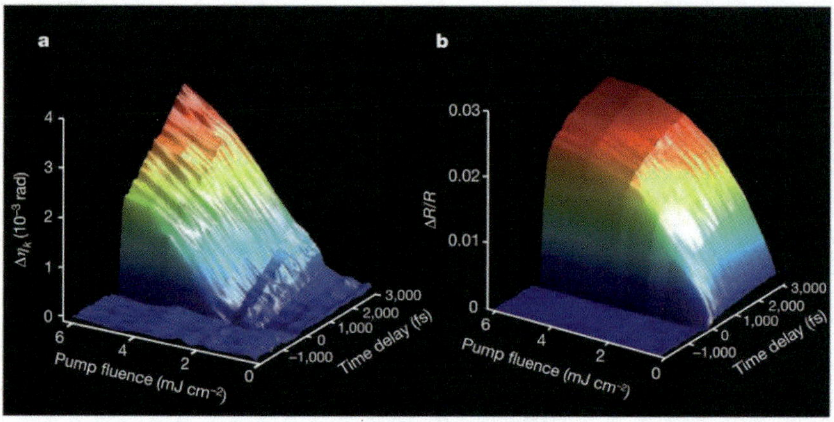


**Figure 2.** A three-dimensional view demonstrating the distinct femtosecond spin and charge dynamics as well as photo-excitation threshold behavior observed experimentally in Ref. [2]. (a) Time-resolved ellipticity change and (b) differential optical reflectivity as function of pump fluence. Measurements were taken under the same experimental conditions. A photo-excitation threshold is seen for emergence of femtosecond magnetization, whereas the charge dynamics exhibits no threshold. Magnetic field B=0.5T and temperature T=30 K.

**Acknowledgments** The authors would like to thank the THALES program NANOPHOS for financial support.

# References

[1] J. Y. Bigot, M. Vomir and E. Beaurepaire, "Coherent ultrafast magnetism induced by femtosecond laser pulses" Nature Physics **5**, 515 (2009).\

[2] T. Li, A. Patz, L. Mouchliadis, J. Yan, T. A. Lograsso, I. E. Perakis, and J. Wang, "Femtosecond switching of magnetism via strongly correlated spin-charge quantum excitations" Nature **496**, 69 (2013).

[3] E. Dagotto, T. Hotta and A. Moreo, "Colossal magnetoresistant materials: the key role of phase separation", Physics Reports **344**, 1 (2001).

# Investigation of non-thermal process in the dynamics of photo-induced FMR in (Ga,Mn)As

**T. Matsuda and H. Munekata**

Imaging science and engineering laboratory, Tokyo Institute of Technology

4259-J3-15 Nagatsuta, Midori-ku, Yokohama 226-8503, Japan

mazda@isl.titech.ac.jp

**Abstract** We report experimental results which indicate the contribution of non-thermal process in photo-induced ferromagnetic resonance (FMR) in (Ga,Mn)As excited by weak, fs laser pulses.

## Introduction

In metallic systems, it has been recognized that randomization of ordered spin systems by the intense laser irradiation (mJ/cm$^2$) is the fundamental process to realize ultra-fast magnetic excitation [1,2]. In (Ga,Mn)As, photo-induced FMR (ph-FMR) could be triggered by a few $\mu$J/cm$^2$ laser pulse without an external magnetic field [3,4]. There have been two different arguments to account for the basic mechanism: the ultra-fast spin randomization (thermal process) [4,5], and excitation of electronic states associated with ordered spins (non-thermal process) [3]. We report here that, in the time-resolved magneto-optical (TRMO) measurements for the weak excitation regime ($\leq$ 10 $\mu$J/cm$^2$), signals due to the ultra-fast randomization have been hardly observed, and the onset of ph-FMR strongly depends on excitation wavelength. These results suggest significant contribution of non-thermal effect.

## Experimental method

TRMO measurements were carried out at 10 K with one-color pump and probe technique. The light source and the sample were a mode-locked Ti:sapphire laser (a pulse width of 150 fs with repetition rate 76 MHz) and a 100 nm-thick $Ga_{0.98}Mn_{0.02}As$ grown on GaAs(001), respectively. Experimental setup has been detailed in Ref.3. The wavelength has been varied in the range $\lambda$ = 750 - 900 nm, whereas the polarization plane was fixed at the [010] GaAs axis for both pump



J.-Y. Bigot et al. (eds), *Ultrafast Magnetism I*, Springer Proceedings in Physics 159
DOI 10.1007/978-3-319-07743-7_2

and probe beams. The fluence of the probe beam was fixed at 84 nJ/cm$^2$, whereas that of the pump ($P^*$) was varied between 0.34 $\mu$J/cm$^2$ and 10 $\mu$J/cm$^2$.

## Results and discussion

Characteristic oscillatory signals due to ph-FMR have been observed for the wavelengths of $\lambda$ = 880 nm or shorter. We have found a striking difference concerning *the onset of oscillation* between the data obtained with $\lambda$ < 820 nm, the region S, and those with $\lambda$ > 820 nm, the region L. TRMO data obtained with $P^*$ = 1.7 $\mu$J/cm$^2$ are shown in Figs.1(a) and (b). Oscillation starts immediately after the excitation in the region L, whereas it is accompanied with noticeable time delay in the region S. Note that, in the ultrafast time regime ($\leqq$ 2 ps), signals due to ultrafast demagnetization have hardly been observed (Fig.1(b)).

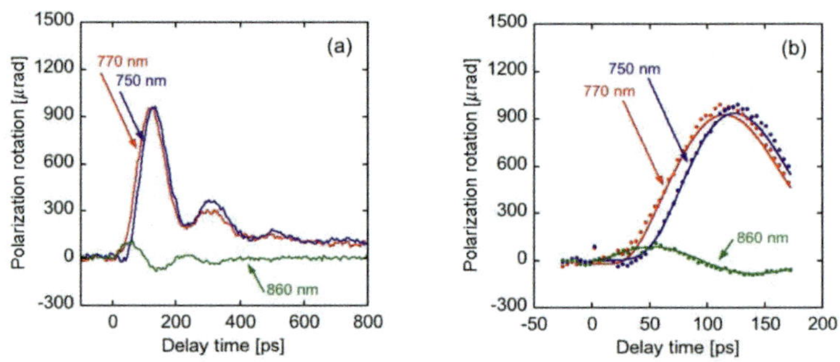


**Fig. 1.** Photo-induced FMR in (Ga,Mn)As measured at long time scale (a), and short time scale (b). Lines and dots in (b) are experimental and calculated TRMO data, respectively.

Magnetization dynamics was calculated numerically by solving LLG equation. We have added a new *delay term* into the dynamics of an effective field that was developed earlier [3]. Before excitation, the effective field $\boldsymbol{H}$ and the magnetization $\boldsymbol{M}$ both lie along the in-plane [100] direction; with certain time delay $\tau_g$ after the excitation at $t$ = 0 ps, $\boldsymbol{H}$ rotates toward the out-of-plane [001] direction and relax back to the [100] direction, with its time constant $\tau_1$ and $\tau_2$, respectively. During this event, $\boldsymbol{M}$ precesses around $\boldsymbol{H}$ with natural damping. With this scenario, the angle $\theta(t)$ of $\boldsymbol{H}$ with respect to the [100] axis was formulated as follows:

$$\theta(t) = \mathrm{erf}(t - \tau_g)\exp(-t/\tau_1)\{1 - \exp(-t/\tau_2)\} \qquad 1$$

Here, the delay term is expressed by the error function erf($t$) with the retardation time $\tau_g$. Our approach has reproduced successfully the experimental data, from which $\tau_g$ is extracted and summarized in Fig. 2 as a function of excitation photon energy and wavelength. Reflecting the instantaneous oscillations, $\tau_g$ = 0 in the region L. This fact indicates the presence of electronic states which allow direct access to the spin subsystem (Fig.3). In the region S, the $\tau_g$ value increases with increasing photon energy. This fact suggests that we enter into the region in which another states, presumably the states associated with a host semiconductor, is first excited, and the excess energy is then transferred to the spin subsystem (Fig.3). A dip observed at 790 nm is presumably due to enhanced LO phonon scattering [6].

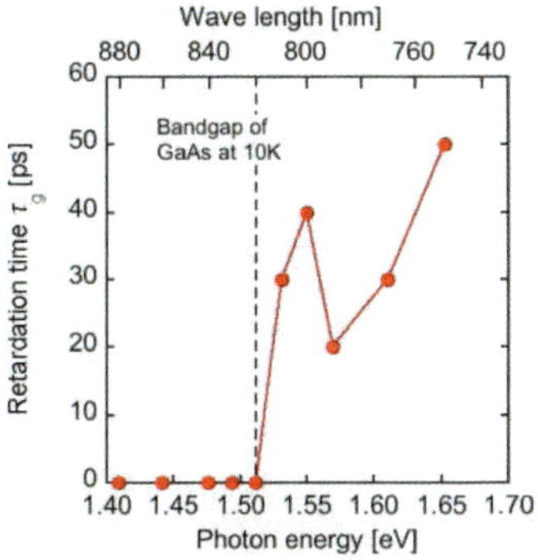


**Fig. 2.** Photon energy and wavelength dependence of retardation time $\tau_g$.

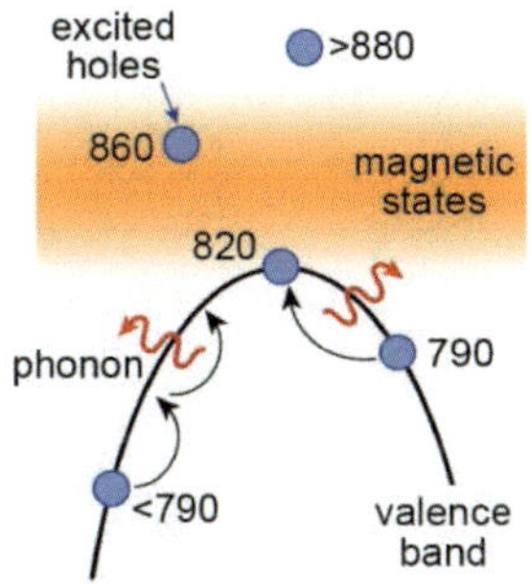


**Fig. 3.** Schematic illustration of the excitation process for the ph-FMR. Values in the figure indicate the excitation wavelength.

## Conclusions

We have found the presence of two different excitation channels for the photoinduced ferromagnetic resonance in (Ga,Mn)As. In the region L ($820 < \lambda \leq 880$ nm), magnetization precession occurs instantaneously with pulsed laser excitation. In the region S ($\lambda < 820$ nm), the precession is accompanied with retardation whose retardation time varies with the wavelength.

## References

[1] E. Beaurepaire, J. –C. Merle, A. Daunois and J.-Y. Bigot, "Ultrafast Spin Dynamics in Ferromagnetic Ni" Phys. Rev. Lett. **76**, 4250 - 4253 (1996).

[2] I. Radu, *et al.*, "Transient ferromagnetic-like state mediating ultrafast reversal of antiferromagnetically coupled spins", Nature **472**, 205 - 208 (2011).

[3] Y. Hashimoto, S. Kobayashi and H. Munekata, "Photoinduced Precession of Magnetization in Ferromagnetic (Ga,Mn)As" Phys. Rev. Lett. **100**, 067202 (2008).

[4] D. M. Wang, et al., "Light-induced magnetic precession in (Ga,Mn)As slabs: Hybrid standing-wave Damon-Eshbach modes", Phys. Rev. B **75**, 233308 (2007).

[5] E. Rozkotova *et al.*, "Light-induced magnetization precession in GaMnAs" Appl. Phys. Lett. **92**, 122507 (2008).

[6] J. S. Blakemore, "Semiconducting and other major properties of gallium arsenide", J. Appl. Phys. **53**, R123 - R180 (1982).

# Time Resolved Spectroscopy in Narrow Gap MOVPE Grown Ferromagnetic Semiconductors

**G.A. Khodaparast[a*], M. Bhowmick[a], C. Feeser[b], B.W. Wessels[b], D. Saha[c], G.D. Sanders[c], C.J. Stanton[c]**

[a] Department of Physics, Virginia Tech, Blacksburg, VA 24061 USA

[b] Department of Materials Science and Engineering, Northwestern University, Evanston, IL 60208 USA

[c] Department of Physics, University of Florida, Gainesville, FL 32611 USA

* khoda@vt.edu

**Abstract** a We report on time resolved experiments that provide insight into the time scales and the nature of the interactions in ferromagnetic InMnAs and InMnSb. Theoretical calculations are performed using an 8 band **k·p** model including non-parabolicity, band-mixing, and the interaction of magnetic Mn impurities with itinerant electrons and holes.

## Introduction

Carrier-induced ferromagnetism in magnetic III-V semiconductors has opened up new opportunities for device applications, as well as fundamental studies in material systems in which itinerant carriers interact with the localized spins of magnetic impurities. A low temperature MBE technique is nearly always used to prepare thin ferromagnetic films, although MOVPE, an alternative technique, allows single phase magnetic InMnAs and InMnSb compounds to be deposited at 500° C, much higher than that used in MBE. Films with hole densities of 1018 cm-3 can have $T_c$ above room temperature [1].

In this work, we perform time resolved differential transmission (TRDT) studies to obtain insight into the dynamics in MOVPE grown ferromagnetic films on the picosecond time scale. To understand the effects of ferromagnetic order on the electronic structure and subsequent relaxation dynamics, we calculate the electronic structure for bulk InMnAs and InMnSb. By calculating the electronic band structure, we can determine where photoexcited carriers are generated by the pump pulse and which regions of the electronic structure are sampled by the probe pulse. The calculations are based on an 8 band **k·p** model which includes the conduction and valence band mixing [2]. We use **k** = 0 Bloch basis states for the conduction bands, heavy-holes, light holes and split-off holes for a total of 8 states in-

 
J.-Y. Bigot et al. (eds), *Ultrafast Magnetism I*, Springer Proceedings in Physics 159
DOI 10.1007/978-3-319-07743-7_3

cluding spin. In our model, we include the effects of the spontaneous magnetization of the Mn ions and the sp-d coupling of this magnetization to the electrons and holes [2]. Figure 1 shows an example of the calculations for ferromagnetic InMnSb at 77 K in the absence of an externally applied magnetic field.

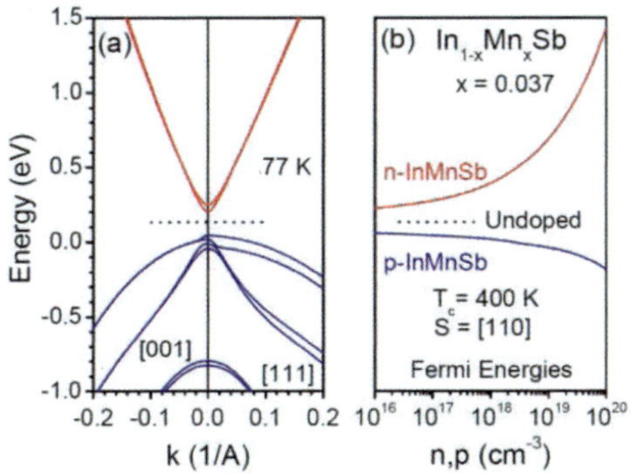


**Fig. 1.** The band structure and the Fermi energies as a function of carrier density for InMnSb at 77 K as a function of n and p carrier concentrations. The Fermi energy for the undoped semiconductors is indicated by the dotted line. The Curie temperature is taken to be 400 K.

## Experimental Approach

By probing the dynamical behavior of the nonequilibrium carriers created by intense laser pulses, we gain valuable information about the band structure and different scattering mechanisms. Using TRDT in the MID-IR, we achieved tunability of carrier dynamics and relaxation times with characteristics unobtainable in MBE grown ferromagnetic structures [2]. The MOVPE grown InMnAs structure is an 800 nm thick film with a Mn content of 4%, and the InMnSb film is a 200 nm film with a Mn content of 3.7%. Both samples, demonstrated $T_c$ above room temperature. The laser pulses were tuned in NIR and MIR using different sources with repetition rates of 1 KHz and pulse durations of ~ 100 fs.

As shown in Fig. 2 for InMnSb, we observe the sensitivity and tunability of the carrier dynamics to the initial excitation. The initial increase in the differential transmission is due to free carrier Drude absorption where the fast component of the temporal evolution is attributed to the relaxation of hot electrons and the slower component is due to electron-hole recombination. Exploiting the selection rules for interband transitions, spin-polarized carriers are created using circularly polarized pump beams. By monitoring the transmission of a weak delayed probe pulse that has the same circular polarization (SCP) or opposite circular polarization (OCP) as the pump pulse, as shown in Fig. 3, the optical polarization can be extracted. Figure 3b shows the exponential fit to the SCP-OCP in the TRDT signal which gives a spin relaxation time of ~ 6.0 ps, which is much higher than the reported time scale for other narrow gap ferromagnetic semiconductors. This fact is due to much higher hole mobilities in these material systems.

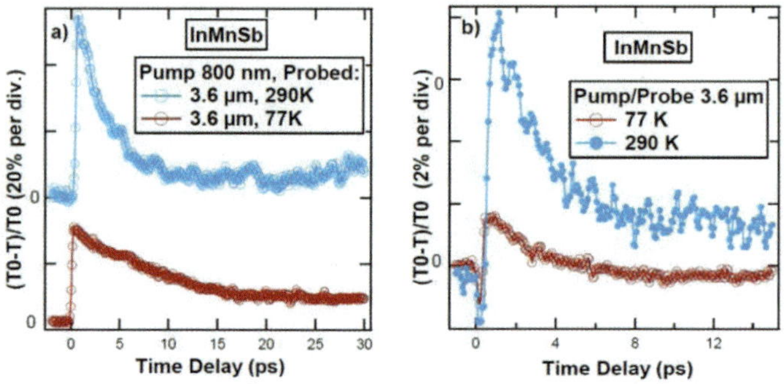


**Fig. 2.** TRDT a) in a non-degenerate scheme, b) when the pump and probe are the same wavelength. Both the magnitude of the TRDT and the time scale of the relaxations vary by tuning the initial excitation wavelength.

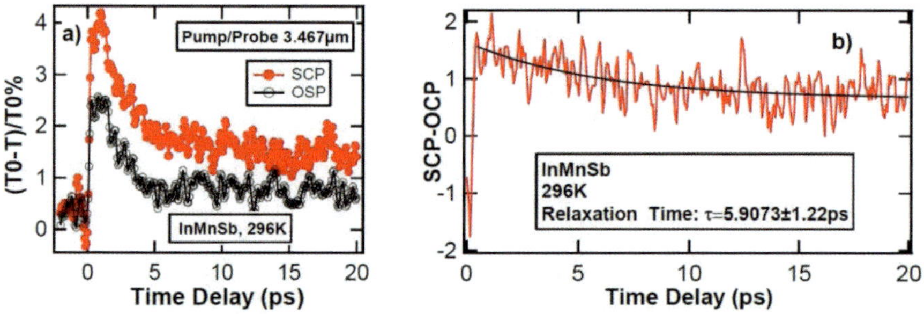


**Fig. 3.** a) Spin Polarized Time Resolved Differential Transmission b) Exponential fit to the SCP-OCP, to extract the spin relaxation time.

## References


[1] C. E. Feeser, L. Lari, V. K. Lazarov, J. A. Peters, and B. W. Wessels "Structural and magnetic properties of epitaxial In1-xMnxSb semiconductor alloys with x > 0.08" J. Vac. Sci. Technol. B **30**, 032801 (2012).

[2] M. Bhowmick, T. R. Merritt, and G. A. Khodaparast, Bruce W. Wessels, Stephen A. McGill, D. Saha, X. Pan, G. D. Sanders, and C. J. Stanton , "Time-resolved differential transmission in MOVPE-grown ferromagnetic InMnAs" Phys. Rev. B **85**, 125313 (2012).

# Magnetization evolution in semiconductor heterostructures after laser excitation

**O. Morandi, G. Manfredi, P.-A. Hervieux**

Institut de Physique et Chimie des Matériaux de Strasbourg

23, rue du Loess, F-67034 Strasbourg, France.

omar.morandi@ipcms.unistra.fr

**Abstract** The magnetization dynamics in semiconductor heterostucutures after short laser irradiation is investigated. By using ab-initio and phenomenological models, we reproduce the observed ultrafast quenching and Landau precession of the total magnetization of the samples.

## Introduction

Ultrafast spin dynamics in diluted magnetic semiconductors and metallic ferromagnetic films is an innovative topic which requires different areas of expertise both in analytical methods and numerical simulations. The possibility of testing the optical response of a semiconductor and ferromagnetic film by performing time-resolved pump-probe experiments opens prospective to probe the many-body electron dynamics in confined systems. Ferromagnetism in nanostructures can be modified at the femtosecond time scale by employing laser pulses [1].

In this contribution, we focus on a specific class of nanometric magnetic devices, the Diluted Magnetic Semiconductors (DMS). They are fabricated by doping semiconductor heterosctuctures containing quantum wells with a small concentration of magnetic ions.

## Ultrafast demagnetization and Landau damping

Various theories concerning the coherent manipulation of magnetization by strong off-resonant light have been put forth for III-Mn-V ferromagnetic semiconductors and for paramagnetic II-Mn-VI materials. We study the ultrafast demagnetization process in a semiconductor quantum well containing a 2D hole gas. The theoretical model used in our simulations is described in Refs. [2-4]. It is based on a first principle many-body approach and the main physically relevant quantities (in par-



J.-Y. Bigot et al. (eds), *Ultrafast Magnetism I*, Springer Proceedings in Physics 159
DOI 10.1007/978-3-319-07743-7_4

ticular the holes and ions spin densities) are obtained from a Boltzmann-Green hierarchy. The miniband structure of the 2D quantum well is evaluated self-consistently during the time evolution of the system. Originating by the strong spin-orbit interaction present in this kind of wide-gap semiconductors, the hole spin relaxation is mainly due to the Elliot-Yafet mechanism. We consider a DMS sample composed of a $Ga_{0.925}Mn_{0.075}As$ layer deposited on a GaAs buffer layer and a semi-insulating GaAs substrate. The laser pulse excitation is assumed to take place at $t = 0$. Details on the heterosctructure used in the calculations are given in Ref. [2]. The main results of the simulations are depicted in Fig. 1, where we display the behavior of the differential magnetization δM defined as $\frac{M_{tot}(t) - M_{tot}(0)}{M_{tot}(0)}$ where $M_{tot}(t)$ is the total magnetization of the sample and $t$ the time. Our model is able to reproduce the initial ultrafast damping of the total magnetization and the subsequent motion toward the equilibrium (left panel). The characteristic time $t_m$ and δM are strongly affected by the hole background density and by the sample thickness. This is illustrated in Fig. 1 (right panel) where δM is depicted as a function of the hole density $n^h$.

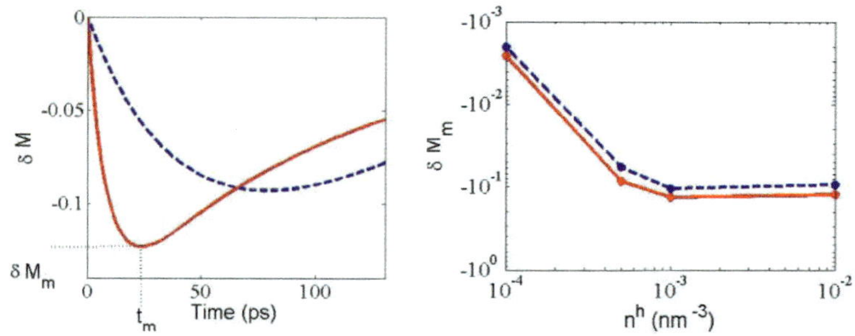


**Fig. 1.** Left panel: time evolution of the differential magnetization δM. Layer thickness: 6 nm (continuous curve) and 4 nm (dashed curves) for $n^h=10^{-3}$ nm$^{-3}$. Right panel: δM as a function of the hole density.

In what follows, we discuss some results concerning the Landau spin precession in magnetic semiconductors. We consider a sample composed of a 12 nm $Zn_{0.77}Cd_{0.23}Se$/ZnSe single quantum well, containing a uniform concentration (5%) of $Mn^{2+}$ ions. An external magnetic field **B** (equal to 0.1 T in our simulations) is applied along the confining direction ($z$ axis). At equilibrium, the sample displays a net magnetization directed along **B.** It results from the strong anti-ferromagnetic coupling between ions and *sp* electrons that overcomes the *d-d* super-exchange interaction between the *d*-electrons of the magnetic ions. The magnetism is mediated by the presence of the particles (in our simulations we consider a hole concentration equal to $n^h = 10^{11}$ cm$^{-2}$) trapped in the quantum well. Concerning the set-up of the irradiation apparatus, we consider a Voigt geometry, where the direction of the light propagation ($x$-axis) is orthogonal to the magnetic field. At $t = 0$, a femtosecond circularly polarized optical pulse impacts the device creating a hot quasi-thermal distribution of spin-polarized particles in the quantum well. Our simulations are based on the theoretical model presented in Ref. [5]. It is a semi-phenomenological approach derived in the framework of the Landau-Lifshitz-Bloch theory.

Figure 1 (left panel) shows the evolution of the longitudinal and out-of-plane components of the ions spin. The spin of the confined particle precesses with a period of the order of 300 fs, and relaxes with a characteristic time of around 2 ps. Our simulations are in good agreement with the experiments discussed in Ref. [6]. Since the spin-flip exchange mechanism between ions and electrons is particularly efficient, the Landau relaxation process, mediated by the 2D electron gas, becomes the dominant relaxation mechanism of the ions spin. In Fig. 2 (right panel) we depict the time evolution of the ion magnetization. The simulation shows that after a first high oscillating evolution (during which the electron spin becomes aligned with the total mean magnetization via the Landau decay), the magnetization start to precess with a period of nearly 50 picoseconds.

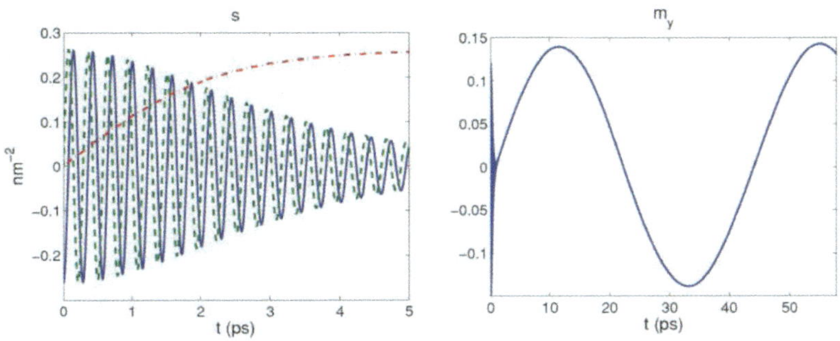


**Fig. 2.** Left panel: Time evolution of the x- (continuous blue curve), y- (dashed green curve) z- (dot-dashed red curve) projection of the total polarization of the trapped particles. Right panel: Time evolution of the transverse *y*-component of the ion magnetization.

## References


[1] E. Beaurepaire, J.-C. Merle, A. Daunois, and J.-Y. Bigot, "Ultrafast Spin Dynamics in Ferromagnetic Nickel", Phys. Rev. Lett. **76**, 4250 (1996).

[2] O. Morandi, P.-A. Hervieux and G. Manfredi, "Time dependent model for Diluted Magnetic semiconductors including band structure and confinement effects", Phys. Rev. B **81**, 155309 (2010).

[3] O. Morandi, P.-A. Hervieux and G. Manfredi, "Ultrafast Magnetization Dynamics in Diluted Magnetic Semiconductors", New J. of Phys **11**, 073010 (2009).

[4] O. Morandi and P.-A. Hervieux, "Third-order many-body perturbation theory applied to Kondo-type dynamics in diluted magnetic semiconductors", Phys. Rev. B **81**, 195215 (2010).

[5] O. Morandi, "Spin evolution in a two-dimensional electron gas after laser excitation", Phys. Rev. B **83**, 224428 (2011).

[6] D. Awschalom and N. Samarth, "Spin Dynamics and Quantum Transport in Magnetic Semiconductor Quantum Structures", J. Magn. Magn. Mater. **200**, 130 (1999).

# Phase and spin relaxation dynamics in high-quality single GaN/AlGaN quantum well

**M. Gallart, M. Ziegler, B. Hönerlage, P. Gilliot**

IPCMS–DON, UMR 7504, CNRS–ULP, 23 rue du Loess, BP 43, 67034 Strasbourg Cedex 2, France

**E. Feltin, J.-F. Carlin, R. Butté, N. Grandjean**

ICMP, École Polytechnique Fédérale de Lausanne (EPFL), CH-1015 Lausanne, Switzerland

Pierre.Gilliot@ipcms.unistra.fr

**Abstract** We report on the first time measurement, using four-wave mixing, of the dephasing time of excitons in a high-quality GaN/Al0.05Ga0.95N quantum well. The comparison with the exciton-spin lifetime, determined via pump-probe spectroscopy using polarized laser pulses, shows that spin relaxation occurs in the motional narrowing regime.

GaN and related materials, because of their weak spin-orbit coupling and large exciton binding energy (~ 17 meV and ~ 26meV in bulk GaN, respectively), are promising candidates for spintronic applications that involve the control and manipulation of the exciton spin in GaN quantum dots, even at high temperature [1]. While some previous experimental works have shown that the spin polarization of excitons in GaN/AlGaN quantum wells (QWs) lasts for a few picoseconds [2], very few data about the dephasing time (T2) in such systems are available. Because spin relaxation in semiconductors is connected to wave-vector relaxation via the spin-orbit interaction, a full understanding of the spin dynamics requires a simultaneous measurement of both time constants.

In the present work, we report on the measurement of the T2 time of excitons in a high-quality single wurtzite GaN/Al0.05Ga0.95N QW, grown by metal organic vapor phase epitaxy, by means of time-resolved four wave-mixing experiments. The exciton-spin lifetime $\tau S$ has also been separately determined via pump-probe spectroscopy using polarized laser pulses. Both experiments have been performed as a function of exciton density and temperature T.

For T > 80 K, the thermal escape of excitons toward the AlGaN barrier leads to a dramatic enhancement of the spin-relaxation rate. On the contrary in the 10 to 80 K temperature range, the spin-relaxation rate remains constant and independent of the dephasing rate that increases linearly with temperature. Moreover, the

 
J.-Y. Bigot et al. (eds), *Ultrafast Magnetism I*, Springer Proceedings in Physics 159
DOI 10.1007/978-3-319-07743-7_5

dephasing rate always remains larger than the spin-relaxation rate in this temperature range. We then conclude that spin relaxation occurs in the motional narrowing regime: scattering events undergone by the excitons are able to hinder spin relaxation up to 80 K beyond which the thermal energy becomes large enough for excitons to escape from the QW. Such measurements demonstrate that GaN-based heterostructures can reach a degree of control that was previously mostly restricted to conventional III-V semiconductors and more specifically to the arsenide family. Moreover, it opens new ways for manipulating spins and charge carriers in nanostructures like polaritonic microcavities.

## References

[1] D. Lagarde, A. Balocchi, H. Carrère, P. Renucci, T. Amand, X. Marie, S. Founta, and H. Mariette, "Room-temperature optical orientation of the exciton spin in cubic GaN/AlN quantum dots" Phys. Rev. B **77**, 041304(R) (2008).

[2] J. Besbas, A. Gadalla, M. Gallart, O. Crégut, B. Hönerlage, P. Gilliot, E. Feltin, J.-F. Carlin, R. Butté, and N. Grandjean, "Spin relaxation of free excitons in narrow GaN/AlxGa1−xN quantum wells" Phys. Rev. B **82**, 195302 (2010).

# Experimental observations of optical spin transfer and spin-orbit torques in magnetic semiconductors

**P. Němec[1], E. Rozkotová[1], N. Tesařová[1], T. Janda[1], D. Butkovičová[1], F. Trojánek[1], P. Malý[1], V. Novák[2], J. Zemen[2], K. Olejník[2], and T. Jungwirth[2]**

[1]Charles University in Prague, Faculty of Mathematics and Physics, Ke Karlovu 3, 121 16 Prague 2, Czech Republic

[2]Institute of Physics ASCR, v.v.i., Cukrovarnická 10, 162 53 Prague 6, Czech Republic

nemec@karlov.mff.cuni.cz

**Abstract** The spin transfer and spin-orbit torques can be used for electrically driven magnetization dynamics. Here, we report on the experimental observation of the optical counterparts of these torques which lead to a sub-picosecond tilt of magnetization and, consequently, to the magnetization precession in (Ga,Mn)As.

The spin transfer torque (STT) is a non-relativistic phenomenon where angular momentum of spin polarized carriers electrically injected into a ferromagnet from an external polarizer is transferred to the magnetization. This effect has opened a new research field of electrically driven magnetization dynamics and plays an important role in the development of a new generation of memory devices. In the absence of an external polarizer a distinct phenomenon can occur in which carriers in a magnet under applied electric field develop a non-equilibrium spin polarization due to the relativistic spin-orbit coupling, resulting in a current induced spin-orbit torque (SOT). Here we show that there exist optical counterparts of STT and SOT – optical spin transfer torque (OSTT) [1] and optical spin-orbit torque (OSOT) [2], respectively.

Ferromagnetic semiconductor (Ga,Mn)As is a favorable material for observing the optical spin torques because the direct-gap GaAs host allows the generation of high density non-equilibrium photo-carriers and the carrier spins interact with ferromagnetic moments on Mn via strong exchange coupling. When the ferromagnetic Mn moments are excited, this can be sensitively detected by probe laser pulses due to large magneto-optical (MO) signals. We show how the optical torques can be identified in the measured time-resolved MO data. Moreover, this MO experiment can be also used for a determination of all micromagnetic parameters [3].

Absorption of laser pulse increases locally the sample temperature and, consequently, it changes the magnetic anisotropy that can lead to a precession of magnetization – see Fig. 1(b). Therefore, for an experimental identification of the opti-

 
J.-Y. Bigot et al. (eds), *Ultrafast Magnetism I*, Springer Proceedings in Physics 159
DOI 10.1007/978-3-319-07743-7_6

cal torques [Fig. 1(a) and (c)] it is necessary to separate their influence from the thermally-induced dynamics. In Fig. 2(a) we show the measured MO signal in a sample where the thermally-induced dynamics was suppressed by a piezostressor [1] – clearly, there is a phase shift of π when the light helicity is changed, which is a signature of the angular momentum transfer [1]. In the case of OSOT, the experimental separation from the thermal effect is even more challenging because both these effects do not depend on the polarization of pump pulse [2]. To achieve this, we used a new technique which enables a 3D magnetization trajectory reconstruction from the measured MO data without any numerical modeling [4]. In the magnetization trajectory OSOT manifests itself as an abrupt tilt of the magnetization while the thermally-induced dynamic corresponds to a much slower onset of the precession – see Fig. 2(b). In addition, we revealed that OSOT can be separated also by a suppression of the thermally-induced easy axis reorientation by a magnetic anisotropy control using the external magnetic field or doping by Mn [2].

The pump-pulse-induced precession of magnetization can be used also for the determination of micromagnetic parameters – see Fig. 3 [3]. In particular, the anisotropy fields and Gilbert damping coefficient can be evaluated from the dependence of the precession frequency and the damping rate on the external magnetic field, respectively. Moreover, the spin stiffness can be obtained from the mutual spacing of the spin-wave resonances in the measured dynamical MO data [3].

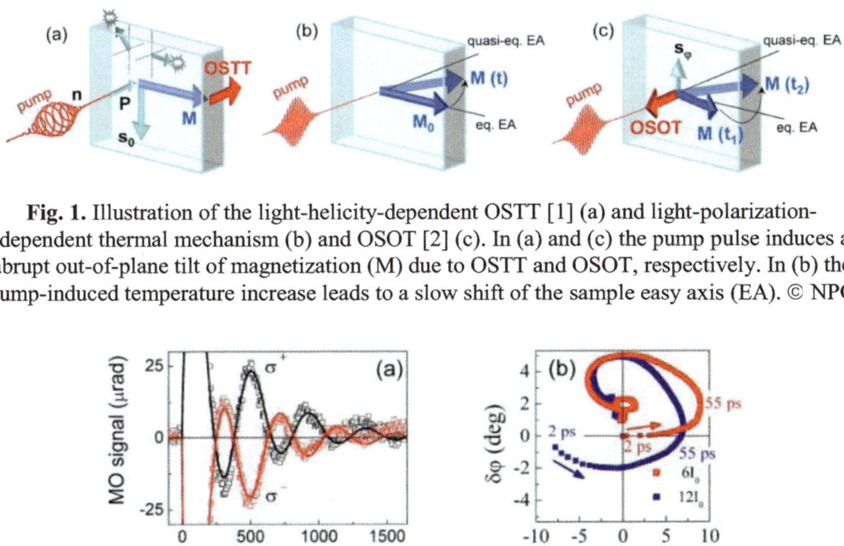


**Fig. 1.** Illustration of the light-helicity-dependent OSTT [1] (a) and light-polarization-independent thermal mechanism (b) and OSOT [2] (c). In (a) and (c) the pump pulse induces an abrupt out-of-plane tilt of magnetization (M) due to OSTT and OSOT, respectively. In (b) the pump-induced temperature increase leads to a slow shift of the sample easy axis (EA). © NPG

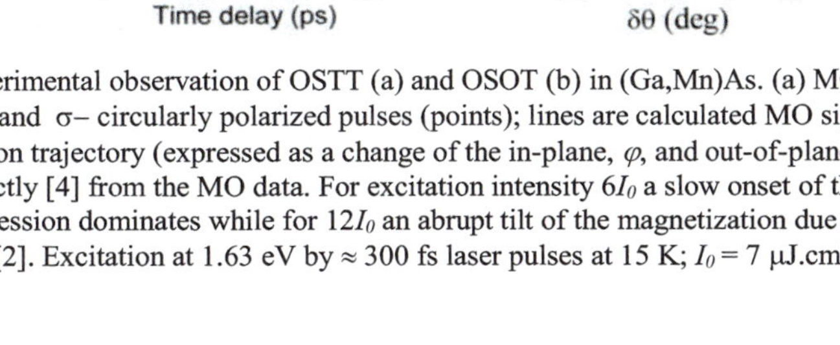


**Fig. 2.** Experimental observation of OSTT (a) and OSOT (b) in (Ga,Mn)As. (a) MO signal induced by σ+ and σ− circularly polarized pulses (points); lines are calculated MO signals [1]. (b) Magnetization trajectory (expressed as a change of the in-plane, $\varphi$, and out-of-plane, $\theta$, angles) obtained directly [4] from the MO data. For excitation intensity $6I_0$ a slow onset of the thermally-induced precession dominates while for $12I_0$ an abrupt tilt of the magnetization due to OSOT also occurs [2]. Excitation at 1.63 eV by ≈ 300 fs laser pulses at 15 K; $I_0 = 7$ μJ.cm$^{-2}$. © NPG

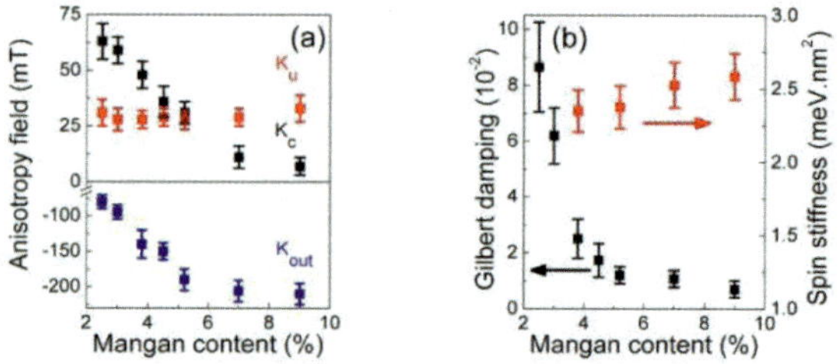


**Fig. 3.** Micromagnetic parameters of optimized epilayers of ferromagnetic (Ga,Mn)As determined from a single set of time-resolved magneto-optical data [3]. © NPG

**Acknowledgments** We acknowledge support from the Grant Agency of the Czech Republic Grant No. P204/12/0853, from EU ERC Advanced Grant No. 268066 and FP7-215368 SemiSpinNet, from the Ministry of Education of the Czech Republic Grants No. LM2011026, from the Charles University in Prague Grant No. SVV-2012-265306 and 443011, and from the Academy of Sciences of the Czech Republic Preamium Academiae.

## References

[1] P. Němec, E. Rozkotová, N. Tesařová, F. Trojánek, E. De Ranieri, K. Olejník, J. Zemen, V. Novák, M. Cukr, P. Malý and T. Jungwirth, "Experimental observation of the optical spin transfer torque" Nature Physics **8**, 411 - 415 (2012).

[2] N. Tesařová, P. Němec, E. Rozkotová, J. Zemen, T. Janda, D. Butkovičová, F. Trojánek, K. Olejník, V. Novák, P. Malý, and T. Jungwirth, "Experimental observation of the optical spin-orbit torque" Nature Photonics **7**, 492-498 (2013).

[3] P. Němec, V. Novák, N. Tesařová, E. Rozkotová, H. Reichlová, D. Butkovičová, F. Trojánek, K. Olejník, P. Malý, R. P. Campion, B. L. Gallagher, J. Sinova, and T. Jungwirth, "Controlling semiconducting and ferromagnetic properties in individually optimized Mn-doped GaAs epilayers" Nature Communications **4**, 1422 (2013).

[4] N. Tesařová, P. Němec, E. Rozkotová , J. Šubrt, H. Reichlová, D. Butkovičová, F. Trojánek, P. Malý, V. Novák, and T. Jungwirth, "Direct measurement of the three-dimensional magnetization vector trajectory in GaMnAs by a magneto-optical pump-and-probe method" Appl. Phys. Lett. **100**, 102403 (2012).

# Laser-induced spin dynamics in ferromagnetic (In,Mn)As at magnetic fields up to 7 T

**R.R. Subkhangulov[1], H. Munekata[2], Th. Rasing[1] and A.V. Kimel[1]**

[1]Radboud University Nijmegen, Institute for Molecules and Materials (IMM) Heyendaalseweg 135, 6525 AJ Nijmegen, the Netherlands

[2]Imaging Science and Engineering Laboratory, Tokyo Institute of Technology, Yokohama,Kanagawa, Japan

r.subkhangulov@science.ru.nl

**Abstract** Laser-induced spin dynamics in (In,Mn)As is studied in magnetic fields up to 7 T. It is shown that a laser pulse can effectively excite homogenous spin precession in this compound at the frequency of the ferromagnetic resonance. Laser excitation of this resonance appears to be very ineffective if the applied magnetic field is below 1.5 T. Our analysis shows that the damping of the laser-induced spin precession is a function of magnetic field and reaches very high values below 1.5 T.

## Introduction

Soon after the discovery, very promising experimental results on optical control of magnetism in both (In,Mn)As and (Ga,Mn)As was achieved with low-power continuous wave excitation [1, 2].These observations raised questions about the feasibility to manipulate their intrinsic magnetic properties, such as magnetization and magnetic anisotropy, in an ultrafast, i.e. (sub)picosecond way by means of femtosecond laser pulses. To date, the possibility of triggering spin precession by light was demonstrated for (Ga,Mn)As in a number of works [3-5]. To the best of our knowledge no successful demonstration of ultrafast optical excitation of ferromagnetic resonance (FMR) in (In,Mn)As has been reported so far. Here we present magneto-optical studies of the laser-induced magnetization dynamics in (In,Mn)As and reveal that laser induced FMR effective damping is high and strongly depends on magnetic field.

 
J.-Y. Bigot et al. (eds), *Ultrafast Magnetism I*, Springer Proceedings in Physics 159
DOI 10.1007/978-3-319-07743-7_7

## Results

(In,Mn)As is semiconductor sample with a thickness of 15nm grown on a 500 nm buffer layer of $AlSb_{0.9}As_{0.1}$ deposited on a GaAs substrate (001). The band-gap is $E_g$~0.4 eV at low temperatures [6, 7]. The SQUID measurements have confirmed the existence of ferromagnetism with a Curie temperature of about $T_C \approx 50$ K and revealed the presence of uniaxial magnetic anisotropy with the easy-axis along [110] in the plane of the sample. (In,Mn)As has p-type conductivity. In (In,Mn)As, it is believed that valence-band holes mediate ferromagnetic exchange interaction between the spins of $Mn^{2+}$ ions [8, 9]. For time-resolved magneto-optical studies we employed 60 fs laser pulses with a repetition rate of 1 kHz and a pump-probe technique. The wavelength of the probe was set to 800 nm, while the wavelength of the pump was 800 nm, which corresponds to an energy higher than that of band-gap.

Figure 1 shows the dynamics of the magneto-optical Kerr rotation measured at different temperatures for pump fluence of $0.25 mJ/cm^2$ at magnetic fields 0.2T and 4T. The laser excitation leads to an ultrafast decrease of the magneto-optical signal. Such a dynamics is explained [10] in terms of subpicosecond laser-induced demagnetization within 100fs (Fig. 1A). The data also disclose that the pulsed excitation triggers oscillations of the magneto-optical signal (Fig. 1B). It is seen that the oscillations are strongly damped and can only be excited in magnetic fields higher than 1T. The mechanism of laser induced FMR was checked to be due to ultrafast quenching of magnetic anisotropy.

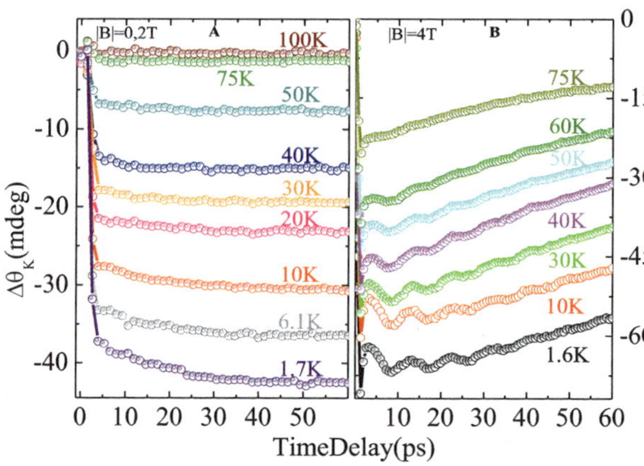


**Fig. 1.** (A) Laser induced dynamics of the magneto-optical Kerr rotation measured at different temperatures for a pump fluence of $0.25 mJ/cm^2$ at B=0.2T and (B) at B=4T.

It is seen that the amplitude of oscillations decreases with increasing bath temperature and vanishes just above the Curie temperature. The decrease of the amplitude is accompanied by a slight decrease of the frequency from 113 GHz at 1.6 K down to 109 GHz at 75 K. The temperature dependence of oscillations amplitude indicate that the observed oscillations originate from optically induced FMR of $Mn^{2+}$ - spins in (In,Mn)As.

In order to substantiate that the origin of the oscillations is indeed ferromagnetic resonance of $Mn^{2+}$-spins, measurements in magnetic fields up to 7 T at 10K were performed. It was found that effective damping (defined as $\gamma = 1/(f \cdot \tau)$, where $\tau$ and $f$ are decay constant of the oscillations and frequency of the oscillations respectively) is a diminishing function of magnetic field and saturated at a magnetic field of 2T.

We argue that the reason of elevated effective magnetic damping is due to the presence of paramagnetic inhomogeneous regions, which enhance two magnon scattering [11]. A magnetic field aligns these paramagnetic regions and erases the boundaries between those regions, which results in a decreasing effective damping with magnetic field. The presence of the paramagnetic contribution even below the Curie temperature was confirmed in SQUID and in magneto-optical measurements.

To conclude, we performed systematic studies of the laser-induced magnetization dynamics in (In,Mn)As for temperatures down to 1.6 K and in magnetic fields up to 7 T. Laser induced ferromagnetic spin precession was observed and it was found that the magnitude of the effective damping of the magnetic oscillations in (In,Mn)As is five times larger than that in (Ga,Mn)As. A strong magnetic field dependence of the effective damping was discovered, which is attributed to the compositional randomness resulting in a presence of paramagnetic regions even below the Curie temperature. The latter hamper ultrafast laser excitation of the ferromagnetic resonance in this ferromagnetic semiconductor and can explain the observed magnetic field dependence of the effective damping.

**Acknowledgments** A Toonen, A. van Roij, and A. van Etteger are acknowledged for indispensable technical support during the experiments. J. Becker is acknowledged for fruitful discussion and suggestion. This work was supported by the European Research Council under the European Union's Seventh Framework Program (FP7/2007-2013)/ERC Grant Agreement No. 257280 (Femtomagnetism).

## References

[1] S. Koshihara, A. Oiwa, M. Hirasawa, S. Katsumoto, Y. Iye, C. Urano, H. Takagi, and H. Munekata "Ferromagnetic Order Induced by Photogenerated Carriers in Magnetic III-V Semiconductor Heterostructures of (In,Mn)As/GaSb" Phys Rev Lett **78**, 4617 (1997).
[2] A. Oiwa, Y. Mitsumori, R. Moriya, T. Słupinski, and H. Munekata, " Effect of Optical Spin Injection on Ferromagnetically Coupled Mn Spins in the III-V Magnetic Alloy Semiconductor (Ga,Mn)As" Phys Rev Lett **88**, 137202 (2002).
[3] D. M. Wang, Y. H. Ren, X. Liu, J. K. Furdyna, M. Grimsditch, and R. Merlin, "Light-induced magnetic precession in (Ga,Mn)As slabs: Hybrid standing-wave Damon-Eshbach modes" Phys Rev B **75**, 233308 (2007).

[4] P. Nemec et al. "The essential role of carefully optimized synthesis for elucidating intrinsic material properties of (Ga,Mn)As" Nat. Commun. **4**, 1422 (2013).

[5] Y. Hashimoto, S. Kobayashi, and H. Munekata, "Photoinduced Precession of Magnetization in Ferromagnetic (Ga,Mn)As" Phys. Rev. Lett. **100**, 067202 (2008).

[6] H. Munekata, H. Ohno, S. von Molnar, Armin Segmüller, L. L. Chang, and L. Esaki, "Diluted magnetic III-V semiconductors" Phys. Rev. Lett. **63**, 1849 (1989).

[7] T. Schallenberg and H. Munekata, "Preparation of ferromagnetic (In,Mn)As with a high Curie temperature of 90K" Appl. Phys. Lett. **89**, 042507 (2006).

[8] F. Matsukuraa, H. Ohno, and T. Dietl, III-V Ferromagnetic Semiconductors (Elsevier, Amsterdam), Vol. 14, Handbook of Magnetic Materials

[9] T. Jungwirth, Jairo Sinova, J. Mašek, J. Kučera, and A. H. MacDonald, " Theory of ferromagnetic (III,Mn)V semiconductors" Rev. Mod. Phys. **78**, 809 (2006).

[10] J. G. Wang et al., J. Phys. Cond. Mat 18 (2006).

[11] M. Sparks, Ferromagnetic-relaxation theory (McGraw-Hill, New York, 1964), McGraw-Hill advanced physics monograph series.

# Evolving magnetization dynamics in $Mn_{3-x}Ga$

**J.M. Wikberg, I. Razdolski, A. Kirilyuk, Th. Rasing**

Institute for Molecules and Materials, Radboud University Nijmegen, Heyendaalseweg 135
6525 AJ Nijmegen, The Netherlands

**J. Sadowski**

MAX-lab, Lund University, P.O. Box 118, SE-221 00 Lund, Sweden

**M. Ottoson**

Department of Materials Chemistry, Uppsala University, P.O. Box 538, SE-751 21 Uppsala, Sweden

**Y. Wei, P. Svedlindh**

Department of Engineering Sciences, Uppsala University, P.O. Box 534, SE-751 21 Uppsala, Sweden

m.wikberg@science.ru.nl

Magnetic materials with high magnetic coercivity ($H_c$), magnetic anisotropy ($K_u$) and low Gilbert damping ($\alpha$) are of great importance for future spintronics devices. For instance, for spin-transfer-torque (STT) memory devices low $\alpha$ and high $K_u$ are desired. Such a combination of material properties includes a built in contradiction since both $K_u$ and $\alpha$ are dependent on the spin-orbit interaction (SO); a high $K_u$ material is expected to have a high α. However, recent experimental investigations of $Mn_{3-x}Ga$, a material exhibiting high Curie temperature ($T_c$) > 700 K and (out-of-plane) perpendicular anisotropy, have shown that such a contradictorily low α and high $K_u$ material exists [1]. Depending on Mn-concentration, $Mn_{3-x}Ga$ can form a $L1_0$ (x = 1-2) or a $D0_{22}$ (x<1) structure, which can be epitaxially grown on ordinary semiconductors such as GaAs or oxides, e.g. MgO, at moderate temperatures. Theoretical predictions [2] together with experimental results [3], of low $\alpha = 0.0003$ (0.001), high perpendicular anisotropy $K_u$ = 26 (20) Merg/cm$^3$, moderate saturation magnetization (Ms) of 845 (305) emu/cm3 and high spin polarization of 71% (88%) in $L1_0$ ($D0_{22}$) materials make $Mn_{3-x}Ga$ an excellent candidate for future spintronic applications.

Epitaxial films of $Mn_{3-x}Ga$ (x = 0-2) have been grown with molecular beam epitaxy (MBE) on GaAs substrates with a growth temperature of 150° C. The

 
J.-Y. Bigot et al. (eds), *Ultrafast Magnetism I*, Springer Proceedings in Physics 159
DOI 10.1007/978-3-319-07743-7_8

growth was monitored with a reflection high-energy electron diffraction (RHEED) system and a short (5 minute) post growth annealing was performed at 250° C. The structural properties of the films have been investigated with x-ray diffraction, while the magnetic properties were investigated with SQUID and VSM magnetometry as well as time-resolved magneto-optical Kerr effect (TRMOKE) in a conventional all-optical pump-probe set-up. The latter allowed time-resolved measurements of the magnetization dynamics as a function of the pump fluence ($F_0$) and magnetic field ($H$) strength.

Our structural measurements confirm the epitaxial film structure and that the nominal Mn-concentrations were reached. The magnetic measurements show that all films exhibit high $T_c > 700$ K and an order of magnitude variation of $M_s$ (42 – 440 emu/cm$^3$) as well as an anisotropy field ($H_a$) from 40 to ~100 kOe as x changed from 0 to 2. Because of the high Ha values and limitations of the used equipment (unable to reach saturation in the magnetic hard axis direction) $H_a$ had to be estimated at times and $K_u$ was evaluated with the relationship:

$$K_u = \frac{M_S H_a}{2} + 2\pi M_S^2 \qquad 1$$

yielding values between 0.85 and 21 Merg/cm$^3$.

Strong influence of the Mn concentration and optical pump beam parameters on the magnetization dynamics was found in the studied samples. The TRMOKE results display characteristic behavior for pump-probe measurements [4] with an ultrafast demagnetization in the first few hundreds of femtoseconds followed by a slow (sub-nanosecond) recovery. The oscillatory dynamics of the transient MOKE signal was proven to be of magnetic nature and corresponds to the uniform precession of the magnetization vector around its new equilibrium with a frequency up to 200 GHz. We observe and analyze the dependence of frequency, amplitude and damping of the oscillations on *Fp* as well as *H*. From fits of the experimental data the spin precession frequency and lifetime can be obtained and together with the expressions derived from the Landau–Lifshitz–Gilbert equation values for $\alpha$ were extracted.

Hence, we demonstrate the evolvement of $\alpha$ and $K_u$ with changes in Mn concentration over the ferro-/ferrimagnetic phase transition. The results will deepen the understanding of the role of SO interaction on $K_u$ and $\alpha$ in $Mn_{3-x}Ga$ and give valuable insights in designing materials with desired properties.

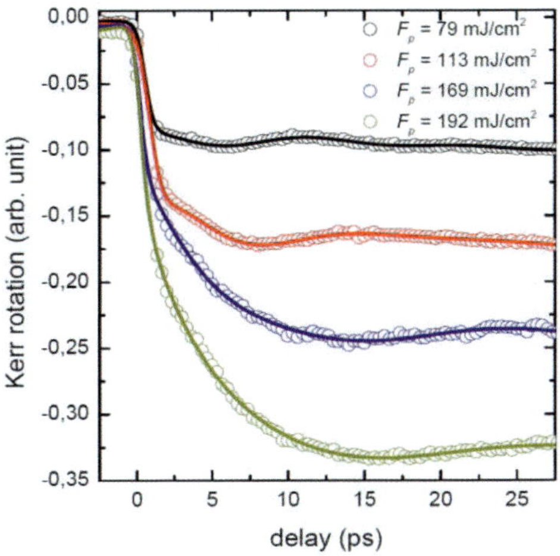


**Fig. 1.** Results from TRMOKE studies on a MnGa (x = 2) sample with $H = 10.2$ kOe at an angle of 78° with respect to the normal of the sample plane and different pump fluences. Open symbols give the experimental results while the solid lines are fits to the experimental results using an exponentially damped sine function.

**Acknowledgments** The authors would like to thank the Swedish Research Council for financial support.

# References

[1] S. Mizukami, et al, "Long-Lived Ultrafast Spin Precession in Manganese Alloys Films with a Large Perpendicular Magnetic Anisotropy" Phys. Rev. Lett. **106**, 117201 (2011);

[2] A. Sakuma, 3 Electronic structures and magnetism of CuAu-type MnNi and MnGa" J. Magn. Magn. Mater. **187**, 105 (1998).

[3] L. J. Zhu , D. Pan , S. H. Nie , J. Lu , and J. H. Zhao, "Tailoring magnetism of multifunctional MnxGa films with giant perpendicular anisotropy" Appl. Phys. Lett. **102**, 132403 (2013);

[4] A. Kirilyuk, A. V. Kimel, and Th. Rasing, "Ultrafast optical manipulation of magnetic order" Rev. Mod. Phys. **82**, 2731 (2010);

# Part II
# Ultrafast Magnetism Dynamics in Metals

# Electronic Scattering Dynamics and Ultrafast Magnetization Dynamics

**M. Aeschlimann, D. Steil, M. Cinchetti, H.C. Schneider**

Department of Physics and Research Center OPTIMAS, University of Kaiserslautern, D-67663 Kaiserslautern, Germany

ma@physik.uni-kl.de

**Abstract**: A dynamical model for Elliott-Yafet type scattering of carriers and its importance for the demagnetization dynamics after ultrafast optical excitation is reviewed. It is pointed out that the demagnetization in 3d ferromagnets as well as recent experimental results on "novel" materials are still in need of a microscopic explanation.

## Introduction

Ultrafast demagnetization dynamics in itinerant ferromagnets has been well established, and explained phenomenologically on the basis of a three-temperature model. More recently, demagnetization has been observed in half-metallic ferromagnets, and there is an enormous current interest in the optically induced magnetization dynamics of systems with different sublattices and magnetic alloys.

There are a number of different theoretical ideas as well as specific models that are contenders for such a microscopic explanation. Different approaches typically treat one aspect of the problem particularly well: coherent dynamics during during the exitation pulse, elementary magnetic excitations, spin-flip scattering transitions, and transport [1-5].

In the following, we will concentrate on the dynamical description of spin-flip scattering transitions, i.e., the dynamical Elliott-Yafet mechanism [3,4]. A treatment beased on dynamical distribution functions is valid even at ultrashort timescales and away from quasi-equilibrium. In the Elliott-Yafet mechanism, the spin expectation value of the carriers is changed by scattering transitions due to a spin-diagonal interaction mechanism, such as the carrier-carrier Coulomb or the carrier-phonon interaction. The spin-orbit coupling is responsible for the spin-mixed character of the single-particle states, which act as the initial and final states for scattering transitions. Here, we focus on carrier dynamics in Heusler alloys.

 
J.-Y. Bigot et al. (eds), *Ultrafast Magnetism I*, Springer Proceedings in Physics 159
DOI 10.1007/978-3-319-07743-7_9

## Dynamical carrier scattering in Heusler compounds

A dynamical model for the numerical calculation of the band and momentum resolved distribution carrier distribution functions $f_s(\vec{k},t)$ has been described in Ref. [2] and applied to half-metallic Heusler compounds of the $Co_2Mn_{1-x}Fe_xSi$ material system [6]. The model describes carrier-carrier Coulomb scattering at the level of Boltzmann scattering integrals, includes a simplified spherically symmetric band structure and a constant spin-mixing parameter. It is therefore well suited to compare the influence of different band structures on the spin-dependent carrier scattering after ultrafast excitation. In Ref. [6], it was found that a fast demagnetization is possible by spin-flip transitions after optical excitation, but these transitions follow different scattering pathways for $Co_2FeSi$ (CFS) and $Co_2MnSi$ (CMS) because of the different line-ups of the minority band-gap with the Fermi energy. In particular, it was found that hole scattering dynamics plays an important role, which makes the line-up of the minority band less important a difference between the two compounds than one would anticipate.

For Heusler alloys, sample quality may be an issue, and the interpretation of magneto-optical measurements of the magnetization dynamics is made more complicated by the possible existence of defects [7]. Müller et al. [7] discussed the fast magnetization dynamics in the case of CMS with regard to possible Co antisite defect states (DO3-disorder, Co on Mn positions) close to the Fermi level. These defect states were predicted for CMS by ab-initio calculations [8].

Here we analyze the influence of defects on the magnetization dynamics in CMS in the framework of our dynamical calculation of carrier distribution functions. To this end, we introduce in the calculation the presence of defects by adding a defect band with a flat dispersion 150 meV below the Fermi energy, similar to Ref. [9].

Figure 1 (left panel) shows the magnetization dynamics for CMS without defects, as well as two different defect concentrations. The case of 100% Co antisites is only presented to elucidate trends of the model because at this defect concentration the overall band structure should be changed considerably. For both cases with defects we obtain a more pronounced quenching and a faster demagnetization by up to a factor of two. The reason for this change in dynamics can be traced back to the minority band line-up in the right panel of Fig. 2. The dynamical calculations show an increase of occupation of the minority band 5 by about 1/3 in the band structure with defects compared to the defect-free case after optical excitation. The excited carriers subsequently flip their spin and contribute to the demagnetization. This increase of occupation in band 5 can be explained only if the defect band serves as an intermediate state for two-photon pumping into the minority band 5 from lower lying minority bands. A few such multi-photon pathways are added as green arrows in Fig. 1.

The present model calculation shows that already a small concentration of impurities may increase the quenching and speed up the demagnetization dynamics

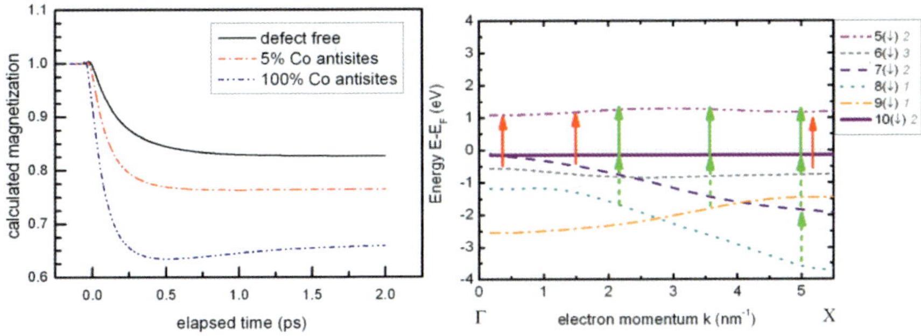


**Fig. 1.** Computed demagnetization dynamics for the Heusler compound (left) and dispersion of selected bands including transitions driven by the optical excitation. The vertical band (solid line) close to zero energy is the defect band, single (red) arrows indicate minority-spin transitions without defects, multiple stacked (green) arrows indicate new transitions due to the defect band.

due to a higher number of minority carrier states below the Fermi energy, if the impurities form a continuous band close to an unoccupied band above the Fermi level.

## References


[1] J.-Y. Bigot, M. Vomir, and E. Beaurepaire, "Coherent ultrafast magnetism induced by femtosecond laser pulses", Nature Physics **5**, 515 (2009).

[2] U. Atxitia, O. Chubykalo-Fesenko, J. Walowski, A. Mann, and M. Münzenberg, "Evidence for thermal mechanisms in laser-induced femtosecond spin dynamics", Phys. Rev. B **81**, 174401 (2010).

[3] M. Krauß, T. Roth, S. Alebrand, D. Steil, M. Cinchetti, M. Aeschlimann, and H. C. Schneider, "Ultrafast demagnetization of ferromagnetic transition metals: The role of the Coulomb interaction", Phys. Rev. B **80**, 180407 (2009).

[4] B. Koopmans, G. Malinowski, F. Dalla Longa, D. Steiauf, M. Fähnle, T. Roth, M. Cinchetti, and M. Aeschlimann, "Explaining the paradoxical diversity of ultrafast laser-induced demagnetization", Nature Materials **9**, 259 (2009).

[5] M. Battiato, K. Carva, and P. M. Oppeneer, "Superdiffusive Spin Transport as a Mechanism of Ultrafast Demagnetization", Phys. Rev. Lett. **105**, 027203 (2010).

[6] D. Steil, S. Alebrand, T. Roth, M. Krauß, T. Kubota, M. Oogane, Y. Ando, H. C. Schneider, M. Aeschlimann, and M. Cinchetti, "Band-Structure-Dependent Demagnetization in the Heusler Alloy $Co_2Mn_{1-x}Fe_xSi$", Phys. Rev. Lett. **105**, 217202 (2010).

[7] G. M. Müller, J. Walowski, M. Djordjevic, G.-X. Miao, A. Gupta, A. V. Ramos, K. Gehrke, V. Moshnyaga, K. Samwer, J. Schmalhorst, A. Thomas, A. Hütten, G. Reiss, J. Moodera, and M. Münzenberg, "Spin polarization in half-metals probed by femtosecond spin excitation", Nature Materials **8**, 56 (2009).

[8] S. Picozzi, A. Continenza, and A. J. Freeman, "Role of structural defects on the half-metallic character of $Co_2MnGe$ and $Co_2MnSi$ Heusler alloys", Phys. Rev. B **69**, 094423 (2004).

[9] S. Picozzi, and A. J. Freeman, "Polarization reduction in half-metallic Heusler alloys: the effect of point defects and interfaces with semiconductors", J. Phys.: Condens. Matter **19**, 315215 (2007).

# Influence of the magnetization compensation point on the all-optical magnetization switching

**L. Le Guyader[1,2], I. Radu[2], A. Eschenlohr[2], S. El Moussaoui[1], M. Buzzi[1], I. Razdolski[3], R. Medapalli3, M. Savoini[3], Ch. Stamm[2], R. Mitzner[2], K. Holldack[2], T. Kachel[2], A. Tsukamoto[4], A. Itoh[4], A. Kirilyuk[3], Th. Rasing[3], F. Nolting[1], A.V. Kimel[3]**

[1]Swiss Light Source, Paul Scherrer Institut, Villigen, Switzerland

[2]Helmholtz-Zentrum Berlin für Materialien und Energie GmbH, Berlin, Germany

[3]Radboud University Nijmegen, Institute for Molecules and Materials, The Netherlands

[4]College of Science and Technology, Nihon University, Chiba, Japan

loic.le_guyader@helmholtz-berlin.de

**Abstract** Combining femtosecond transmission measurements with picosecond time-resolved photo-emission electron microscopy, both using x-ray magnetic circular dichroism, new insights into the all-optical magnetization switching mechanism in GdFe based rare-earth transition metal ferrimagnetic alloys is provided, with emphasis on the role played by the magnetization compensation temperature $T_M$ of the alloy.

## Introduction

Understanding ultrafast all-optical magnetization switching (AOS), i.e. the permanent reversal of the magnetization by the sole action of a femtosecond laser pulse in the absence of any applied magnetic field, is a challenging issue that could have tremendous impact for the magnetic recording industry. While a qualitative agreement between spin atomistic simulations and experiments exists [1], the exact role played by the magnetization compensation point $T_M$ in the ultrafast demagnetization [2] and switching [3] in ferrimagnetic rare-earth transition-metal alloys is still not completely clear.



J.-Y. Bigot et al. (eds), *Ultrafast Magnetism I*, Springer Proceedings in Physics 159
DOI 10.1007/978-3-319-07743-7_10

## Results

By combining femtosecond X-ray transmission measurements with picosecond time-resolved photo-emission electron microscopy (PEEM), both using X-ray magnetic circular dichroism, we report on new insights into the AOS mechanism in GdFe based ferrimagnetic alloys. In agreement with previous experiments and theoretical predictions, AOS is seen below and above $T_M$, and in particular against a 0.18 T magnetic field. However, at temperatures far from $T_M$, no AOS is observable. Collapse of the reversed domain could be ruled out using time-resolved XMCD PEEM imaging. Static imaging of the magnetic domain configuration after AOS reveals that no domain wall (DW) motion occurs within the 100 nm spatial resolution, ruling out a nucleation and growth switching mechanism favored by the DW velocity divergence at TM. Furthermore, investigation of the formation speed of the transient ferromagnetic-like state as a function of $T_M$ shows very pronounced variations. These results provide evidence that the $T_M$ is somehow a more important condition for the formation of the transient ferromagnetic-like state [4] and the occurrence of AOS than initially thought.

**Acknowledgments** This work was partially supported by the European Research Council under the European Union's Seventh Framework Programme (FP7/2007- 2013) ERC Grant agreements No. NMP3-SL-2008-214469 (UltraMagnetron) and No. 214810 (FANTOMAS).

## References

[1] T. A. Ostler, J. Barker, R. F. L. Evans, R. Chantrell, U. Atxitia, O. Chubykalo-Fesenko, S. El Moussaoui, L. Le Guyader, E. Mengotti, L. J. Heyderman, F. Nolting, A. Tsukamoto, A. Itoh, D. Afanasiev, B. A. Ivanov, A. M. Kalashnikova, K. Vahaplar, J. Mentink, A. Kirilyuk,Th. Rasing and A. V. Kimel, "Ultrafast Heating as a Sufficient Stimulus for Magnetization Reversal in a Ferrimagnet" Nat. Commun. **3**, 666 (2012).

[2] R. Medapalli, I. Razdolski, M. Savoini, A. R. Khorsand, A. Kirilyuk, A. V. Kimel, Th. Rasing, A. M. Kalashnikova, A. Tsukamoto, A. Itoh, "Efficiency of ultrafast laser-induced demagnetization in $Gd_xFe_{100-x-y}Co_y$ alloys" Phys. Rev. B **86**, 054442 (2012).

[3] K. Vahaplar, A. M. Kalashnikova, A. V. Kimel, D. Hinzke, U. Nowak, R. Chantrell, A. Tsukamoto, A. Itoh, A. Kirilyuk, Th. Rasing, "Ultrafast Path for Optical Magnetization Reversal via a Strongly Nonequilibrium State" Phys. Rev. Lett. **103,** 117201 (2009).

[4] I. Radu, K. Vahaplar, C. Stamm, T. Kachel, N. Pontius, H. A. Dürr, T. A. Ostler, J. Barker, R. F. L. Evans, R. W. Chantrell, A. Tsukamoto, A. Itoh, A. Kirilyuk, Th. Rasing, A. V. Kimel, "Transient ferromagnetic-like state mediating ultrafast reversal of antiferromagnetically coupled spins" Nature **472**, 205–208 (2001).

# Element-specific probing of ultrafast magnetization dynamics in the visible spectral range

**M. Savoini,[1] A.R. Khorsand,[1] A. Kirilyuk,[1] A.V. Kimel,[1] A. Tsukamoto,[2] A. Itoh,[2] and Th. Rasing[1]**

[1]Radboud University Nijmegen, Institute for Molecules and Materials, Nijmegen, The Netherlands

[2]College of Science and Technology, Nihon University, 7-24-1 Funabashi, Chiba, Japan

m.savoini@science.ru.nl

**Abstract** Femtosecond laser excitation of thin magnetic films consisting of multiple magnetic sublattices triggers ultrafast spin dynamics and even magnetization reversal driven by the exchange interaction between the sublattices. To explore these intriguing phenomena requires element specific studies.We demonstrate that element specific probing of ultrafast spin dynamics can also be realized with visible light and does not require sophisticated X-ray facilities.

A challenging and intriguing topic in today's condensed matter physics is the non-equilibrium spin dynamics on sub-ps timescales. Spin dynamics is understood very well at timescales of magnetic resonances (GHz), during which the alignment of spins is defined by the exchange interaction, i.e., the strongest force in magnetism. On the other hand, little is known about their coupling when magnetic sublattices are almost instantaneously brought out of equilibrium with, for example, an optical laser pulse with a duration that corresponds to the exchange interaction, i.e., 10-100 fs. In this regime of non-adiabatic spin-dynamics, novel and counter-intuitive phenomena can emerge [1,2].

In order to track the magnetic response of each sublattice individually, an element-specific probe is necessary. For this purpose, photons in the soft x-ray regime are used [1-3]. Techniques utilizing high energy photons however, have important limitations. There are only very few femtosecond x-ray or High-Harmonic Generation facilities in the world, strongly limiting the amount of experiments which can be performed. As a consequence, only a few element specific studies are reported until now.

We demonstrate that element-specific probing of ultrafast spin dynamics can also be realized in the visible spectral range and does not require sophisticated high photon-energy facilities. We present the requirements a magnetic material should meet to make such an optical study feasible, and use a TbFe-alloy as our

 
J.-Y. Bigot et al. (eds), *Ultrafast Magnetism I*, Springer Proceedings in Physics 159
DOI 10.1007/978-3-319-07743-7_11

model sample. We demonstrate that in this material one can study the dynamics of the Tb- and Fe-spins individually, by choosing the wavelength of light above and below 600 nm, respectively, as presented in Fig. 1. This creates a vista of opportunities for the investigation of ultrafast element-specific magnetization dynamics. For instance, one can easily extend the conclusions of this work to other optical probes making, for instance, magneto-optical imaging or Raman scattering on magnons element specific.

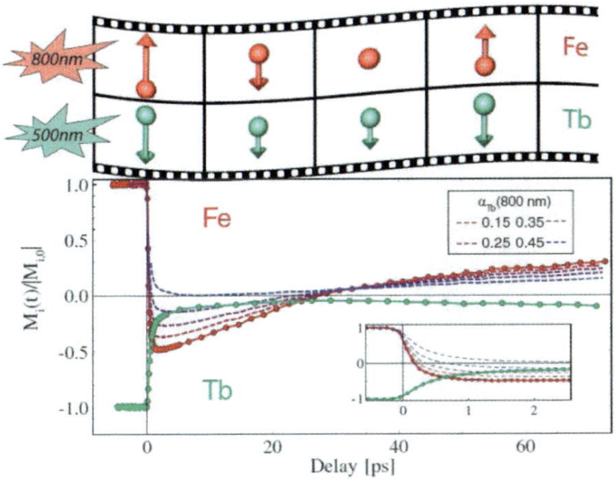


**Fig. 1.**Ultrafast sublattice magnetization dynamics of TbFe at $\lambda_{Fe}$ = 800 nm and $\lambda_{Tb}$ = 500 nm, corresponding to dominantly the Fe and Tb sublattices, respectively. A long lasting ferromagnetic state is excited, with a fluence of $F$ = 8 mJ/cm$^2$[4].

**Acknowledgments** We thank A. van Roij and A. Toonen for technicalsupport. This research has received funding fromStichtingvoorFundamenteelOnderzoek der Materie(FOM), De NederlandseOrganisatievoorWetenschappelijkOnderzoek (NWO), and EC FP7 ITN GrantNo. 214810 (FANTOMAS) and ERC Grant No. 257280 (Femtomagnetism).

## References

[1] C. Stamm et al., “Femtosecond modification of electron localization and transfer of angular momentum in nickel” Nat. Mater. **6**, 740 (2007).
[2] I. Radu et al., “Transient ferromagnetic-like state mediating ultrafast reversal of antiferromagnetically coupled spins” Nature **472**, 205 (2011).
[3] C. La-O-Vorakiat et al., “Ultrafast Demagnetization Dynamics at the M Edges of Magnetic Elements Observed Using a Tabletop High-Harmonic Soft X-Ray Source” Phys. Rev. Lett. **103**, 257402(2009).
[4] A.R. Khorsand et al., “Element-specific probing of ultrafast spin dynamics in multisublattice magnets with visible light” Phys. Rev. Lett. **110**, 107205 (2013).

# Ultrafast non-local spin dynamics in metallic bilayers by linear and non-linear magneto-optics

**A. Melnikov[1], A. Alekhin[1], D. Bürstel[2], D. Diesing[2], T.O. Wehling[3], I. Rungger[4], M. Stamenova[4], S. Sanvito[4], and U. Bovensiepen[5]**

[1]Fritz-Haber-Institut der MPG, Phys. Chemie, Faradayweg 4-6, 14195 Berlin, Germany

[2]Universität Duisburg-Essen, Fakultät für Chemie, Universitätsstr. 5, 45117 Essen, Germany

[3]Universität Bremen, Theor. Physik, Otto-Hahn-Allee 1 (NW1), 28359 Bremen, Germany

[4]School of Physics and CRANN, Trinity College Dublin, Dublin 2, Ireland

[5]Universität Duisburg-Essen, Fakultät für Physik, Lotharstraße 1, 47048 Duisburg, Germany

melnikov@fhi-berlin.mpg.de

**Abstract** We make a step towards the understanding of spin dynamics induced by spin-polarized hot carriers in metals. Exciting the Fe layer of Au/Fe/MgO(001) structures with femtosecond laser pulses, we demonstrate the ultrafast spin transport from Fe into Au using time-resolved MOKE and mSHG for depth-sensitive detection of the transient magnetization.

Ultrafast spin dynamics is the key for development of data storage and spintronics devices. Modern all-optical techniques provide ultimate time resolution of sub-10 fs for the related studies. However, recent findings on laser-induced magnetization dynamics [1] still challenge the understanding of ultrafast magnetism. With this contribution, we make a step towards the understanding of ultrafast spin dynamics in metallic multi-layers, which is (i) spatially non-uniform due to interfaces and strong absorption of the pump pulse and (ii) non-local due to highly mobile hot carriers (HC) that are spin-polarized if excited in a ferromagnet.

The transient magnetization **M** was monitored by the magneto-optical Kerr effect (MOKE) and magneto-induced second harmonic (SH) generation (mSHG) using the 800 nm, 14 fs output of a cavity-dumped Ti:sapphire oscillator. The SH intensity $I^{2\omega}(\mathbf{M},t) \propto |\mathbf{E}_{\mathbf{even}}+\mathbf{E}_{\mathbf{odd}}|^2$, where $t$ is the pump-probe delay, $E_{even}(t)$ is independent of **M** and $E_{\mathrm{odd}}(t) \propto M(t)$. The interface spin dynamics is analyzed by $\Delta_{odd}(t) \approx E_{odd}(t)/E_{odd}(t_0) - 1 \approx M(t)/M_0 - 1$ defined from $I^{2\omega}(\pm M,t)$ and $I^{2\omega}(\pm M_0,t_0)$; $t_0$ represents the absence of excitation [2]. The bulk **M** is probed by $\Delta_{MR,\,ML} \equiv \theta(t)/\theta_0 - 1 \approx M(t)/M_0 - 1$, where θ is the MOKE rotation (MR) or ellipticity (ME). The results obtained in epitaxial Au/Fe/MgO(001) [3] serving as a model system, show pronounced dependence of $\Delta_{MR}$ and $\Delta_{odd}$ on the Fe thickness $d_{Fe}$ (Fig.1 a, b). This is explained by an increase of the Fe/Au interface contribution with reducing

 
J.-Y. Bigot et al. (eds), *Ultrafast Magnetism I*, Springer Proceedings in Physics 159
DOI 10.1007/978-3-319-07743-7_12

$d_{Fe}$ due to absorption (at 800 nm the light penetration depth in Fe $\lambda_{\omega}^{Fe}\approx 20$ nm), which is corroborated by the transient reflectivity and $E_{even}(t)$ (not shown).

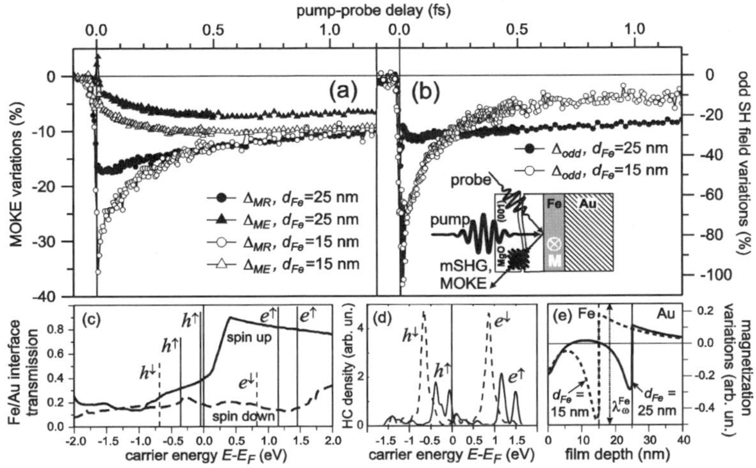


**Fig. 1.** A Pump-induced variations of (a) MOKE rotation and ellipticity, and (b) odd SH field $E_{odd}$ measured in the longitudinal, p-probe and transversal, p-P (inset) geometries, respectively. Experiments were performed for p-polarized pump with the fluence on the order of 1 mJ/cm$^2$ in Au/Fe/MgO(001) bi-layers with layer thicknesses $d_{Au}$=60 nm and $d_{Fe}$=25 and 15 nm, as indicated. (c) Average spin-dependent transmission probability of the Fe/Au interface per available single-electron channel in Fe, calculated from first principles with the Smeagol code [4]. (d) Calculated density of hot carriers [3] excited in Fe by 1.5 eV photons; peak positions are marked in Panel (c) by vertical lines. (e) Variations of the magnetization $\Delta M$ after the excitation for 15 and 25 nm Fe films in contact with bulk Au; the model accounts only for the ballistic transport; calculations are based on the data from Panels (c) and (d), and Table 1.

The complete description of the observed transients requires an advanced model like that one developed for the superdiffusive HC transport in Ni on Al [5]. Here we focus on the fastest (within our 20 fs resolution) timescale of the $\Delta_{MR}$ and $\Delta_{odd}$ break-down (Fig.1 a, b) and model the ballistic HC transport. The related HC parameters are summarized in Table 1 and Fig. 1 c, d. Owing to the largest $\lambda_{Fe}$, smallest $\lambda_{Au}$, and largest transmission of Fe/Au interface (Fig. 1 c), majority electrons $e^{\uparrow}$ dominate the ballistic spin transport. After the ballistic regime has passed at delays $t>10$ fs, i.e. $t>\tau_{Fe}$ (see Table 1), $M$ is reduced in Fe and increased in Au near the Fe/Au interface (Fig. 1 e). To explain the different behavior of $\Delta_{MR}(t)$ and $\Delta_{ME}(t)$ (Fig. 1 a) with this model, we make following assumptions: (i) Owing to the large spin-orbit coupling, the transiently magnetized Au provides a MOKE response. (ii) Since at 1.5 eV photon energy Au has high reflectivity but low absorption (|Re ε|>>|Im ε|), it contributes much stronger to MR (defined by Re ε) than to ME (defined by Im ε). (iii) The Au contribution to MR has opposite sign with respect to the Fe one, which can be explained by a non-resonant character of the Au optical response (in contrast to the resonant case of Fe). This together with $\Delta M(t>0)$ (Fig. 1 e) explains the strong break-down of $\Delta_{MR}$, which increases with reducing $d_{Fe}$. Due to similar reasons the reduction of $M$ at the Fe side of the Fe/Au interface and the build-up of $M$ at the Au side after the excitation (Fig. 1 e) leads to strong break-down of $\Delta_{odd}$ (Fig. 1 b), which increases with reducing $d_{Fe}$. This effect however is more pronounced than the break-down of $\Delta_{MR}$ due to the interface sensitivity of mSHG. The following dynamics is driven by the superdiffusive HC transport [5] and broadens the $\Delta M$ distribution across the Fe film and the Fe/Au interface, which leads to the recovery of $\Delta_{odd}$ and $\Delta_{MR}$ and further reduction of $\Delta_{ME}$.

In conclusion, we have demonstrated the ultrafast spin transport across the (buried) Fe/Au interface and shown that the combination of time-resolved MOKE and mSHG is a powerful tool to study the non-local spin dynamics in metals.

**Table 1.** Calculated [3] ballistic velocities of the excited HC in Fe and Au, $v_{Fe}$ and $v_{Au}$, lifetimes $\tau_{Fe}$ and $\tau_{Au}$ estimated from Refs. [6], and resulting ballistic propagation lengths $\lambda_{Fe}$ and $\lambda_{Au}$.

| HC | $v_{Fe}$ (nm/fs) | $\tau_{Fe}$ (fs) | $\lambda_{Fe}$ (nm) | $v_{Au}$ (nm/fs) | $\tau_{Au}$ (fs) | $\lambda_{Au}$ (nm) |
|---|---|---|---|---|---|---|
| $e^{\uparrow}$ | 0.43 | 8 | 3.4 | 0.95 | 15 | 14 |
| $h^{\uparrow}$ | 0.08 | ~8 | ~0.6 | 0.77 | 200 | 154 |
| $e^{\downarrow}$ | 0.30 | 2 | 0.6 | 1.17 | 40 | 47 |
| $h^{\downarrow}$ | 0.21 | ~2 | ~0.4 | 0.94 | 80 | 75 |

**Acknowledgments** The authors thank A.I. Lichtenstein and P.M. Oppeneer for fruitful discussions. Funding by the DFG (ME 3570/1, Sfb 616) and by the EU 7-th framework program (CRONOS) is gratefully acknowledged.

## References

[1] A. Kirilyuk, A.V. Kimel, and T. Rasing, "Ultrafast optical manipulation of magnetic order" Rev. Mod. Phys. **82**, 2731 (2010).

[2] A. Melnikov, I. Radu, U. Bovensiepen, O. Krupin, K. Starke, E. Matthias, and M. Wolf, "Coherent optical phonons and parametrically coupled magnons induced by femtosecond laser excitation of the Gd(0001) surface" Phys. Rev. Lett. **91**, 227403 (2003).

[3] A. Melnikov, et al., "Ultrafast transport of laser-excited spin polarized carriers in Au/Fe/MgO(001)" Phys. Rev. Lett. **107**, 076601 (2011).

[4] J.M. Soler, et al., "The SIESTA method for *ab initio* order-N materials simulation" J. Phys.: Condens. Matter **14**, 2745 (2002); A.R. Rocha, V.M. Garcia-Suarez, S. Bailey, C. Lambert, J. Ferrer, and S. Sanvito, "Spin and molecular electronics in atomically generated orbital landscapes" Phys. Rev. B **73**, 085414 (2006).

[5] M. Battiato, K. Carva, and P.M. Oppeneer, "Superdiffusive Spin Transport as a Mechanism of Ultrafast Demagnetization" Phys. Rev. Lett. **105**, 027203 (2010).

[6] V.P. Zhukov, E.V. Chulkov, and P.M. Echenique, "Lifetimes and inelastic mean free path of low-energy excited electrons in Fe, Ni, Pt, and Au: Ab initio GW+T calculations" Phys. Rev. B **73**, 125105 (2006).

[7] V.P. Zhukov, E.V. Chulkov, and P.M. Echenique, "Lifetimes of excited electrons in Fe and Ni: first-principles GW and the T-matrix theory" Phys. Rev. Lett. **93**, 096401 (2004); J.J. Quinn, "Range of excited electrons in metals" Phys. Rev. **126**, 1453 (1962).

# Balance of angular momentum and magnetization switching in ferrimagnetic alloys

**Andrei Kirilyuk**

Radboud University Nijmegen, Institute for Molecules and Materials, Heyendaalseweg 135, 6525 AJ Nijmegen, The Netherlands

A.Kirilyuk@science.ru.nl

**Abstract** Here we consider the mechanism responsible for the ultrafast heat-induced switching of the magnetization in ferrimagnetic rare-earth – transition metal films. This switching is found to follow a counterintuitive pathway via a ferromagnetic, which crucially depends on the angular momentum balance of the two opposite sublattices.

Many peculiarities of magnetization dynamics are related to the fact that a certain amount of angular momentum is associated with magnetic moments. In fact, the dynamics of the magnetic moments is described by a mechanical relation between the change of angular momentum and the torque applied to the system, expressed as the Landau-Lifshitz equation. The law of the conservation of angular momentum is therefore an important issue to be considered when sudden changes of the magnetization are applied, such as under the influence of an ultrashort laser pulse [1].

While the basic possibilities for direct laser manipulation of magnetization have been indicated a long time ago, only recently was it possible to apply such control in magnetically ordered materials [2,3].

The question was immediately triggered whether one could use the same mechanism for practical switching of the magnetization in e.g. recording media. A seemingly straightforward answer came very soon afterwards, with a direct demonstration of all-optical light helicity-dependent magnetic recording in thin films of metallic GdFeCo alloys [4]. In spite of the fact that the switching was clearly reproducible and robust, the exact process and mechanism of it remained elusive for a long time. The most obvious explanation via the inverse Faraday effect could only very qualitatively account for the observed features.

Single-shot time-resolved magnetic imaging revealed more details of the process, demonstrating that the switching occurs via a strong non-equilibrium state [5], which could not be described with the usual macrospin-like approach, and required the use of atomistic level spin dynamics calculations. Such calculations, though in a qualitative agreement with most of the observed features [6], failed to explain some essential points quantitatively. For example, the pulse width required

 
J.-Y. Bigot et al. (eds), *Ultrafast Magnetism I*, Springer Proceedings in Physics 159
DOI 10.1007/978-3-319-07743-7_13

to reverse the magnetization was calculated to be at least 250 fs under the most favorable conditions, while experimentally, pulses as short as 40 fs were used. Optical coherence was invoked, though it seemed very unlikely that it would last over 100 fs.

The solution to the problem was found in the multi-sublattice nature of the samples. XMCD experiments showed that the until now used single-component approach is invalid for these ferrimagnetic alloys. By using element resolved femtosecond visible–x-ray pump–probe experiments it has been demonstrated that the RE and TM sublattices may have very different dynamics, to the point that the FeCo sublattice reverses before that of Gd, creating a transient ferromagnetic state [7]. This state was directly reproduced with atomistic simulations, even though all exchange constants were assumed to be unchanged by the laser pulse, and only ultrafast heating was taken into account. Most intriguingly, this transient state was shown to be followed by a reversal of the second sublattice without any additional stimulus or external reference (magnetic field), in that way creating a fully switched magnetic domain [8]. This very unusual process, that actually happens against the very strong exchange interaction that couples the sublattices, was shown to be polarization-independent, thus ruling out the opto-magnetic interaction such as the inverse Faraday effect or anything equivalent, as the driving mechanism. What was it then?

It appeared that the transient state is due to two reasons: (i) considerably different demagnetization times of the RE and TM sublattices, and (ii) angular-momentum conservation and its transfer between the sublattices. First of all, it is the very different magnetic origin of the two sublattices that leads to considerably different demagnetization rates at the time scale of the first few hundreds of femtoseconds [9]. At this time scale, the sublattices are effectively decoupled because the electronic bath temperature exceeds TC by far. Later, when the electrons cool down, the angular-momentum transfer to the hot electron bath reduces, and the exchange coupling becomes effective again, preserving the total angular momentum of the two sublattices. The conservation of the angular momentum thus first pushes the TM sublattice through zero, to help further demagnetization of the RE system, as the latter has not yet achieved a thermal equilibrium with the environment because of its slow demagnetisation. Then, on the longer time scale, the RE sublattice is restored back in the direction against the TM one, to save the exchange energy. This latter process is not angular-momentum conserving. However, it happens much more slowly, on a time scale of several picoseconds, and the conservation is much less strict then. Thus, each laser pulse with an intensity above a certain threshold results in magnetic switching [8].

What is the origin of the helicity dependence in the all-optical switching? A careful study [10] showed that it is the helicity-dependent absorption in the RE–TM magnetic layer, caused by the circular magnetic dichroism, that exactly matches the helicity-dependent intensity window as observed in switching experiments [6]. It happened that also multiple pulse effects, such as observed in the original work [4], could also be explained by the dichroism effects [10]. Thus, the effect is purely thermal and only depends on the total amount of energy absorbed in the sample.

To summarize, though originally this research was triggered by the observation of the inverse Faraday effect in magnetically ordered media [2,3], the all-optical switching of the magnetization in RE–TM alloys was found to be due to a completely different effect: a combination of ultrafast laser-induced demagnetization with the angular-momentum conservation on a sub-picosecond time scale. In this new mechanism, the helicity happened to play a secondary role, via the dichroism-induced difference in absorption. Therefore, the research leading to this conclusion provided us with invaluable information about the behavior of magnets away from their thermodynamic equilibrium.

**Acknowledgments** This research has received funding from Nederlandse Organisatie voor Wetenschappelijk Onderzoek (NWO), Stichting voor Fundamenteel Onderzoek der Materie (FOM) and EC FP7 contributions under grants NMP3-SL-2008-214469 (UltraMagnetron), N 214810 (FANTOMAS), NMP-2011-281043 (FEMTOSPIN) and NMP3-LA-2010-246102 (IFOX).

## References

[1] A. Kirilyuk, A.V. Kimel, and Th. Rasing, "Laser-induced magnetization dynamics and reversal in ferrimagnetic alloys" Rep. Prog. Phys. **76**, 026501 (2013).

[2] A.V. Kimel, A. Kirilyuk, P.A. Usachev, R.V. Pisarev, A.M. Balbashov, and Th.Rasing, "Ultrafast nonthermal control of magnetization by instantaneous photomagnetic pulses" Nature **435** 655 (2005).

[3] F. Hansteen, A.V. Kimel, A. Kirilyuk and Th. Rasing, "Femtosecond photomagnetic switching of spins in ferrimagnetic garnet films" Phys. Rev. Lett. **95** 047402 (2005).

[4] C.D. Stanciu, F. Hansteen, A.V. Kimel, A. Tsukamoto, A. Itoh, A. Kirilyuk and Th. Rasing "All optical magnetic recording with circularly polarized light" Phys. Rev. Lett. **99** 047601 (2007).

[5] K. Vahaplar, A.M. Kalashnikova, A.V. Kimel, D. Hinzke, U. Nowak, R.W. Chantrell, A. Tsukamoto, A. Itoh, A. Kirilyuk and Th. Rasing "Ultrafast path for optical magnetization reversal via a strongly non-equilibrium state" Phys. Rev. Lett. **103**, 117201 (2009).

[6] K. Vahaplar et al., "All-optical magnetization reversal by circularly polarized laser pulses: Experiment and multiscale modeling" Phys. Rev. B **85**, 104402 (2012).

[7] I. Radu et al., "Transient ferromagnetic-like state mediating ultrafast reversal of antiferromagnetically coupled spins" Nature **472**, 205 (2011).

[8] T.A. Ostler et al., "Ultrafast heating as a sufficient stimulus for magnetization reversal in a ferrimagnet" Nature Commun. **3**, 666 (2012).

[9] J.H. Mentink, J. Hellsvik, D.V. Afanasiev, B.A. Ivanov, A. Kirilyuk, A.V. Kimel, O. Eriksson, M.I. Katsnelson and Th. Rasing "Ultrafast spin dynamics in multisublattice magnets" Phys. Rev. Lett. **108**, 057202 (2012).

[10] A.R. Khorsand, M. Savoini, A. Kirilyuk, A.V. Kimel, A. Tsukamoto, A. Itoh and Th. Rasing "Role of magnetic circular dichroism in all-optical magnetic recording" Phys. Rev. Lett. **108**, 127205 (2012).

# Disentangling spin and charge dynamics with magneto-optics

**E. Carpene[1], F. Boschini[2], H. Hedayat[2], C. Piovera[2], C. Dallera[2], E. Puppin[3], M. Mansurova[4], M. Münzenberg[4], X. Zhang[5], A. Gupta[5]**

1 IFN-CNR, Dipartimento di Fisica, Politecnico di Milano, 20133 Milano, Italy

2 Dipartimento di Fisica, Politecnico di Milano, 20133 Milano, Italy

3 CNISM, Dipartimento di Fisica, Politecnico di Milano, 20133 Milano, Italy

4 I. Physikalisches Institut, Georg-August-Universität Göttingen, 37077 Göttingen, Germany

5 Department of Chemistry, University of Alabama, Tuscaloosa, Alabama 35487, USA

ettore.carpene@polimi.it

**Abstract** The magneto-optical response of Fe and CrO2 epitaxial films has been investigated by pump-probe polarimetry, showing that charge and spin dynamics can be unambiguously disentangled. Both diagonal and off-diagonal terms of the dielectric tensor can be retrieved, providing the complete dynamical characterization of magnetic and optical properties in a ferromagnet.

## Introduction

Time-resolved magneto-optical Kerr effect (TR-MOKE) is a well-established technique to investigate the spin dynamics in ferromagnetic layers. Thanks to the pump-probe method, the technique can easily achieve time resolution of a few tens of femtosecond [1], making it a unique tool to explore ultrafast spin dynamics. The magnetic information is extracted from the so-called Kerr angle $\Theta = \theta + i\varepsilon$, a complex quantity that incorporates the light polarization rotation, $\theta$, and ellipticity, $\varepsilon$, both related to the magnetic state of the sample. A long-lasting controversy is whether optical contributions alter the measurement and to what extent the technique is reliable in deriving the true spin dynamics in ferromagnets [2-4]. Here, we present a detailed analysis of the TR-MOKE signals on two representative cases of ferromagnetic samples showing that genuine magnetic information can be disentangled from non-magnetic one.

 
J.-Y. Bigot et al. (eds), *Ultrafast Magnetism I*, Springer Proceedings in Physics 159
DOI 10.1007/978-3-319-07743-7_14

## Results and Discussion

We have investigated the optically-induced spin dynamics in Fe(001) and CrO2 (100) epitaxial layers using films with thickness (100 nm and 300 nm, respectively) larger than the optical penetration depth in order to avoid artifacts due to light reflection at the film-substrate interface. Transient reflectivity, Kerr ellipticity and rotation have been measured for p and s polarizations of the probing beam using an amplified Ti:sapphire laser system and the standard pump-probe method. Fig. 1 reports the magneto-optical measurements on CrO2: panel (a) shows the time-resolved Kerr rotations ($\theta_s$ and $\theta_p$) and ellipticities ($\varepsilon_s$ and $\varepsilon_p$) as a function of pump-probe delay; panel (b) shows the normalized Kerr angles ($|\Theta_s|$ and $|\Theta_p|$). It should be noticed how the dynamics of rotation and ellipticity differ from each other (especially at zero delay), raising a fundamental question: which one truly describe the spin dynamics? On the other hand, the normalized moduli ($|\Theta_s|$ and $|\Theta_p|$) are very similar to each other (it can be proven that the ratio $|\Theta_s|/|\Theta_p|=\sqrt{R_p/R_s}$, with $R_{p,s}$ being the transient reflectivities and $R_p/R_s$<1.06 in our experiment, see Ref. [5]).

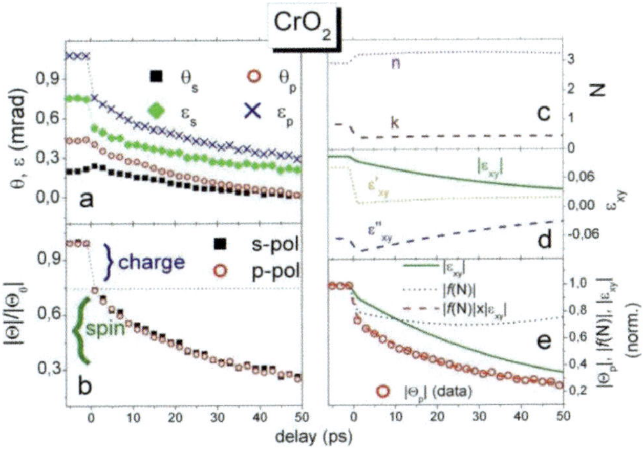


**Fig. 1.** A (a): time-resolved Kerr rotations ($\theta_s$ and $\theta_p$) and ellipticities ($\varepsilon_s$ and $\varepsilon_p$) of $CrO_2$ vs. pump-probe delay. (b): normalized Kerr angle ($|\Theta_s|$ and $|\Theta_p|$). (c) and (d): dynamics of refractive index $N = n + ik$ and off-diagonal parts $\varepsilon_{xy} = \varepsilon'_{xy} + i\varepsilon''_{xy}$ of the dielectric tensor. (e): comparison between extracted and measured $|\Theta p|$.

According to the analytical expressions [5,6] the Kerr angles $\Theta_{s,p}$ can be written as the product $f(N)_{s,p} \times \varepsilon_{xy}$ and consequently $|\Theta_{s,p}|=|f(N)_{s,p}| \times |\varepsilon_{xy}|$ ($\varepsilon_{xy}$ is the off-diagonal element of the dielectric tensor carrying all the magnetic information, $f(N)$ is a complex function of the refractive index N which reflects the charge dynamics). This expression clearly reveals that the Kerr angle might be affected by charge evolution, i.e. $f(N)$. In fact, the steep change of $|\Theta s,p|$ at zero delay is attributed to the dynamics of photo-excited hot carriers, while the subsequent dynamics is determined by the spin evolution. In order to provide a quantitative proof, we have used the four measured rotations and ellipticities curves shown in Fig. 1(a) to retrieve the dynamics of the refractive index and off-diagonal elements of the dielectric tensor ($N = n+ik$ and $\varepsilon_{xy} = \varepsilon'_{xy} + i\varepsilon''_{xy}$, respectively). By employing a numerical procedure to invert the analytical formulas connecting $\theta_s$, $\theta_p$, $\varepsilon_s$, $\varepsilon_p$ to $N$ and $\varepsilon_{xy}$ we have unambiguously deduced the four quantities $n$, $k$, $\varepsilon'_{xy}$ and $\varepsilon''_{xy}$. The results are reported in Fig. 1 (c)-(d). Both $n$ and $k$ show a steep change across zero pump-probe delay and a rather flat temporal evolution afterward [7]. A similar behavior is expected for $f(N)$. The off-diagonal dielectric element displays a prompt variation with the pump arrival, but its modulus $|\varepsilon_{xy}|$ smoothly changes at zero pump-probe delay and therefore it is not significantly affected by charge carriers. Thus, its dynamics is dominated by the spins (i.e. the magnetization). Fig. 1 (e) reports the evolution of $|f(N)|$ and $|\varepsilon_{xy}|$ as extracted from inversion procedure. Notice the steep change of $|f(N)|$ at zero delay and the flat dynamics afterwards. Besides, the quantity $|f(N)|\times|\varepsilon_{xy}|$ excellently match the experimental Kerr angle $|\Theta|$, providing proof that the prompt variation of the Kerr angle at zero pump-probe delay is determined by charge dynamics (i.e. $|f(N)|$), while the subsequent evolution genuinely represents the magnetization.

## Conclusions

In conclusion, we have proven that only a detailed magneto-optical analysis clarifies to what extent the TR-MOKE technique can reliably reveal spin dynamics. Our experiment shows that exhaustive magneto-optical information is accomplished only by measuring both real and imaginary parts of the complex Kerr signal. This allows one to retrieve the complete dynamical characterization of spin and charges in a ferromagnet, unambiguously disentangling their evolutions after an intense optical excitation.

**Acknowledgments** Fondazione Cariplo is gratefully acknowledged for financial support.

## References

[1] E. Beaurepaire, J.-C. Merle, A. Daunois, and J.-Y. Bigot, "Ultrafast Spin Dynamics in Ferromagnetic Nickel" Phys. Rev. Lett. **76**, 4250 (1996).
[2] B. Koopmans, M. van Kampen, J. T. Kohlhepp, and W. J. M. de Jonge, "Ultrafast Magneto-Optics in Nickel: Magnetism or Optics?" Phys. Rev. Lett. **85**, 844 (2000).
[3] L. Guidoni, E. Beaurepaire, and J.-Y. Bigot, "Magneto-optics in the Ultrafast Regime: Thermalization of Spin Populations in Ferromagnetic Films" Phys. Rev. Lett. **89**, 017401 (2002).

[4] G. P. Zhang, W. Hübner, Georgios Lefkidis Yihua Bai & T. F. George, "Paradigm of the time-resolved magneto-optical Kerr effect for femtosecond magnetism" Nat. Phys. **5**, 499 (2009).
[5] E. Carpene, F. Boschini, H. Hedayat, C. Piovera, C. Dallera, E. Puppin, M. Mansurova, M. Münzenberg, X. Zhang, and A. Gupta "Measurement of the magneto-optical response of Fe and CrO2 epitaxial films by pump-probe spectroscopy: Evidence for spin-charge separation" Phys. Rev. B **87**, 174437 (2013).
[6] S. Visnovsky, "Magneto-optical, longitudinal and transversal Kerr birefringence effects in orthorhombic crystals" Czech. J. Phys. B **34**, 969 (1984).
[7] From the dynamics of N, recalling that $N = \sqrt{\varepsilon_{xx}}$ , the time evolution of the diagonal dielectric element $\varepsilon_{xx}$ can be easily obtained.

# Laser-Induced Spin Dynamics in amorphous NdFeCo

**J. Becker[1], I. Razdolski[1], A. Tsukamoto[2], A. Itoh[2], A. Kirilyuk[1], A.V. Kimel[1], and Th. Rasing[1]**

[1]Radboud University, IMM, Nijmegen 6525 ED, The Netherlands

[2]Department of Electronics and Computer Science, Nihon University, 7-24-1 Narashino-dai Funabashi, Chiba 274-8501, Japan

j.becker@science.ru.nl

**Abstract** We studied magnetization dynamics in the amorphous light rare earth-transition metal compound NdFeCo. We report a two step demagnetization process. Furthermore, an unusual overshooting behaviour is observed that can be linked to the particular magnetic structure of the sample.

## Introduction

Rare earth-transition metal (RE-TM) alloys provide a wide playground for investigating magnetization dynamics at ultrafast timescales. Recently, a number of fascinating results have been reported for GdFeCo alloys with an anti-parallel alignment of the two spin subsystems [1]. However, current research is mainly focused on ferrimagnetic heavy rare RE-TM compounds. From a fundamental point of view, magnetization dynamics in the light RE-TM compound NdFeCo is of interest because of an unusually complex configuration of the magnetic moments of the Nd and FeCo magnetic sublattices.

Although the spins of both TM elements and Nd couple antiferromagnetically, the system is ferromagnetic due to the large orbital moment of Nd that dominates the RE magnetization [2].

## Two-Step Demagnetization

$Nd_{30}Fe_{60}Co_{10}$ amorphous thin films were studied by means of a time-resolved pump-probe technique exploiting the magneto optical Kerr effect. Magnetization dynamics were investigated as a function of temperature, pump fluence and applied magnetic field.

 
J.-Y. Bigot et al. (eds), *Ultrafast Magnetism I*, Springer Proceedings in Physics 159
DOI 10.1007/978-3-319-07743-7_15

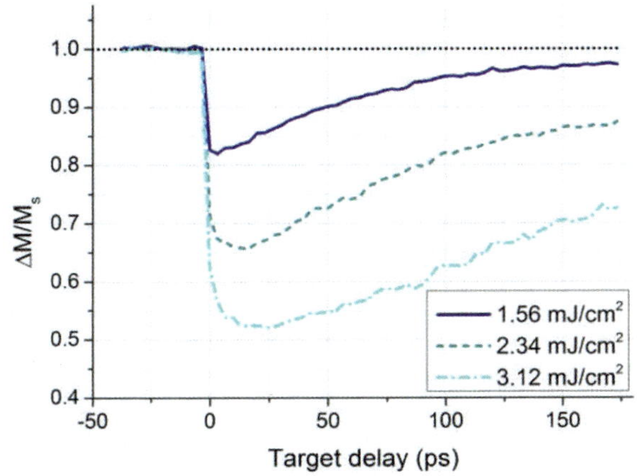


**Fig. 1.** Laser induced dynamics of the magneto-optical signal measured at different fluences.

At room temperature we observe two demagnetization regimes, an ultrafast demagnetization (< 1ps) followed by a slower demagnetization (~ 30 ps) of lower amplitude (fig.1). This second demagnetization step takes up to tens of picoseconds depending on the sample temperature and the pump fluence. Increasing temperature or pump fluence will increase the demagnetization time of this slower process whereas it completely disappears for low temperatures and low fluences.

Similar behavior has been observed in pure TM at high fluences by Koopmans et al. [3] and in RE-TM such as TbFe [4] GdFeCo [5]. Mekonnen et al. showed that within a 4-temperature model this behaviour can be explained by the contribution of the RE magnetic sublattice which is known to display slower magnetization dynamics than the TM [6].

## Overshooting Signal

Another interesting signature appears for certain measurement conditions (fig.2). After the ultrafast demagnetization, the magnetization signal sharply increases up to a value above the initial saturation magnetization. It then slowly relaxes back to the initial value. This overshooting behavior strongly depends on applied field, pump fluence and temperature. Increasing the field will decrease the overshooting amplitude and eventually suppress it. Increasing temperature or fluence will shift the maximum towards longer delay times and eventually also suppress the overshooting.

Only measurements at fields comparable to the anisotropy field - which increases with decreasing temperature [7] - seem to show this overshooting behavior, hinting towards a link between the two. Interestingly, the anisotropy in this amorphous sample has a profound effect on the magnetic structure. Random local ion anisotropy prevents a completely co-linear alignment of the magnetic moments of the rare-earth and transition metal elements [7]. Both magnetic sublattices show a fan-like distribution of magnetic moments, the FeCo sublattice having a much narrower cone opening than the Nd sublattice.

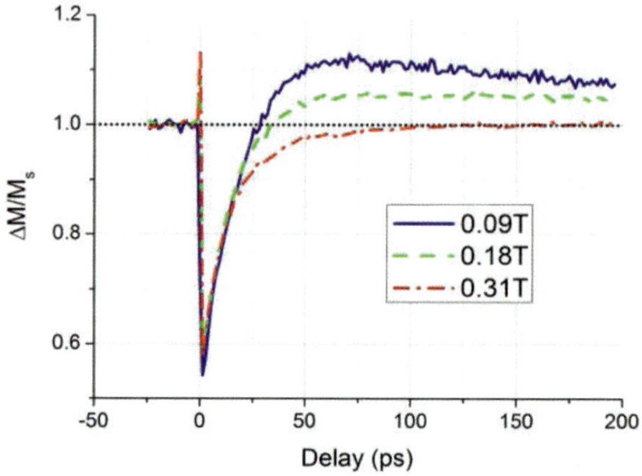


**Fig. 2.** Laser induced dynamics of the magneto-optical signal at different magnetic fields.

**Acknowledgments** This research was partially supported by the European Research Council ERC Grant Agreement No. 257280 (Femtomagnetism), the European Union's Seventh Framework Program (FP7/2007-2013) FEMTOSPIN, the Nederlandse Organisatie voor Wetenschappelijk Onderzoek (NWO) and the Stichting voor Fundamenteel Onderzoek der Materie (FOM).

## References

[1] I. Radu, K. Vahaplar, C. Stamm, T. Kachel, N. Pontius, H. A. Dürr, T. A. Ostler, J. Backer, R. F. Evans, R.W. Chantrell, A. Tsukamoto, A. Itoh, A. Kirilyuk, Th. Rasing, A. V. Kimel, "Transient ferromagnetic-like state mediating ultrafast reversal of antiferromagnetically coupled spins" Nature **472**, 205 (2011).

[2] R.J. Gambino, T.R. McGuire, "Magneto-optic properties of Nd-Fe-Co amorphous alloys" J. Appl. Phys. **57**, 3906 (1985).

[3] B. Koopmans, G. Malinowski, F. Dalla Longa, D. Steiauf, M. Fahnle, T. Roth, M. Cinchetti, M. Aeschlimann, "Explaining the paradoxical diversity of ultrafast laser-induced demagnetization" Nature Mater. **9**, 259-265 ( 2010).

[4] J.-W. Kim, K.-D. Lee, J.-W. Jeong, S.-C. Shin, "Ultrafast spin demagnetization by nonthermal electrons of tbfe alloy film" Appl. Phys. Lett. **94**, 192506 (2009).

[5] A. Mekonnen, A. R. Khorsand, M. Cormier, A. V. Kimel, A. Kirilyuk, A. Hrabec, L. Ranno, A. Tsukamoto, A. Itoh, Th. Rasing, "Role of the inter-sublattice exchange coupling in short-laser-pulse-induced demagnetization dynamics of GdCo and GdCoFe alloys" Phys. Rev. B **87**, 180406(R) (2013).

[6] A. Vaterlaus, T. Beutler, F. Meier, "Spin-lattice relaxation time of ferromagnetic gadolinium determined with time-resolved spin-polarized photoemission" Phys. Rev. Lett. **67**, 3314-3317 (1991).

[7] R. C. Taylor, T. R. McGuire, J. M. D. Coey, A. Gangulee "Magnetic properties of amorphous neodymium-transition-metal films" J. Appl. Phys. **49**, 2885-2893 (1978).

# Probing ultrafast spin moment change of bcc iron in crystal-momentum space: A proposal

**M.S. Si[1,2], J.Y. Li[1], D.S. Xue[1], and G.P. Zhang[2]**

[1]Key Laboratory for Magnetism and Magnetic Material of the Ministry of Education, Lanzhou University, Lanzhou 730000, China

[2]Department of Physics, Indiana State University, Terre Haute, Indiana 47809, USA

gpzhang@indstate.edu

**Abstract** A new method is proposed to study the ultrafast spin change in the crystal-momentum space. Our first-principles results reveal the hot spin spots in the first Brillouin zone of bcc Fe. This can be tested experimentally.

## Introduction

Since ultrafast quenching magnetization induced by a femtosecond laser pulse was first reported in 1996 [1], lots of studies [2-4] have been devoted to its underlying magnetization mechanism. However, up to now no consensus has been reached. One of the main reasons is that such ultrafast process involves various interactions among photons, phonons and magnons. It is convenient to describe these excitations in the momentum space. Therefore, an access to the dynamics of crystal-momentum-dependent ultrafast spin change will provide new information about the origin of femtosecond magnetism (femtomagnetism) at the microscopic level.

Very recently, a study of spin-orbit hybridization points in fcc Co has been reported [5], where a magnetic linear dichroism (MLD) combined with two-photon photoemission (2PPE) was used to measure those points along a few selected crystal moment lines. This is a major step in femtomagnetism as it demonstrates a potential means to detect the spin-flip scattering events. However, these hybridization points do not reflect the spin moment change in a sample. This is because experiments measure the spin polarization of those ejected electrons not those electrons in the sample, while one is interested in the spin moment change of those electrons left behind. Thus, we suggested a new method, i.e., optical spin generator, to detect the spin moment change in the momentum space, where those spin moment changes are identified and thus labeled as hot spin spots [6].

 
J.-Y. Bigot et al. (eds), *Ultrafast Magnetism I*, Springer Proceedings in Physics 159
DOI 10.1007/978-3-319-07743-7_16

## Theoretical algorithm and results

Our theory is based on the density functional theory and the dynamical Liouville equation. While the detailed computational descriptions can be found elsewhere [7], here in brief we first self-consistently solve the Kohn-Sham equation to find eigenstates of a system, and then numerically solve the Liouville equation. For bcc Fe, the initial spin moment is 2.173 $\mu_B$ under the optimized lattice constant of 2.834 Å [8]. We compute the crystal-momentum-resolved spin moment change, where the hot spin spots are identified in several planes within the first Brillouin zone (BZ).

Our proposed demagnetization mechanism is schematically shown in the inset of Fig. 1(a). As a femtosecond laser pulse irradiates upon bcc Fe, the demagnetization starts (see Fig. 1(a)). Fig. 1(a) shows that the spin moment first decreases sharply and then reaches its minimum value at ~24 fs. This relaxation time is larger than that of ~11 fs in fcc Ni [7]. After the maximum demagnetization, the spin moment recovers slightly and oscillates around an equilibrium value. This is a typical feature of spin dynamics on a femtosecond timescale. However, much information about the magnetization is missing. In order to obtain more insight into its physical origin, we disperse the ultrafast spin moment change in the momentum space, as below.

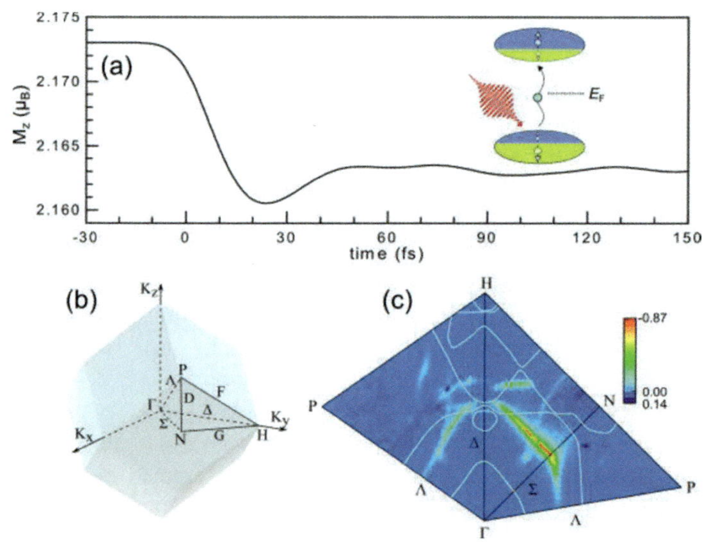


**Fig. 1.** (a) Time evolution of the spin moment for bcc Fe. The laser's parameters are taken as follows: laser amplitude $A_0 = 0.05$ V/Å, laser pulse duration $\tau = 12$ fs, and optimum photon energy $\hbar\omega = 2.01$ eV. Inset: our proposed mechanism of demagnetization in femtomagnetism. (b) The first BZ for the bcc lattice, where high symmetry points and lines are displayed. (c) Time-averaged spin change dispersed in the Γ-H-P, Γ-H-N, and Γ-P-N planes. The white curves denote the Fermi surfaces.

Crystal-momentum-dependent ultrafast spin moment change $\Delta M_{\mathbf{k}}$ is defined as

$$\Delta M_{\mathbf{k}} = M_{\mathbf{k}}(t) - M_{\mathbf{k}}(-\infty)$$

where $M_{\mathbf{k}}(t)$ is the time-averaged spin moment after 90 fs, $M_{\mathbf{k}}(-\infty)$ is the initial spin moment without the influence of laser pulse. To compare our results with the technique of angular-resolved PE [5], we disperse the ultrafast spin moment change in crystal-momentum space. Figure 1(c) shows the crystal-momentum dispersion of ultrafast spin moment change, where the high symmetry points and lines within first BZ of bcc crystal structure as schematically depicted in Fig. 1(b). A few features are interesting. The hot spin spots induced by the laser are localized in the momentum space, only in the Γ-H-P, Γ-H-N, and Γ-P-N planes. We also find the hot spin spots across the Λ, Δ, and Σ lines. An experimental verification of our findings is needed.

## Conclusions

We propose to detect the ultrafast spin change in the crystal momentum space. Our first-principles calculation demonstrates several striking feature of the spin moment change. We find that the spin change is localized in some special planes in the BZ. This can be directly tested experimentally.

**Acknowledgments** This work was supported by the National Basic Research Program of China under No. 2012CB933101 (DSX) and the U.S. Department of Energy under Contract No. DE-FG0206ER46304 (GPZ).

## References

[1] E. Beaurepaire, J.-C. Merle, A. Daunois, and J.-Y. Bigot, "Ultrafast spin dynamics in ferromagnetic nickel" Phys. Rev. Lett. **76**, 4250 (1996).
[2] G. P. Zhang and W. Hübner, "Laser-induced ultrafast demagnetization in ferromagnetic metals" Phys. Rev. Lett. **85**, 3025 (2000).
[3] M. S. Si and G. P. Zhang, "Resolving photon-shortage mystery in femtosecond magnetism" J. Phys.: Condens. Matter **22**, 076005 (2010).
[4] A. Kirilyuk, A. V. Kimel, and T. Rasing, "Ultrafast optical manipulation of magnetic order" Rev. Mol. Phys. **82**, 2731 (2010).
[5] M. Pickel, A. B. Schmidt, F. Giesen, J. Braun, J. Minár, H. Ebert, M. Donath, and M. Weinelt, "Spin-orbit hybridization points in the faced-centered cubic cobalt band structure" Phys. Rev. Lett. **101**, 066402 (2008).
[6] M. S. Si and G. P. Zhang, "Hot spin spots in the laser-induced demagnetization" AIP Advances **2**, 012158 (2012).
[7] G. P. Zhang, Y. Bai, and T. F. George, "Energy- and crystal momentum-resolved study of laser-induced femtosecond magnetism" Phys. Rev. B **80**, 214415 (2009).
[8] Neil W. Ashcroft and N. David Mermin, "Solid state physics", (Rinehart and Winston: Holt 1976), Chap. 4, pp 70.

# Angular dependence of Gilbert damping in ferromagnetic metallic systems

**E. Barati[1], M. Cinal[1], D.M. Edwards[2], and A. Umerski[3]**

[1]Institute of Physical Chemistry of the Polish Academy of Sciences, Warsaw 01-224, Poland

[2]Department of Mathematics, Imperial College, London SW7 2BZ, UK

[3]Department of Mathematics, Open University, Milton Keynes MK7 6AA, UK

ebaratie@gmail.com

**Abstract** The Gilbert damping constant, present in the phenomenological Landau-Lifshitz-Gilbert equation, is calculated within a realistic tight-binding model, for arbitrary magnetization orientation in ferromagnetic systems. The dependence of the Gilbert damping constant on magnetization direction is investigated for bulk Fe, Co and Ni as well as Co films.

## Introduction

Magnetic relaxation in ferromagnetic metallic systems is governed by the Gilbert damping which enters the phenomenological Landau-Lifshitz-Gilbert (LLG) equation [1] describing the dynamics of magnetization in such systems. Thus, Gilbert damping is of crucial importance in understanding spin-dependent transport in magnetic nanostructures as well as their application in spintronic devices. Such damping is also is the mechanism through which energy dissipates in magnetic systems.

In our previous work [2], the Gilbert damping constant has been calculated for bulk cubic ferromagnetic metals with magnetization along [001] axis as well as for (001) fcc Co films with magnetization perpendicular to the surface. Therein, it has been shown that the damping constants obtained for bulk ferromagnets are in very good agreement with ab initio calculations. The damping constant has been found to be considerably enhanced for Co films of few-monolayer (ML) thickness. A further remarkable enhancement is obtained as a result of covering Co films with a Pd cap in Co/Pd bilayers.

In this short report, we employ a realistic nine-band tight-binding model [2] to calculate the Gilbert damping constant for an arbitrary magnetization direction in ferromagnetic metallic systems. The dependence of the Gilbert damping constant on magnetization orientation is investigated for bulk Fe, Co and Ni as well as (001) fcc Co films.

 
J.-Y. Bigot et al. (eds), *Ultrafast Magnetism I*, Springer Proceedings in Physics 159
DOI 10.1007/978-3-319-07743-7_17

## Theory

The Gilbert damping constant for the magnetization **M** oriented along the z axis is given by the Kamberský formula [2,3] including the torque $B_{-}=[S_{-}, H_{SO}]$. The spin-orbit (SO) interaction $H_{SO} = L_zS_z+1/2(L_+S_-+L_-S_+)$ is represented by the operators of orbital and spin angular momenta ( $L_z$ , $L_\pm = L_x \pm iL_y$ and $S_z$ , $S_\pm = S_x \pm iS_y$ ) defined with the fixed *x*, *y*, *z* axes chosen along the [100], [010], [001] axes. For arbitrary direction of magnetization **M**, described by the polar and azimuthal angles, θ, ϕ, electronic states are expressed in the rotated spin state basis $|\uparrow\rangle_r$ and $|\downarrow\rangle_r$ and the operator $B_-$ is replaced with $B'_- = [S'_-, H_{SO}]$ where $S'_- = S_x - IS_y$ (in the rotated frame of reference $Ox'y'z'$ with $z'$ parallel to **M** ). The Gilbert damping constant is then calculated with the matrices of $H_{SO}$ and $B'_-$ in the rotated basis expanded [4] as $|\uparrow\rangle_r = e^{-i\phi/2}\cos(\theta/2)|\uparrow\rangle + e^{i\phi/2}\sin(\theta/2)|\downarrow\rangle$ and $|\downarrow\rangle_r = e^{-i\phi/2}\sin(\theta/2)|\uparrow\rangle + e^{i\phi/2}\cos(\theta/2)|\downarrow\rangle$ in the unrotated basis $|\uparrow\rangle$ and $|\downarrow\rangle$.

## Results: bulk ferromagnetic metals, Co films

The calculated damping constant α for bulk cubic ferromagnets is the same for **M** along different principal axes. Figure 1 shows the dependence of $\alpha$ on the polar angle θ, in the *xz* plane (ϕ = 0), for bulk Fe, Co and Ni. It is symmetric with respect to θ = 45°, as expected. At θ = 45°, we find a maximum damping for Fe and minimum damping for Co. For bulk Ni such minimum is present for electron scattering rates $\Gamma \leq 0.05$ eV and it is replaced by a weakly pronounced maximum for larger Γ. The change of α with the magnetization direction does not exceed 13%, 3%, and 18% for Fe, Co and Ni, respectively, with Γ = 0.01 eV. The damping constant is less sensitive to the magnetization direction for larger Γ.

In Fig. 2(a) we plot the Gilbert damping constant $\alpha$ for (001) fcc Co films with different magnetization directions and film thicknesses. The change of the damping constant due to the magnetization re-orientation from out-of-plane to in-plane is over threefold for the Co monolayer and decays for thicker films. The difference diminishes for larger Co thicknesses as α has the same value for θ = 0° and θ = 90° in bulk Co (cf. Fig. 1).

The dependence of the Gilbert damping on the direction of magnetization is much stronger in thin films than in bulk metals. The damping constant for Γ = 0.01 eV has the maximum at θ close to 45° where its value increases considerably (by nearly 60% for 10-ML Co film) in comparison with in-plane and out-ofplane orientations; Fig. 2(b). Such maximum disappears for $\Gamma \geq 0.1$ eV.

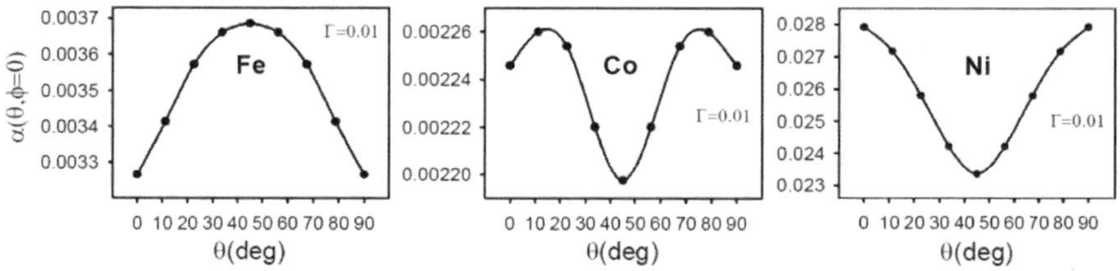


**Fig. 1.** A Gilbert damping constant for bulk bcc Fe, fcc Co, fcc Ni vs polar angle θ, obtained with the scattering rate $\Gamma = 0.01$ eV.

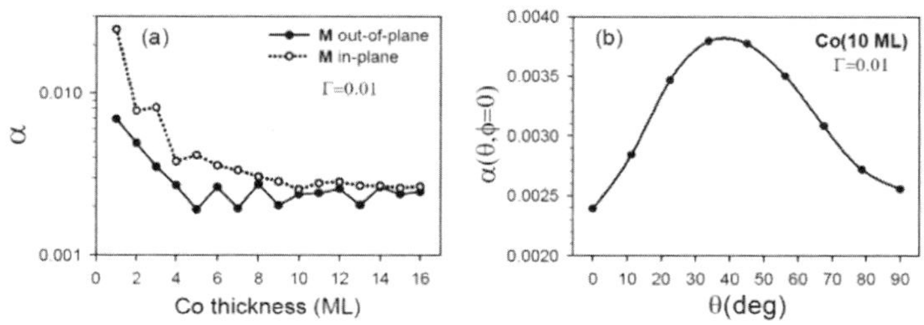


**Fig. 2.** A Gilbert damping constant vs (a) Co film thickness for in-plane and perpendicular magnetization, (b) polar angle θ for Co(10 ML); the scattering rate $\Gamma = 0.01$ eV.

In conclusion, the Gilbert damping constant is found to depend on the magnetization orientation, weakly for bulk ferromagnets but much strongly for ultrathin Co films. The damping is particularly strongly enhanced in Co films with magnetization orientation intermediate between in-plane and out-of-plane directions.

**Acknowledgments** a Two of us (E.B. and M.C.) acknowledge the financial support of the Foundation for Polish Science within the International PhD Projects Programme, cofinanced from European Regional Development Fund within Innovative Economy Operational Programme "Grants for innovation".

## References

[1] D. M. Edwards and O. Wessely, "The quantum-mechanical basis of an extended Landau–Lifshitz–Gilbert equation for a current-carrying ferromagnetic wire" J. Phys.: Condens. Matter **21**, 146002 (2009).

[2] E. Barati, M. Cinal, D. M. Edwards and A. Umerski, "Calculation of Gilbert damping in ferromagnetic films" EPJ Web of Conferences **40**, 18003 (2013).

[3] V. Kamberský, "On ferromagnetic resonance damping in metals" Czech. J. Phys. B **26**, 1366 (1976).

[4] Ch. Li, A.J. Freeman, H. J. F. Jansen, and C. L. Fu, "Magnetic anisotropy in low-dimensional ferromagnetic systems: Fe monolayers on Ag(001), Au(001), and Pd(001) substrates" Phys. Rev. B **42**, 5433 (1990).

# Novel dual-colour architecture for ultrafast spin dynamics measurements in the sub-10 fs regime

**C.S. Gonçalves, A.S. Silva, M. Miranda, F. Silva, P. Oliveira, H. Crespo, and D.S. Schmool**

Department of Physics and Astronomy and IFIMUP-IN, Faculdade de Ciências, Universidade do Porto, Rua Campo Alegre 687, 4169-007 Porto, Portugal.

dschmool@fc.up.pt

**Abstract** We have designed and built an ultrafast pump-probe apparatus based on a dual-color scheme for the measurement of spin dynamics, via the MOKE signal, with pulse durations in both pump and probe lines of <10 fs. Such temporal resolution is obtained using hollow-fiber and chirped mirror pulse compression techniques.

## Introduction

Ultrafast spectroscopies based on the pump and probe methodology are a general tool that can be applied to the study of many ultrafast physical phenomena in materials. Such techniques utilize pulsed lasers, whose temporal width defines the ultimate resolution of measurement. By splitting the optical pulse into two components, a controlled time delay can be introduced between the arrival times of the two parts, at a sample, using a delay stage. As we are concerned with the study of spin dynamics in magnetic materials, we use the measurement of the magneto-optic Kerr effect (MOKE), which allows us to optically probe the state of magnetisation of the sample.

## Dual-colour pump-probe apparatus for magnetization dynamics

Our pump-probe system (Fig. 1) is based on a commercial Ti: Sapphire laser amplifier (Femtolasers Compact Pro CE Phase) delivering sub-30-fs laser pulses (~40 nm bandwidth centred at 800 nm) with 1 mJ of energy at a repetition rate of 1 kHz. These pulses are further post-compressed in a home-built state-of-the-art hollow-fibre chirped-mirror compressor that uses the dispersion-scan (d-scan) technique for the simultaneous measurement of the pulses [1, 2]. This compressor consistently delivers sub-4-fs pulses with 200-300 μJ energy that can be directly

 
J.-Y. Bigot et al. (eds), *Ultrafast Magnetism I*, Springer Proceedings in Physics 159
DOI 10.1007/978-3-319-07743-7_18

carrier-envelope phase stabilized with the same d-scan setup [3]; its output is extremely broadband, spanning from 450 to 1050 nm, with a Fourier-limit of <3 fs. The broadband pulses are sent through an additional chirped mirror compressor covering the bandwidth 500-950 nm and capable of compensating, up to high-order, the material dispersion associated with the air paths and all subsequent optical components (filters, lenses, waveplates and polarisers) traversed by the pulses up to the sample stage. A dichroic mirror is used to produce the pump (reflected) and probe (transmitted) pulses.

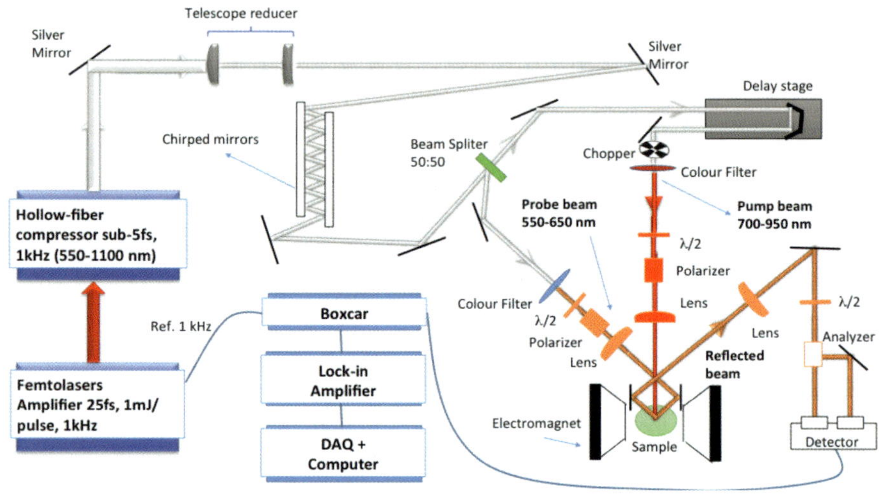


**Fig. 1.** Schematic illustration of our dual-colour pump-probe spectrometer.

Contrary to conventional dual-colour systems, which usually employ one or more optical parametric amplifiers (OPAs) as a secondary source of spectrally tuneable femtosecond pulses, our approach consists in obtaining the dual-colour pump and probe pulses from the same broadband hollow-fibre source. This approach has several advantages. Unlike OPA systems, the pump and probe pulses are very broadband (550-700 nm and 650-950 nm, respectively) and can both support pulse durations below 10 fs, as shown in Fig. 2. The small overlap between the pump and probe spectra can be reduced with optical filters. The pulses have similar energies, enabling us to exchange their roles when studying specific samples that may benefit from a particular combination of central wavelengths for the pump and probe pulses. The set-up permits control of both polarisation and energy of the pulses, allowing the detection of both polarisation rotation and ellipticity. Detection is based on balanced photo-detection in conjunction with boxcar and lock-in techniques. The electromagnet and bipolar power supply can generate magnetic fields of about 1 T in the sample plane.

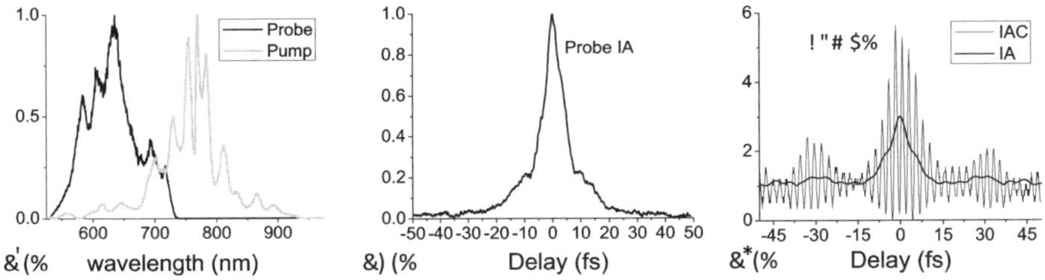


**Fig.2.** (a) Spectra for the pump and probe beams, (b) measured intensity autocorrelation of probe beam and (c) measured interferometric autocorrelation of pump beam and corresponding (calculated) intensity autocorrelation. Retrieved pulse duration (assuming sech2 shapes) is 8 fs for both pulses.

## Conclusions

We have developed a state-of-the-art pump-probe system with sub 10 fs temporal resolution. The full spectrum of the broadband source can be used for both pump and probe pulses with minimal changes to the setup, this is expected to reduce the ultimate temporal resolution to the order of 6 fs. This unique apparatus will be dedicated to the study of ultrafast magnetisation dynamics in low dimensional ferromagnetic systems.

**Acknowledgments** This work was supported by the FCT (Portugal) through grants PTDC/FIS/115561/2009 and PTDC/FIS/115102/2009. CSG acknowledges financial support from FCT via a PhD grant SFRH/BD/61466/2009.

## References

[1] M. Miranda, T. Fordell, C. Arnold, A. L'Huillier, and H. Crespo, "Simultaneous compression and characterization of ultrashort laser pulses using chirped mirrors and glass wedges," Opt. Express **20**, 688-697 (2012).

[2] B. Alonso, M. Miranda, F. Silva, V. Pervak, J. Rauschenberger, J. San Román, Í. J. Sola, and H. Crespo, "Characterization of sub-two-cycle pulses from a hollow-core fiber compressor in the spatiotemporal and spatio-spectral domains" Applied Physics B **112**, 105-114 (2013).

[3] F. Silva, M. Miranda, B. Alonso, J. Rauschenberger, V. Pervak, and H. Crespo, "Simultaneous compression, measurement and phase stabilization of >50 GW, 3.2 fs pulses using a single dispersion-scan setup" Ultrafast Optics, UFO IX, Davos, Switzerland, 2-8 March (2013).

# Spin dynamics in Rare Earth doped Cobalt ferromagnetic films

**L.H.F. Andrade[2,3], M. Vomir[1], J. Kim[1], M. Sanches Piaia[1], A.D. Santos[4], and J.-Y. Bigot[1]**

1-Institut de Physique et Chimie des Matériaux de Strasbourg, UMR 7504 CNRS, Université de Strasbourg, France

2-Departamento de Física, ICEB/UFOP, Ouro Preto, MG, Brasil

3-Centro de Lasers e Aplicações, IPEN-CNEN, São Paulo, SP, Brasil

4-Departamento de Física dos Materiais e Mecânica, IF/USP, São Paulo, SP, Brasil

bigot@unistra.fr

**Abstract** We have investigated the demagnetization ($\tau_M$) and the precession damping ($\eta$) times of the magnetization of Co ferromagnetic films doped with rare earth elements excited by femtosecond laser pulses. Both times are measured as a function of the absorbed laser energy density $E_0$ and doping concentration $c_{RE}$ with rare-earth elements. The results indicate a priori no direct connection between $\tau_M$ and $\eta$ with respect to spin-impurity scattering processes, while both times depend on $E_0$, due to a raise of respectively the electronic and lattice temperatures.

Two key parameters characterizing the ultrafast magnetization dynamics are the timescale of the demagnetization ($\tau_M$) and the dynamics of the precession of the magnetization characterized both by its period $T_{prec}$ and damping time ($\eta$) [1-7]. In the present article we investigate $\tau_M$ and $\eta$ in cobalt films doped with rare earth elements by varying systematically $E_0$ the laser energy density and $c_{RE}$ the rare earth concentration. These two relaxation times do not follow the same behavior upon varying each parameter, showing that the mechanisms involved in the spin-flip scattering process occurring at these timescales are different.

The investigated magnetic films were sputtered at an Ar pressure of 5 mTorr at room temperature on substrates of glass and silicon, by co-sputtering a Co target with rare-earth targets (Tb, Ho, Sm and Dy). The time-resolved magneto-optical measurements were done with 48 fs laser pulses (center wavelength 800 nm) delivered at 5 kHz by an amplified Titanium:Sapphire oscillator in a setup similar to that used in Ref. [1].

Figure 1 shows the time-resolved normalized differential magneto-optical polar Kerr signal $\Delta M/M(t)$ (stars) for a Tb doped Co film, $Co_{89.5}Tb_{10.5}$, together with its transmission $\Delta T/T(t)$ (solid dots), which represents the electron dynamics. The electron and magnetization dynamics with their distinctive features are shown for

 
J.-Y. Bigot et al. (eds), *Ultrafast Magnetism I*, Springer Proceedings in Physics 159
DOI 10.1007/978-3-319-07743-7_19

long pump-probe delays, until 450 ps, Figure 1(a), and for short delays, until 500 fs, Figure 1(b). For long delays the magneto-optical experimental signal, which is representative of the magnetization precession, is fitted to a damped harmonic function which in this case decays with a time constant $\eta$ of 149 ps. Fitting an exponential to the differential Kerr signal at short delays we obtain a demagnetization time $\tau_M$ of 71 ± 5 fs for $Co_{89.5}Tb_{10.5}$.

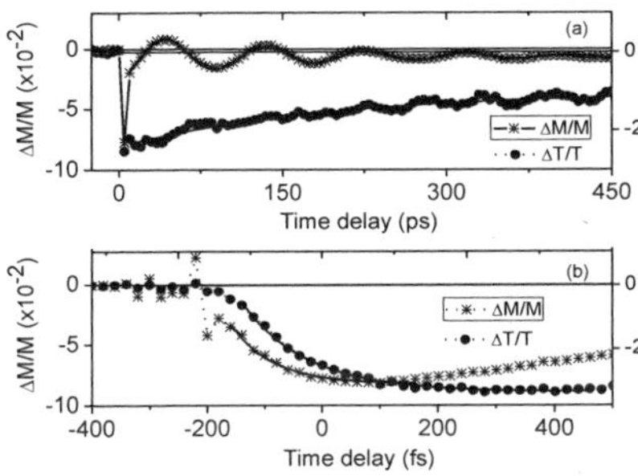


**Fig. 1.** Polar Kerr dynamics and differential transmission measured for a $Co_{89.5}Tb_{10.5}$ film for (a): "long" and (b): "short" pump-probe delay times.

In Figure 2a and 2b $\tau_M$ and $\eta$ are shown for several rare-earth doped Co samples. $\tau_M$ is not significantly affected by the doping concentration and the type of impurities. In contrast, $\eta$ decreases for increasing $c_{RE}$ showing the influence of the spin scattering with magnetic impurities and lattice defects on the precession dynamics as reported before in the case of permalloy thin films [3, 5, 6]. Our results first show that the 5d-4f spins scattering between the transition metal and rare earth spins does not affect the thermalization of the 5d spins non-equilibrium distribution excited by the laser. In contrast this mechanism prevails during the diffusive transport regime of the spins i.e. when the precession and its damping occur.

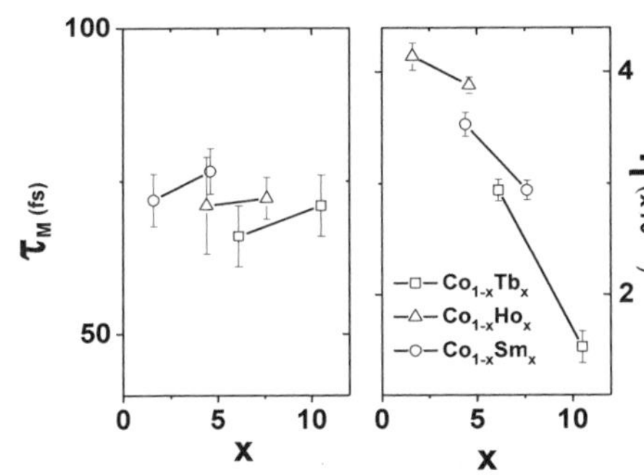


**Fig. 2.** Variation of the thermalization $\tau_M$ and damping $\eta$ times for Co films doped with several concentrations of Sm, Tb and Ho impurities.

The influence of an increase of $E_0$ on the magnetization dynamics of a rare-earth doped Co film is to increase both $\tau_M$ and $\eta$. The increase of $\tau_M$ with $E_0$ is obviously related to the increase of the spin specific heat as the temperature approaches the Curie temperature. On the other hand, the increase of the damping time η with $E_0$ means that there is less "friction" on the precession of the magnetization with increasing excitation. This result disfavours the role of thermal excitations on the precession damping, as has been recently suggested [8]. For example, the relaxation time η changes for $Co_{93.9}Tb_{6.1}$ from 215 ps to 295 ps when increasing $E_0$ by a factor 3. This result can be explained as follows: according to a three bath modelling of the charges, spins and lattice dynamics, an increase of laser intensity manifests by an increasing lattice temperature, which is in quasi equilibrium with the electrons and spins after a few picoseconds [1]. Such increase of temperature reduces the modulus of the magnetization according to the Curie-Weiss law. As a result the friction decreases since it is proportional to M x (dM/dt) in the modified Landau-Lifshitz-Gilbert equation, providing we take into account the changes of the magnetization modulus [1].

In conclusion, we have measured the ultrafast demagnetization time $\tau_M$ and damping time η of the precession in rare-earth doped Co films. $\tau_M$ of rare-earth doped Co films is not affected by the doping concentration, in contrast to η. In addition, η increases when we increase the laser energy density. This is attributed to the reduction of the magnetization modulus as the temperature increases.

**Acknowledgments** We acknowledge the financial support of the European Research Council with the ERC Advanced Grant ATOMAG (No. ERC-2009-AdG-20090325 247452).

## References

[1] M. Vomir, L. H. F. Andrade, L. Guidoni, E. Beaurepaire, J.-Y. Bigot, "Real Space Trajectory of the Ultrafast Magnetization Dynamics in Ferromagnetic Metals" Phys. Rev. Lett. **94**, 237601 (2005).

[2] B. Koopmans, J. J. M. Ruigrok, F. Dalla Longa, W. J. M. de Jonge, "Unifying Ultrafast Magnetization Dynamics" Phys. Rev. Lett. **95**, 267207 (2005).

[3] J. Walowski, G. Müller, M. Djordjevic, M. Münzenberg, M. Kläui, C. A. F. Vaz, and J. A. C. Bland, Phys. Rev. Lett. "Energy Equilibration Processes of Electrons, Magnons, and Phonons at the Femtosecond Time Scale" **101**, 237401 (2008).

[4] J.-Y. Bigot, M. Vomir, E. Beaurepaire, "Coherent ultrafast magnetism induced by femtosecond laser pulses" Nature Physics **5**, 515 (2009).

[5] I. Radu, G. Woltersdorf, M. Kiessling, A. Melnikov, U. Bovensiepen, J.-U. Thiele, C. H. Back, "Laser-Induced Magnetization Dynamics of Lanthanide-Doped Permalloy Thin Films" Phys. Rev. Lett. **102**, 117201 (2009).

[6] W. Bailey, P. Kabos, F. Mancoff, S. Russek, "Control of magnetization dynamics in Ni81Fe19 thin films through the use of rare-earth dopants" IEEE Trans. Magn. **37**, 1749 (2001).

[7] H. Vonesch and J.-Y. Bigot, "Ultrafast spin-photon interaction investigated with coherent magneto-optics" Phys. Rev. B **85**, 180407(R) (2012).

[8] D. Steiauf, M. Fähnle, "Elliott-Yafet mechanism and the discussion of femtosecond magnetization dynamics" Phys. Rev. B **79**, 140401(R) (2009).

# Ultrafast Ferrofluids Magnetization Frameworks

**A. Larionescu, C. Buzduga, C. Ciufudean**

University Stefan cel Mare, 13 University, 720229, Suceava, Romania

cbuzduga@eed.usv.ro

**Abstract** This paper presents the construction of a stand for the study of magnetic materials, especially ferrofluids. This stand is supplied in two ways, with an AC and a DC source.

## Introduction

Ferrofluids are colloidal suspensions in liquid of very fine (» 10 *nm*) magnetic particles and when exposed to magnetic field increase their viscosity (e.g. the magnetic viscosity phenomenon). These liquids were named magneto-rheological fluids [1,2]. Ferrofluids are involved in implementing new devices and technologies and therefore the research of magnetic liquids has a strongly multidisciplinary character. As much as any research work experiments are very extensively and intensively used in order to have a break-through in technological field. We are aware of previous works in this field of knowledge as well as some theoretical models and a few applications [1-3] and we developed in our laboratory an experimental stand for studying ferrofluids. Ferrofluids are a new type of material that preserves the properties of the basic fluid and acquire the magnetic properties of the particles in their own composition, responding instantly to the application of a magnetic field [2-5].

In order to generate the magnetic field around the conductor, we need high current amplitude, obtained in our work through an electrical scheme for continuous current and an electrical scheme for alternating current, for each case being highlighted the hydrostatic profile of the ferrofluid. This paper approaches several aspects of particular problems of ferrofluids statics, related to the shape of the ferrofluids free surface around a vertical electrical conductor of different shapes (round, square, tape and cross), traversed by an electrical current.

 
J.-Y. Bigot et al. (eds), *Ultrafast Magnetism I*, Springer Proceedings in Physics 159
DOI 10.1007/978-3-319-07743-7_20

## Magnetic source

In this experiment, in order to obtain the shape of ferrofluid around a vertical current, we applied three types of signals (ramp, step and impulse) for several types of profiles made of different materials. Materials are processed profiles (round, square, triangle, cross and band), and are made by copper and steel.

The stand presented in Fig. 1, serving to investigate the hydrostatic profile of a ferrofluid surrounding a vertical electrical conductor, uses several bottoms, each of them associated with a vertical wire, and made of a conductive material diamagnetic, paramagnetic or ferromagnetic. This source is reliable and economical being realized in the research lab.

Ferrofluids applications in technical are similar semiconductors applications. The most common applications require study ferrofluids in magnetic field. To study this we design is necessary magnetic field sources. An example of the source is presented below.

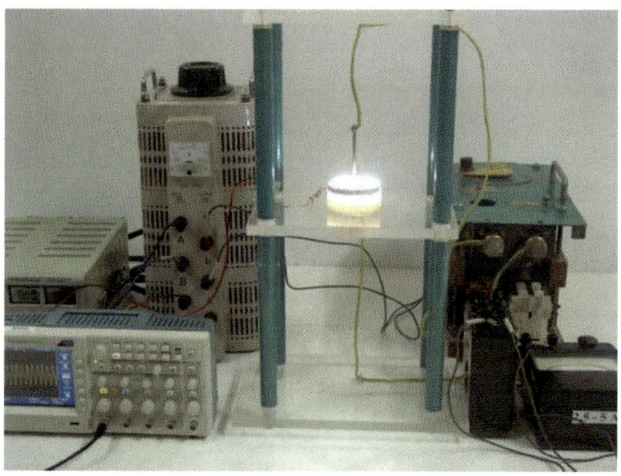

**Fig. 1.** Overview of the stand

Experimental results for the three types of signals is illustrated in Fig. 2, 3 and 4, the images are captured using a high speed camera. Ferrofluid is arranged around the current that passes through metallic conductors in the form of a truncated cone. When excitation signal is step and ramp we noticed that besides the attraction of ferrofluid on the top of conductor, there are also some peaks of various shapes and sizes. The phenomenon is easily visible on alternative current only for high current values. Ferrofluid column height wired around the current that passes through a metal conductor depends by the conductor diameter. The smaller is the diameter of the conductor the greater is ferrofluid column height. We estimate that our experiments are useful in industry for improving sensors sensitivity (i.e. inclinometers, flow meters, accelerometers, etc.); for improving the sound quality of audio loudspeakers, for increasing the capacity of magnetic disks, for computer new memory storage facilities, and sputtering apparatus in the semiconductive industry.

**Fig. 2.** Profile of hydrostatic ferrofluids for impulse signal

**Fig. 3.** Profile of hydrostatic ferrofluids for ramp signal

**Fig. 4.** Profile of hydrostatic ferrofluids for step signal

## References


[1] S. Odenbach, "Magnetoviscous effects in ferrofluids", Springer-Verlag, Heidelberg, (2002).

[2] C. Scherer, A. M. Figueiredo Neto, "Ferrofluids: Properties and Applications", Brazilian Journal of Physics, vol. **35**, no. 3A, pp. 718-727, September, (2005).

[3] R.E. Rosensweig, „Towards ferrofluids with enhanced magnetization". Journal of Magnetism and Magnetic Materials **323**, Elsevier, ScienceDirect, pp. 1191–1197, (2011).

[4] S. Odenbach, "Ferrofluids/magnetically controlled suspensions", Colloids and Surfaces A: Physicochem. Eng. Aspects **217**, Elsevier Science B.V, pp. 171-178, (2003).

[5] S. Odenbach, "Ferrofluids". Springer-Verlag, Berlin, (2002).

# Magnetization Reversal in a Cobalt Nanoparticle

**G. Klughertz, P.-A. Hervieux, and G. Manfredi**

Institut de Physique et Chimie des Matériaux de Strasbourg, CNRS and Université de Strasbourg, 23 rue du Loess, F-67034 Strasbourg, France

klughertz@ipcms.unistra.fr

**Abstract** We study the magnetization reversal of isolated and interacting single-domain cobalt nanoparticles with uniaxial anisotropy under the macrospin approximation. We analyze the consequences of dipolar interaction on the reversal time. The effect of temperature on the probability to switch the magnetization using a linearly chirped pulse to drive the precession is investigated.

The quest for higher density in magnetic data storage implies future use of smaller nanoparticles with high magnetic anisotropy to overcome the superparamagnetic limit. In order to address this problem, we have developed a Fokker-Planck model to simulate the magnetization dynamics of isolated and interacting single-domain magnetic nanoparticles in the mean-field approximation [1]using a static field only.

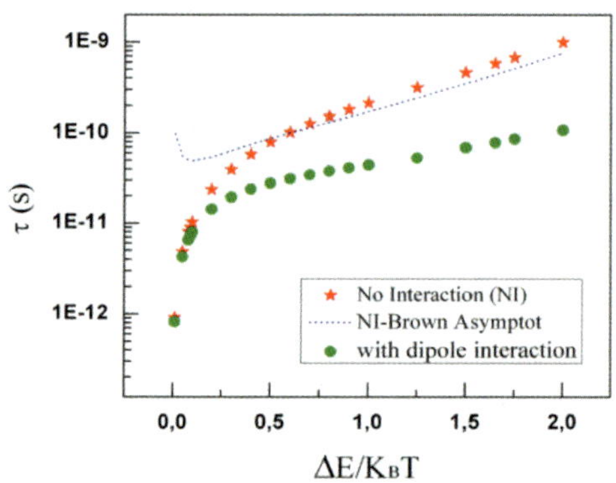


**Fig. 1.** The variation of the reversal time $\tau$ as a function of the inverse temperature $1/T$ for damping constant $\alpha = 1$ : numerical results (red) and Brown asymptote (dashed line) for no interaction, magnetic dipolar interaction for 25nm interparticle distance (green).

Fig. 1 shows that the inclusion of dipolar interactions accelerates the reversal process at low temperatures while it does not affect it at high temperatures. Also, the

 
J.-Y. Bigot et al. (eds), *Ultrafast Magnetism I*, Springer Proceedings in Physics 159
DOI 10.1007/978-3-319-07743-7_21

results for isolated nanoparticles are in agreement with the Brown expression (dashed line).

But an increased anisotropy requires larger fields to reverse the magnetization of the nanoparticle, which is currently difficult to achieve experimentally. To elude this limitation, a microwave field can be combined to the static field [2], and it has been shown that the optimal microwave field should be modulated both in frequency and magnitude [3].

We investigate here the magnetization reversal of an isolated single-domain Co nanoparticle with uniaxial anisotropy under the macrospin approximation, combining a static field with a linearly chirped excitation consisting in a microwave field to drive the precession until the magnetization switches.

We study an isolated single-domain Co nanoparticle with uniaxial anisotropy along $\boldsymbol{e_z}$ in the macrospin approximation ($|\boldsymbol{M}|$ is constant) by solving the Landau-Lifshitz-Gilbert equation:

$$\frac{d\boldsymbol{M}}{dt} = -\frac{\gamma}{(1+\lambda)^2}\left(\boldsymbol{M} \times \boldsymbol{B_{eff}}\right) - \frac{\gamma\lambda}{(1+\lambda)^2 \mu_S}\left(\boldsymbol{M} \times \left(\boldsymbol{M} \times \boldsymbol{B_{eff}}\right)\right) \quad 1$$

where $\gamma$ is the gyromagnetic ratio, $\lambda$ the phenomenological damping parameter, and $\boldsymbol{B_{eff}}$ the effective field acting on the particle.

The effective field comprises the anisotropy field, a static external field collinear to the anisotropy axis, and a microwave field of varying frequency which will constitute the chirped pulse. Two kinds of microwave fields are considered: one of constant amplitude rotating in the $(\boldsymbol{e_x}, \boldsymbol{e_y})$ plane a second one with constant direction along $\boldsymbol{e_x}$ but oscillating amplitude. We have found that the rotating field is twice as effective as the constant-direction field for the magnetization reversal process.

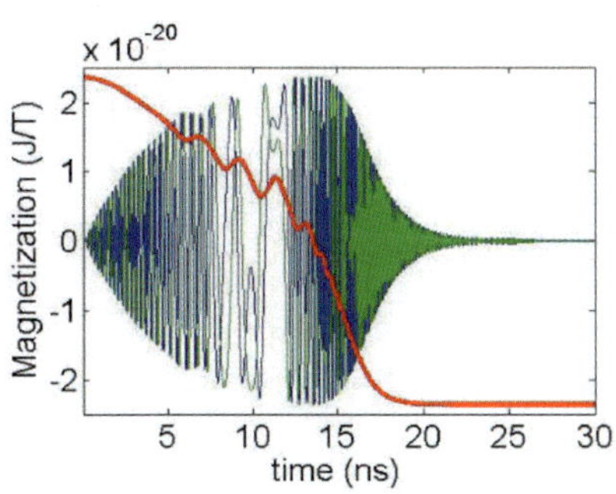


**Fig. 2.** Evolution of the magnetization components ($M_x$, $M_y$ and $M_z$ in green, blue and red respectively).

Fig. 2 shows that the moment is captured and dragged until its complete reversal, which corresponds to a change in the sign of the $M_Z$ component of the magnetic moment.

We now turn to the effect of a finite temperature on the magnetization reversal. One can define a threshold amplitude $\varepsilon_{th}$ of the oscillating microwave field above which the magnetization switches.
In the zero temperature case, the probability to capture and drive the precession of the moment is zero below $\varepsilon_{th}$ and equal to unity above, with a very sharp and step-like transition. At finite temperature, the moment is subjected to thermal fluctuations and it may or may not be captured, depending on its orientation when the chirped pulse is activated. This randomness in the initial conditions creates a finite width of the transition around $\varepsilon_{th}$. We show analytically and numerically that this transition width scales as $\sqrt{T/V}$ where T is the temperature and V is the volume of the nanoparticle.

In summary, we observed that the inclusion of dipolar interactions lead to a faster reversal of the magnetization. We also showed that a microwave field with a varying frequency and very small amplitude (a hundred times smaller than the static field) can efficiently reverse the magnetization, thus reducing the switching field. The capture process is strongly altered by the temperature because of the random amplitude and phase difference between the excitation and the moment.

**Acknowledgments** The authors acknowledge the financial support of EQUIPEX UNION Grant (ANR-10-EQPX-52).

## References

[1] H. Kesserwan, G. Manfredi, J.-Y. Bigot, P.-A. Hervieux, "Magnetization reversal in isolated and interacting single-domain nanoparticles" Phys. Rev. B **84**, 172407 (2011).
[2] C. Thirion, W. Wernsdorfer, D. Mailly, "Switching of magnetization by nonlinear resonance studied in single nanoparticles" Nat. Mater. **2**, 524-527 (2003).
[3] N. Barros, M. Rassam, H. Jirari, H. Kachkachi, "Optimal switching of a nanomagnet assisted by microwaves" Phys. Rev. B **83**, 144418 (2011).

# Ultrafast magnetization dynamics driven by equilibration of temperatures and chemical potentials

**B.Y. Mueller and B. Rethfeld**

Department of Physics and Research Center OPTIMAS, University of Kaiserslautern, Erwin-Schrödinger-Str. 46, 67663 Kaiserslautern, Germany

bmueller@physik.uni-kl.de

**Abstract** Applying spin-resolved Boltzmann collision integrals, we find that ultrafast demagnetization dynamics is driven by both the equilibration of separated temperatures and chemical potentials of spin-up and spin-down electrons. We set up a system of equation as a simplified phenomenological model to describe the demagnetization dynamics governed by these equilibration processes.

It is well known that ultrashort laser pulses irradiating itinerant ferromagnets cause a demagnetization of the material on a sub-picosecond time scale[1, 2]. Several models which are intended to explain this effect microscopically are currently under discussion [2–4]. One of them considers changes of total magnetization induced by spin flips of single electrons during collisions with other particles. Such processes, i.e. collisions accompanied with a spin flip, are known as Elliot Yaffet processes [5].

In this context, the interest of our group on magnetization dynamics arouse. Generally, we are studying electron dynamics in laser excited solids with help of Boltzmann collision integrals [6, 7]. Information on the transient electron distribution functions is of large interest for instance in the field of photoemission [8, 9], electron emission [10] and transport [11].

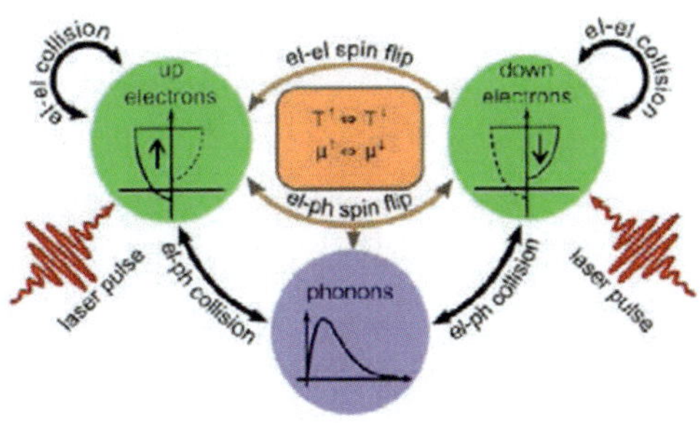


**Fig. 1.** Relaxation dynamics of the majority and minority electron systems as well as the phonons. Figure is taken from Ref. [12]

 
J.-Y. Bigot et al. (eds), *Ultrafast Magnetism I*, Springer Proceedings in Physics 159
DOI 10.1007/978-3-319-07743-7_22

We study in particular moderate excitation densities and resulting effects as the thermalization of excited electrons, the electron phonon coupling and the mutual influence of both. Such microscopic effects due to the excitation of many electrons in a metal may provide a key to understand also magnetization dynamics in laser-excited metals. To that end, we distinguish the electrons, separating them into spin-up and spin-down electrons, and consider in particular collisions in between the subsystems of spin-up electrons, spin-down electrons and phonons. The considered systems and collision processes are sketched schematically in Fig. 1. Following [3, 13], spin flips are considered to occur due to both, electron–electron and electron– phonon collisions.

Details of our calculation are published in Ref. [12]. Assuming the Stoner model, the densities of states are shifted with respect to each other with an energy difference given by the exchange splitting. Both subsystems, however, obey a joint level of chemical potential leading to a larger density of one kind of electrons than the other kind. When exciting the material, energy is brought into both electron systems, leading to increased temperatures. As we have shown in Ref. [12], these temperatures do not equal each other, thus we have to distinguish between the temperature of spin-up electrons $T_e^{\uparrow}$, and the temperature of spin-down electrons $T_e^{\downarrow}$. The equilibration of both temperatures drives an energy exchange between the spin-up and the spin-down system. Moreover, the chemical potential depends in turn on the electron temperature, however, this dependence is not necessarily linear. Thus, when the temperatures $T_e^{\uparrow}$ and $T_e^{\downarrow}$. have equilibrated, the chemical potentials $\mu_e^{\uparrow}$ and $\mu_e^{\downarrow}$ may still differ [17]. The equilibration of both chemical potentials $\mu_e^{\uparrow}$ and $\mu_e^{\downarrow}$ , drives a *particle exchange* between the spin-up and the spin-down system. We have identified this equilibration of temperatures *and* chemical potentials as the driving force for ultrafast demagnetization [12].

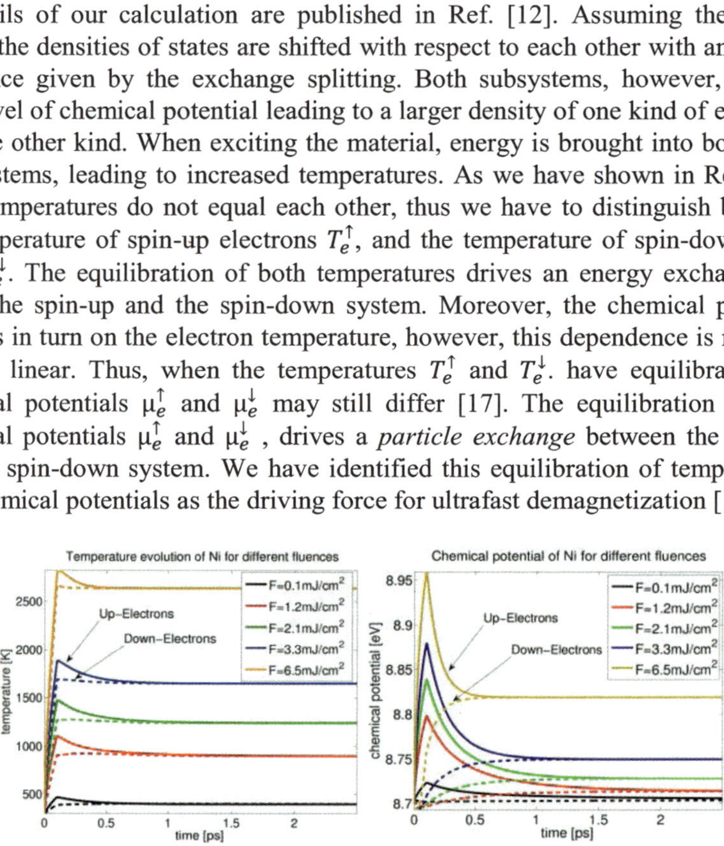


**Fig. 2.** Relaxation dynamics of the spin-up and spin-down electron temperatures (left) and of the respective chemical potentials (right).

In Fig. 2, the equilibration of temperatures and chemical potentials after laser excitation with different absorbed laser fluences is shown for the case of Nickel, considering its density of states [7, 14]. The evolution of the temperatures $T_e^{\uparrow}$ and $T_e^{\downarrow}$ shows the increase during the irradiation (pulse duration 100 fs, constant intensity), the different maximum values of $T_e^{\uparrow}$ and $T_e^{\downarrow}$, respectively, and the subsequent equilibration of these quantities. Note that the energy exchange with the phonons was "switched off" in this calculation. Fig. 2 shows that the time of equilibration of the chemical potentials depends on excitation strength. Since this time is directly connected with the timescale of demagnetization (disregarding the effect of electron–phonon collisions discussed in [12]), this dependence nicely confirms the relation between quenching time an irradiation strength found in Ref. [2, 15].

In order to simplify the kinetic description we develop a model considering the equilibration of temperatures and chemical potentials in a phenomenological way. The main idea is similar to the two-temperature description of electrons and phonons [16] governed by the difference of the respective temperatures. In contrast to the three temperature model introduced in Ref. [2], we, however, do not consider an explicit spin temperature, but solely distinguish between the temperatures of the electronic subsystems $T_e^{\uparrow,\downarrow}$, as well as the chemical potentials $\mu_e^{\uparrow,\downarrow}$. The model implies terms considering their respective equilibration, additionally to the equilibration between electron temperature and phonon temperature. Due to the consideration of *particle exchange* between spin-up and spin-down electrons, we are able to calculate the evolution of magnetization during and after ultrafast excitation. We compare the results of this simplified model with the results of the complete simulation applying spin resolved Boltzmann collision integrals.

**Acknowledgments** Financial support of the Deutsche Forschungsgemeinschaft through the Emmy-Noether and Heisenberg Programm is gratefully acknowledged.

## References

[1] E. Beaurepaire, J.-C. Merle, A. Daunois, and J.-Y. Bigot, "Ultrafast Spin Dynamics in Ferromagnetic Nickel" Phys. Rev. Lett. **76**, 4250 (1996).
[2] B. Koopmans, G. Malinowski, F. Dalla Longa, D. Steiauf, M. Fähnle, T. Roth, M. Cinchetti, and M. Aeschlimann," Explaining the paradoxical diversity of ultrafast laser-induced demagnetization" Nature Materials **9**, 259 (2010).
[3] M. Krauss, M. Aeschlimann, and H. C. Schneider, "Ultrafast Spin Dynamics Including Spin-Orbit Interaction in Semiconductors" Phys. Rev. Lett. **100**, 256601 (2008).
[4] M. Battiato, K. Carva, and P. M. Oppeneer, "Superdiffusive Spin Transport as a Mechanism of Ultrafast Demagnetization" Phys. Rev. Lett.**105**, 027203 (2010).
[5] I. Žutić, J. Fabian, and S. Das Sarma, "Spintronics: Fundamentals and applications" Rev. Mod. Phys. **76**, 323 (2004).
[6] B. Rethfeld, A. Kaiser, M. Vicanek, and G. Simon, "Ultrafast dynamics of nonequilibrium electrons in metals under femtosecond laser irradiation" Phys. Rev. B **65**, 214303 (2002).
[7] B. Y. Mueller and B. Rethfeld, "Relaxation dynamics in laser-excited metals under nonequilibrium conditions" Phys. Rev. B **87**, 035139 (2013).
[8] R. Knorren, K. H. Bennemann, R. Burgermeister, and M. Aeschlimann, "Dynamics of excited electrons in copper and ferromagnetic transition metals: Theory and experiment" Phys. Rev. B **61**, 9427 (2000).

[9] A. Gloskovskii, D. Valdaitsev, S. Nepijkoa, G. Schönhense, and B. Rethfeld, "Coexisting electron emission mechanisms in small metal particles observed in fs-laser excited PEEM" Surface Science **601**, 4706 (2007).
[10] W.Wendelen, B. Y. Mueller, D. Autrique, B. Rethfeld, and A. Bogaerts, "Space charge corrected electron emission from an aluminum surface under non-equilibrium conditions" J. Appl. Phys. **111**, 113110 (2012).
[11] O. Osmani, N. Medvedev, M. Schleberger, and B. Rethfeld, "Energy dissipation in dielectrics after swift heavy-ion impact: A hybrid model" Phys. Rev. B **84**, 214105 (2011).
[12] B. Y. Mueller, T. Roth, M. Cinchetti, M. Aeschlimann, and B. Rethfeld, "Driving force of ultrafast magnetization dynamics" New Journal of Physics **13**, 123010 (2011).
[13] S. Essert and H. C. Schneider, Electron-phonon scattering dynamics in ferromagnetic metals and their influence on ultrafast demagnetization processes" Phys. Rev. B **84**, 224405 (2011).
[14] Z. Lin, L. V. Zhigilei, and V. Celli, "Electron-phonon coupling and electron heat capacity of metals under conditions of strong electron-phonon nonequilibrium" Phys. Rev. B **77**, 075133 (2008).
[15] T. Roth, A. J. Schellekens, S. Alebrand, O. Schmitt, D. Steil, B. Koopmans, M. Cinchetti, and M. Aeschlimann, "Temperature Dependence of Laser-Induced Demagnetization in Ni: A Key for Identifying the Underlying Mechanism" Phys. Rev. X **2**, 021006 (2012).
[16] S. I. Anisimov, B. L. Kapeliovich, and T. L. Perel'man, "Electron emission from metal surfaces exposed to ultrashort laser pulses" Sov. Phys. JETP **39**, 375 (1974).
[17] Note that for a free electron gas, the relation between chemical potential and temperature can be expressed explicitly by the Sommerfeld expansion. For realistic densities of states the relation may be more complicated but can be determined directly from the density of states and the density of electrons, assuming a Fermi distribution of their energy.

# Layer-Specific Probing of Ultrafast Spin Dynamics in Multilayered Magnets with Visible Light

**Yu. Tsema,[1] M. Savoini,[1] A. Kirilyuk,[1] A. Tsukamoto,[2] and Th. Rasing[1]**

[1] Radboud University Nijmegen, Institute for Molecules and Materials, Heyendaalseweg 135, 6525 AJ Nijmegen, The Netherlands

[2] College of Science and Technology, Nihon University, 7-24-1 Funabashi, Chiba, Japan

Y.Tsema@science.ru.nl

**Abstract** In this study GdFeCo-multilayered structures are investigated by means of static magneto-optical measurements and time-resolved pump-probe spectroscopy. Fluence dependent time-resolved measurements demonstrate a complex behavior of the magnetization and the possibility of magnetization reversal. A method to distinguish contributions from different layers is proposed.

It was recently demonstrated that femtosecond laser pulses can manipulate the magnetic order [1] and even result in a full magnetization reversal of metallic samples [2]. The latter was first shown in GdFeCo which is a rare earth (RE) – transition metal (TM) ferrimagnetic alloy that used to be a common material for magneto-optical data storage devices [3]. Due to the different strengths of the exchange interaction of the nonequivalent magnetic sublattices, the magnetic moments of the RE and the TM elements behave in a different way as a function of temperature. This leads to the existence of a so-called magnetization compensation temperature ($T_M$) at which the magnetic moments of both elements are aligned anti-parallel and have the same absolute values. This temperature is a function of the alloy chemical composition and plays a crucial role in its magnetization dynamics. Particularly, the degree of laser induced demagnetization is higher when the system crosses the magnetization compensation temperature due to the laser heating [4]. It was in fact demonstrated that the ultrafast demagnetization efficiency strongly depends on ($T-T_M$). In our study of $Gd_xFe_yCo_{100-x-y}$ samples, where x is between 22% and 27%, the position of $T_M$ is found to be in an interval between 70K and 320K which means that an efficient magnetization reversal of GdFeCo-systems is possible for only a narrow range of concentrations of Gd. However, the fabrication of multilayered GdFeCo-systems could allow to extend this range.

The complexity in the study of any multilayered systems originates from the superposition of the responses coming from the different layers of the system. Be-

 
J.-Y. Bigot et al. (eds), *Ultrafast Magnetism I*, Springer Proceedings in Physics 159
DOI 10.1007/978-3-319-07743-7_23

ing able to address single features may be the key for a deeper understanding of various phenomena. For example, mixing different RE-concentrations in one sample consisting of RE-TM layers could reveal an atypical behavior of the net magnetization dynamics, different from those for each separate alloy.

Figure 1 shows a static magneto-optical characterization for the sample which main layers are $Gd_{27}Fe_{63.9}Co_{9.1}$ and $Gd_{26}Fe_{64.7}Co_{9.3}$, both 10nm-thick, separated by a layer of 5nm of $Si_3N_4$. A typical hysteresis measured at 120K is shown in the inset, where dashed lines indicate the coercive field of $Gd_{26}Fe_{64.7}Co_{9.3}$ and the dotted lines are for $Gd_{27}Fe_{63.9}Co_{9.1}$. The existence of a different magnetization compensation points for each RE-TM layer is observed.

The possibility to distinguish contributions of RE and TM in RE-TM ferromagnetic alloys such as TbFeCo by magneto-optical spectroscopy was recently shown [5]. For our multi-layer sample it appears that multiple scattering effects lead to a layer specific magneto-optical response, depending on the wavelength and angle of incidence. Figure 2 demonstrates the possibility to achieve a 100% sensitivity to the top layer of the structure. This phase-based effect allows to apply time-resolved pump-probe spectroscopy to separate the contributions of the layers and unravel the superposition of layer specific trends.

This method allows us to investigate all-optical switching in a buried layer, extracting information that can lead (with a proper engineering) to the creation of an extended switching window or the creation of synthetic (anti-)ferromagnetic stacks of different layers, opening the way for new discoveries in the exchange interaction dynamics.

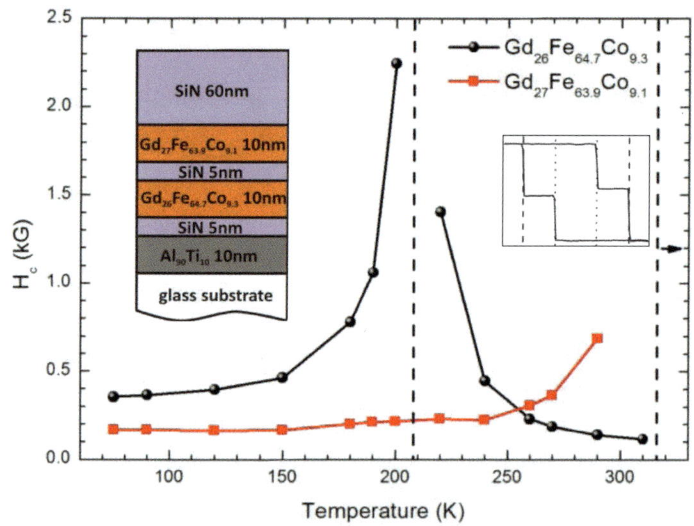


**Fig. 1.** The coercive field $H_C$ vs temperature for the $Gd_xFe_yCo_{100-x-y}$ layers with different concentrations ($x$=26% and 27%). The divergence in the coercivity indicates the magnetization compensation points which are shown as dashed lines. Schematic view of the sample and common hysteresis measured at 120K are shown in the insets.

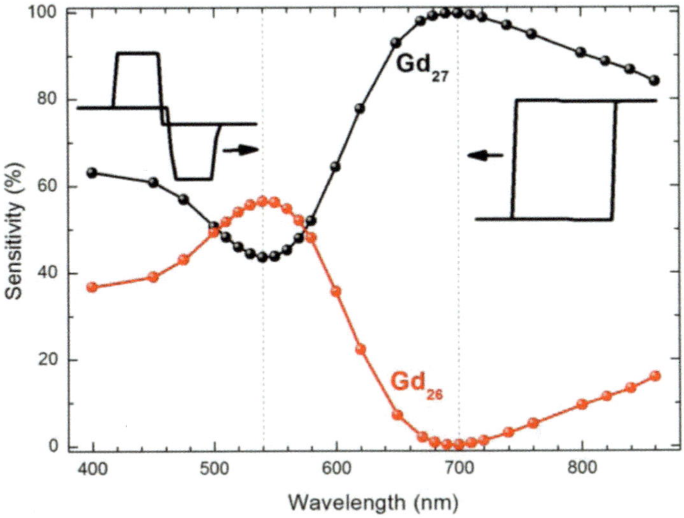


**Fig. 2**. Relative magneto-optical sensitivity to the two magnetic layers as a function of the wavelength. The sensitivity is determined here as Kerr rotation coming from one layer divided by the total Kerr rotation for the whole sample. The sample was illuminated with monochromatic light at 45 degrees of incidence at room temperature. For the chosen conditions maxima of sensitivity are around 540nm and 700nm.

## References


[1] E. Beaurepaire, J.-C. Merle, A. Daunois, and J.-Y. Bigot, "Ultrafast Spin Dynamics in Ferromagnetic Nickel" PRL **76**, 4250 (1996).

[2] C. D. Stanciu, F. Hansteen, A. V. Kimel, A. Kirilyuk, A. Tsukamoto, A. Itoh, and Th. Rasing, "All-Optical Magnetic Recording with Circularly Polarized Light" PRL **99**, 047601 (2007).

[3] M. Mansuripur, "The Physical Principles of Magneto-optical Recording", Cambridge University Press (1995).

[4] R. Medapalli, I. Razdolski, M. Savoini, A. R. Khorsand, A. Kirilyuk, A. V. Kimel, and Th. Rasing, "Efficiency of ultrafast laser-induced demagnetization in GdxFe100−x−yCoy alloys" PRB **86**, 054442 (2012).

[5] A. R. Khorsand, M. Savoini, A. Kirilyuk, A. V. Kimel, A. Tsukamoto, A. Itoh, and Th. Rasing, "Element-Specific Probing of Ultrafast Spin Dynamics in Multisublattice Magnets with Visible Light" PRL **110**, 107205 (2013).

# Precession of the magnetization and breathing motion of assemblies of Co-Pt nanoparticles

**Hasan Kesserwan, Valérie Halté, Jean-Yves Bigot**

Institut de Physique et Chimie des Matériaux de Strasbourg, CNRS, Université de Strasbourg, UMR 7504, France

valerie.halte@ipcms.unistra.fr

**Abstract:** The time resolved reflectivity $\Delta R(t)$ and magneto-optical Kerr response $\Delta\theta(t)$ of assemblies of Co-Pt nanoparticles are studied with femtosecond laser pulses. In ordered arrangements of such superparamagnetic nanoparticles $\Delta R(t)$ displays a low frequency mode (145 ps) characteristic of collective vibrations, while $\Delta\theta(t)$ varies monotonously. In contrast, after the nanoparticles are thermally annealed, $\Delta\theta(t)$ displays a motion of precession and a monotonous variation of $\Delta R(t)$ both characteristic of a ferromagnetic phase and disordered arrangement.

Magnetic nano-objects are promising for making fast data recording devices [1]. In that context, the study of their ultrafast magneto-optical properties is of prior importance [2]. In particular, two dynamical regimes are interesting to study: when the nano-objects are either superparamagnetic or ferromagnetic and when dipolar magnetic interactions take place between them [3-5]. In this work, we have studied the magnetization dynamics of close-packed assemblies of Co(core)-Pt(shell) nanoparticles. By comparing different annealing conditions of the nanoparticles it is shown that both their spatial organization and their structural phase have a strong influence on their ultrafast dynamics.

The native nanoparticles have been elaborated by redox transmetalation reactions [6]. They result in a 5 nm cobalt core and a 1.5 nm platinum shell and are superparamagnetic at room temperature with a blocking temperature $T_B$=66 K as shown in the figure 1a). They are assembled into pellets by cold pressing under 160 Pa. The ultrafast dynamics of the reflectivity and the magnetization are measured, in a pump-probe configuration, using a regenerative amplified Titanium: Sapphire laser, cadenced at 5kHz and delivering pulses with 130 fs duration. The pump, centered at 400 nm, is generated by frequency doubling in a BBO crystal whereas the probe is centered at the fundamental wavelength (800 nm). A polarization bridge allows us to measure simultaneously the normalized reflectivity $\Delta R/R(t)$ and the Kerr rotation $\Delta\theta/\theta(t)$ using lock-in amplifiers and photodiodes. First, a mild local laser annealing in the region of pump and probe spatial overlap is performed. The maximum density of energy for this laser annealing is 3 mJ/cm² and corresponds to a transient increase of the lattice temperature of 370 K. For

 
J.-Y. Bigot et al. (eds), *Ultrafast Magnetism I*, Springer Proceedings in Physics 159
DOI 10.1007/978-3-319-07743-7_24

higher laser annealing temperatures systematic photo-ionization processes take place leading to sample damage. Second, we performed annealing in an oven, allowing higher temperatures, up to 700 K.

First, we study the consequences of the laser annealing. The differential reflectivity measured on the annealed zone is plotted on the figure 1b) for a density of excitation of 0.8 mJ/cm² (empty squares). After the pump pulse arrival, the signal increases corresponding to a heating of electrons until a maximum is reached after 160 fs corresponding to the thermalization time of the electrons. Then, the electrons cool down to the lattice via electrons-phonons interactions in a few picoseconds. At longer delays, oscillations show up with a period $T_{vib}$=146 ps. They correspond to collective vibrations due to nanoparticles re-arrangement locally induced by the laser annealing. A similar behavior has been demonstrated on 3D self-organized supra-crystals of Co nanoparticles [7]. Let us add that we observe a heat diffusion time of 170 ps and a damping time of the oscillations of 157 ps. It shows that the collective vibrations disappear as soon as the heat diffuses away from the probed area. In figure 1b), the main features of the magnetization dynamics (solid triangles) are also represented. First the ultrafast demagnetization takes place, followed by the spins-phonons relaxation in 360 fs and later a long-standing monotonous heat diffusion occurs. This monotonous re-magnetization is consistent with superparamagnetic fluctuations without specific magnetic order at room temperature.

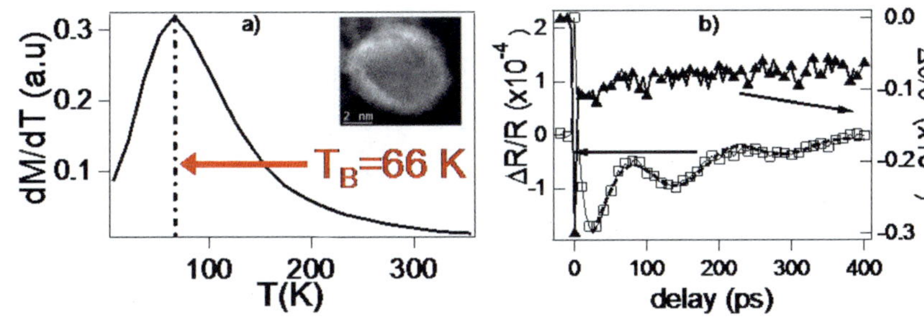


**Fig. 1.** Dynamics of an ordered assembly of superparamagnetic Co-Pt nanoparticles. a) Blocking temperature of the native nanoparticles extracted from ZFC-FC SQUID measurements. Inset: TEM image of a single core-shell CoPt nanoparticle. b) Dynamics of the differential reflectivity (empty squares) and of the magneto-optical Kerr rotation (solid triangles).

Secondly, we proceed to the thermal annealing of the pellet in an oven at 650 K for one hour. The nanoparticles undergo a phase transition resulting in a ferromagnetic behavior at 300 K with a mean diameter of 8 nm, as shown in the figure 2a). The analysis of electron microscopy images of these nanoparticles [8] show that their structure evolves from a Co-Pt core-shell into homogenous CoPt and $CoPt_3$ phases, corresponding to an enrichment of the Co core in platinum. As spin-orbit interactions are rather strong in such alloy, it reinforces the magneto-crystalline anisotropy such as the thermal energy $k_BT$ is not sufficient to allow magnetization fluctuations over the anisotropy barrier. It results in a ferromagnetic behavior of the nanoparticles at room temperature as confirmed by ZFC-FC measurements with a blocking temperature of 347 K (figure 2a)) and an open hysteresis loop at 300K with a coercive field of 50 Oe. The figure 2b) shows the dynamics of $\Delta\theta/\theta(t)$ (solid triangles) of the annealed sample with an external magnetic field of 1.5 kOe. At long pump-probe temporal delays, it is drastically different from the native case. The oscillations correspond to the precession behavior of the magnetization, projected on the polar axis as observed in ferromagnetic cobalt nanoparticles [9]. The period of these oscillations is 80 ps. This result is a confirmation of the observations made by electron microscopy (CoPt phase). Moreover, the existence of the magnetization precession indicates that the nanoparticles assembly is homogeneous upon thermal annealing. Note that, as compared to fig. 1b), the reflectivity now behaves monotonously at long time delays in fig. 2b). This is due to the high temperature annealing which results in a disordered assembly of the nanoparticles, suppressing their coherent collective vibrations observed with the mild laser annealing.

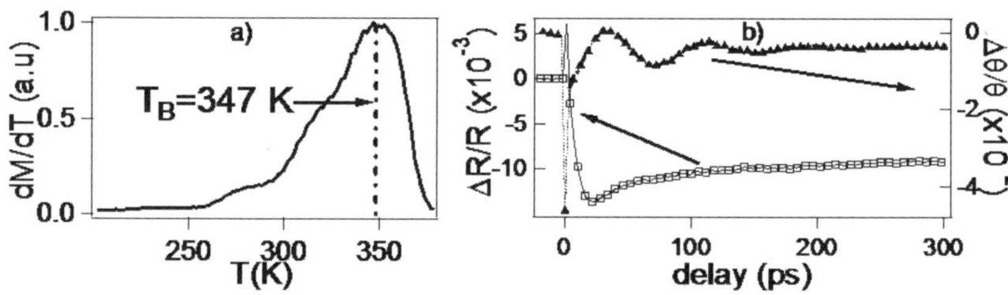


**Fig. 2.** Dynamics of a disordered assembly of annealed ferromagnetic Co-Pt nanoparticles. a) Blocking temperature extracted from ZFC-FC SQUID measurements. b) Dynamics of the differential reflectivity (empty squares) and of the magneto-optical Kerr rotation (solid triangles) of thermally annealed nanoparticles for a density of excitation of 0.45 mJ/cm².

**Acknowledgments:** The authors thank the European Research Council for financial support (ERC grant "ATOMAG" ERC-2009-AdG-20090325 #247452).

## References

[1] D. Weller, A. Moser, L. Folks, M. E. Best, W. Lee, M. F. Toney, M. Schwickert, J.-U. Thiele, M. F. Doerner, "High $K_u$ materials approach to 100 Gbits/in$^2$" IEEE Trans. Magn. **36**, 10–15 (2000).

[2] J.-Y. Bigot, M. Vomir, "Ultrafast magnetization dynamics of nanostructures" Annalen der Physik **525,** No. 1–2, 2–30 (2013).

[3] A. Laraoui, J. Vénuat, V. Halté, M. Albrecht, E. Beaurepaire , J.-Y. Bigot, "Study of individual ferromagnetic disks with femtosecond optical pulses" J. Appl. Phys. **101**, 09C105/1-3 (2007).

[4] J. L. Dormann, D. Fiorani, E. Tronc, "Magnetic Relaxation in Fine-Particle Systems" Adv. in Chem. Phys. 98, 283 (1997).

[5] H. Kesserwan, G. Manfredi, J.-Y. Bigot, P.-A. Hervieux, "Magnetization reversal in isolated and interacting single-domain nanoparticles" Phys. Rev. B **84**, 172407 (2011).

[6] J-I. Park, and J. Cheon, "Synthesis of "Solid Solution" and "Core-Shell" Type Cobalt–Platinum Magnetic Nanoparticles via Transmetalation Reactions" J. Am. Chem. Soc. **123**, 5743-5746 (2001).

[7] I. Lisiecki, V. Halté, C. Petit, M.-P. Piléni, J.-Y. Bigot, "Vibration Dynamics of Supra-Crystals of Cobalt Nanocrystals Studied With Femtosecond Laser Pulses" Adv. Mat. **20**, 4176-4179 (2008).

[8] J.-Y. Bigot, H. Kesserwan, V. Halté, O. Ersen, M. S. Moldovan, T. H. Kim, J-tak. Jang, J. Cheon, "Magnetic Properties of Annealed Core–Shell CoPt Nanoparticles" Nano letters **12**, 1189 (2012).

[9] L. H. F. Andrade, A. Laraoui, M. Vomir, D. Muller, J.-P. Stoquert, C. Estournès, E. Beaurepaire, J.-Y. Bigot, "Damped Precession of the Magnetization Vector of Superparamagnetic Nanoparticles Excited by Femtosecond Optical Pulses" Phys. Rev. Lett. **97**, 127401 (2006).

# Laser heated ferromagnetic simulations

**Raghuveer Chimata[1], Jonathan Chico[1], Anders Bergman[1], Lars Berqvist[2], Biplab Sanyal[1], and Olle Eriksson[1]**

[1]Department of Physics and Astronomy, Uppsala University, Box 516, 75120 Uppsala, Sweden
[2]Department of Materials Science and Engineering, KTH Royal Institute of Technology, Brinellv. 23, Stockholm, Sweden

Raghuveer.Chimata@physics.uu.se

**Abstract** In this work, we show a model of ferromagnetic material heated by a laser pulse. The laser creates a pattern of circles on the ferromagnetic materials with hot regions heated up to 3000K and cold regions at 100K and 400K. In our model the Landau-Lifshitz-Gilbert equation for a macrospin and spin temperature is passed through stochastic field. We show that the difference of magnon dispersion in the cold regions of the material.

## Introduction

Recent interest in advanced magnetic materials for data storage and data processing on short timescales drags a huge interest to study in a detailed atomistic description of magnetic materials. During ultrafast demagnetization processes, angular momentum will be transferred between the electrons, spins and phonons in the systems to conserver it[1]. The process of demagnetization will take place in sub-picosecond scales[2,3] and remagnetization will take place in picosecond scales[4]. The dynamics of individual local moments can be studied using atomistic spin dynamics codes.

Atomistic spin dynamics simulations have provided insight into the local spin as well as the macro spin of the magnetic materials. Owing to computational limitations, work so far has been limited to nanometer scales and the laser dot is very big. Therefor, to study such type of systems has to use massively parallel computing structures. For this purpose we used parallelized Uppsala atomistic spin dynamics ( UppASD ) code[5].

## Method

To study laser induced ferromagnetic materials we used a two-step method ap-

 
J.-Y. Bigot et al. (eds), *Ultrafast Magnetism I*, Springer Proceedings in Physics 159
DOI 10.1007/978-3-319-07743-7_25

proach. In first step we obtained the Heisenberg exchange parameters and local moments from DFT codes, while a second step we used atomistic spin dynamics. The microscopic equations of local moments, which evolve in external fields, are expressed as follows,

$$\frac{dm_i}{dt} = -\Upsilon m_i \times [B_i + b_i(t)] - \Upsilon \frac{\alpha}{m} m_i \times (m_i \times [B_i + b_i(t)]) \qquad 1$$

In the above equation $\Upsilon$ is the electron gyromagnetic ratio and α is the damping parameter. The stochastic field, $b_i(t)$, enters into the effective field in both the precessional term and the damping term.

## Micromagnetic simulations

The simulations of the impact of a change in the site dependent laser spot were carried out using the UppASD code. A 2D super cell size of bccFe 300×300×1 has been used. The laser pulse was used to heat the material in specific regions and simulated heat regions reach 3000 K while cold regions are at 0K and 400 K. The simulations are run up to 10ps. The evolution of local moments is calculated and represented in form of figures. Figure1, the hot region is at 3000 K and the cold region is at 0K. The evolution in this model shows there is no magnon dispersion taking the place due to huge landscape between the hot region and cold region. In figure 2, the dynamics was shown but in this case the hot regions are at 3000 K and cold region is 400 K. When simulations carried out clearly that there is magnon dispersion in the slab.

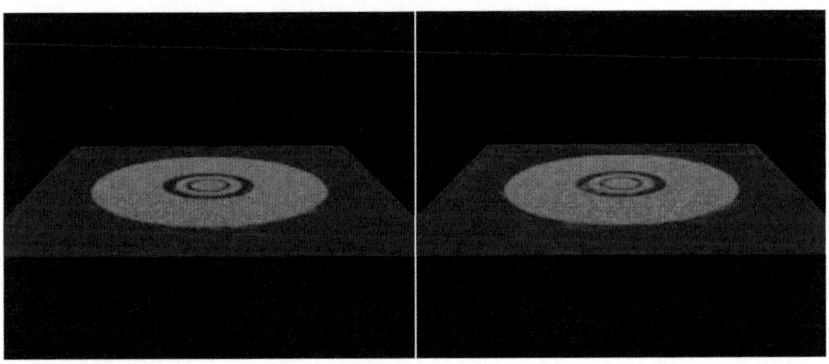

**Fig. 1.** Evolution of magnetic moment of bcc Fe. The region with blue vectors is at 0 K and while the green and red vectors indicate hot regions of 3000 K.

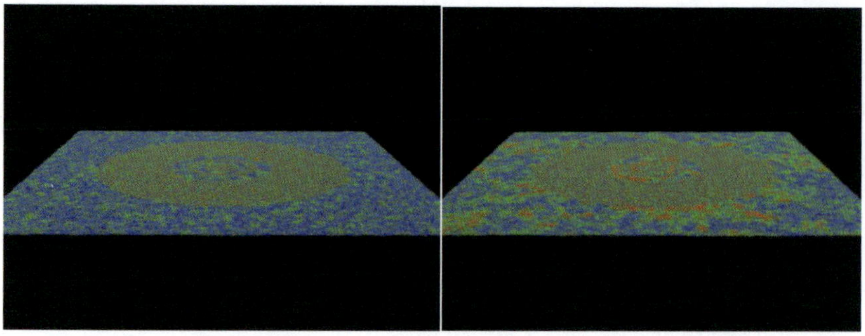

**Fig. 2.** Evolution of magnetic moment of bcc Fe are shown. The region with blue vectors is at 400 K and while the green and red vectors indicate hot regions of 3000 K.

By studying the above simulations can create a new texture material by using two different curie temperature materials by considering consider a cold region has a high Tc material while low temperature has a low Tc material. It promotes the new type of dynamics and creation of new artificial materials.

**Acknowledgement** We gratefully acknowledge financial support from the Swedish Research Council (VR). O. E. is in addition grateful to the ERC (project 247062—ASD) and KAW foundation for support. Support from eSSENCE and SeRC are acknowledged. We also acknowledge Swedish National Infrastructure for Computing (SNIC) for the allocation of time in high-performance supercomputers.

## References

[1] B. Koopmans, J. J. M. Ruigrok, F. Dalla Longa, and W.J.M. de Jonge, "Unifying Ultrafast Magnetization Dynamics", Phys. Rev. Lett. **95**, 267207 (2005).

[2] E. Beaurepaire, J. C.Merle, A.Daunois, and J.-Y. Bigot, Phys. Rev. Lett. "Ultrafast Spin Dynamics in Ferromagnetic Nickel", **76**, 4250 (1996).

[3] J.-Y. Bigot, "Femtosecond magneto-optical processes in metals", C. R. Acad. Sci. Ser. IV **2**, 1483-1504 (2001).

[4] R. Chimata, A. Bergman, L. Bergqvist, B. Sanyal, and O. Eriksson, Phys. Rev. Lett. **109**, 157201 (2012).

[5] B. Skubic, J. Hellsvik, L. Nordström, and O. Eriksson, J. Phys. Condens. Matter **20**, 315203 (2008); http://www.physics.uu.se/en/page/UppASD.

# Part III
# Spin Waves Dynamics

# Excitation and control of spin wave by light pulses

**Takuya Satoh[1,2], Yuki Terui[1], Rai Moriya[1], Boris A. Ivanov[1,3], Kazuya Ando[4], Eiji Saitoh[4,5,6,7], Tsutomu Shimura[1], and Kazuo Kuroda[1]**

[1] Institute of Industrial Science, The University of Tokyo, Tokyo 153-8505, Japan

[2] PRESTO, Japan Science and Technology Agency, Tokyo 102-0076, Japan

[3] Institute of Magnetism, NASU, 03142 Kiev, Ukraine,

[4] Institute for Materials Research, Tohoku University, Sendai 980-8577, Japan,

[5] CREST, Japan Science and Technology Agency, Tokyo 102-0075, Japan,

[6] The Advanced Science Research Center, Japan Atomic Agency, Tokai 319-1195, Japan,

[7] WPI-AIMR, Tohoku University, Sendai 980-8577, Japan;

tsatoh@iis.u-tokyo.ac.jp

**Abstract** Time- and space-resolved spin wave excitation by circularly polarized pulses was investigated in a garnet ferrite. Propagation of the spin wave was controlled by spatially shaping of the light pulses.

## Introduction

Magnetization dynamics has attracted great deal of interest in recent years. In particular, the propagation characteristics of spin waves have been extensively studied because of their importance as the basis of "magnonics," which aims to use spin wave packets instead of electric currents as the information transmitter [1].
Recently, a novel method of magnetization control has been reported. The method employs the inverse Faraday effect (IFE), where a circularly polarized light pulse nonthermally generates an effective magnetic field [2]. The IFE has been described as impulsive stimulated Raman scattering. Thus, ultrafast magnetic switching is expected via the IFE. So far, there has been limited discussion on local spin oscillations. Non-local spin dynamics, i.e. propagating spin waves, excited by this impulsive excitation method had not been observed. By using this method, 2D propagation can be observed in a non-contact manner, and broadband excitation can be realized, unlike resonant excitation with a microwave field.
Yttrium iron garnet has often been used as a sample in magnonics because of its intrinsically low magnetic damping [3]. In an experiment involving photoin-duced spin waves, Bi-doped rare-earth iron garnet was found to be more suitable than yt-

 
J.-Y. Bigot et al. (eds), *Ultrafast Magnetism I*, Springer Proceedings in Physics 159
DOI 10.1007/978-3-319-07743-7_26

trium iron garnet due to its high magneto-optical susceptibility. By doping the sample with bismuth and magnetic rare-earth ions, the Gilbert damping coefficient increases. Even so, spin waves can propagate over distances of hundreds of micrometers, and they can be spatially resolved.

Here we report spin wave excitation via the IFE in a ferrite garnet. 2D propagation of the spin waves was investigated by a pump-probe experiment. Direction of the spin wave propagation was controlled by spatially shaping of the light pulses.

## Excitation and control of spin wave

The composition of the garnet ferrite was $Gd_{4/3}Yb_{2/3}BiFe_5O_{12}$, and the thickness was 110 μm. An in-plane magnetic field of 1 kOe was applied to saturate the magnetization. We employed the magneto-optical Faraday effect to measure space- and time-resolved spin waves. Circularly polarized pump pulses with a wavelength of 1400 nm were focused in a circular spot (50 μm in diameter) on the sample surface. Polarization rotation of the probe light was recorded for each pump position. All measurements were performed at room temperature, which is below the Curie temperature ($T_c$ = 573 K).

Figure 1(a) shows time-resolved polarization rotation when the pump spot was scanned in the directions parallel (x axis) to the magnetic field. Propagation of a spin wave packet was clearly observed. Figure 1(b) shows a spatiotemporal map of the spin waves when the pump spot was scanned in the x axis, whereas the posi-tion of the probe spot was fixed. We found that the wavelength of the spin wave was 200–300 μm, and the group velocity was about 100 km/s. It is clearly seen that a spin wave packet propagated from the pump spot, and the phase of the wave packet moved toward the pump position from the outside, which is a characteristic of backward volume magnetostatic waves. Its dispersion of the lowest-order mode is shown in Fig. 1(c) where $\varphi$ is the propagation angle with respect to the x axis.

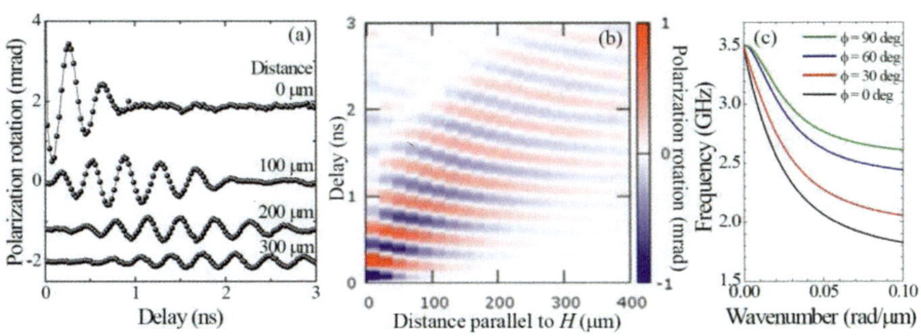


**Fig. 1.** (a) Time-resolved polarization rotation when the pump spot was scanned in the directions parallel to the magnetic field. (b) Spatiotemporal maps of spin waves. (c) The dispersion of the lowest-order mode of the backward volume magnetostatic wave.

In order to control the direction of spin wave emission, we made an elliptical pump spot shape on the sample surface. When the major axis of the ellipse was (a) perpendicular and (b) parallel to the x axis, a spin wave propagated perpendicular or parallel to the *x* axis, respectively (see Figure 2). In this way, we achieved directional control of the spin wave emission [4].

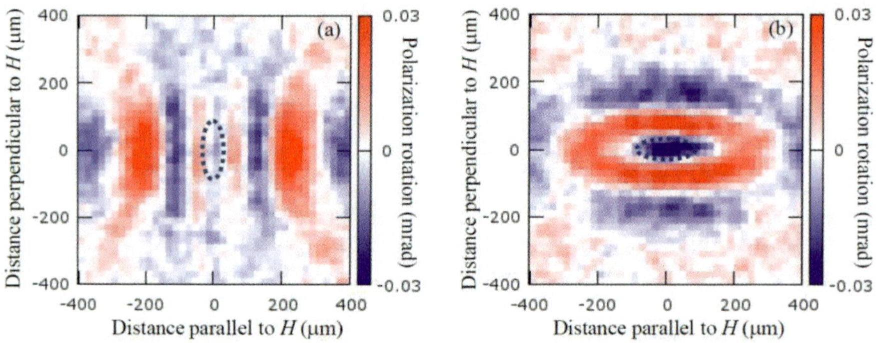


**Fig. 2.** 2D maps of spin wave emission at time delay 1.5 ns. Major axis of the ellipse was (a) perpendicular and (b) parallel to the magnetic field, respectively.

## Conclusion

The spin wave is excited and directed only by the spatially shaped light pulse. We expect that our technique will allow much higher design flexibility in magnonics, and that this technique will pave the way for a new field of "opto-magnonics" for manipulating spin waves with light spots of various shapes.

**Acknowledgments** This work was supported by JST-PRESTO and KAKENHI (23104706).

## References

[1] S. O. Demokritov, A. N. Slavin, eds. "Magnonics: From Fundamentals to Applications", (Springer, 2013).
[2] A. V. Kimel et al., "Ultrafast non-thermal control of magnetization by instantaneous photomagnetic Pulses", Nature **435**, 655-657 (2005).
[3] Y. Kajiwara et al., "Transmission of electrical signals by spinwave interconversion in a magnetic insulator", Nature **464**, 262-266 (2010).
[4] T. Satoh et al., "Directional control of spin wave emission by spatially shaped light", Nature Photonics **6**, 662-666 (2012).

# k-vector distribution of magneto-static spin waves excited by micro-fabricated antenna structures

**H.G. Bauer, J.-Y. Chauleau, G. Woltersdorf, C.H. Back**

Fakultät für Physik, Universität Regensburg, Universitätsstrasse 31, 93040 Regensburg, Germany

christian.back@ur.de

**Abstract** In this article we examine the k-vector distribution of magneto-static spin waves which are excited by microfabricated antenna structures. Changing the repetition of the antenna structures from a simple coplanar wave guide like structure to a meandering type antenna the k-vector distribution can be adjusted to a more or less peaked distribution representing the pitch of the meander. However, intrinsic damping of the magnetic system counteracts this effort.

## Introduction

Spin wave Doppler shift experiments have played an important role in determining the spin polarization of the conduction electrons in ferromagnetic conductors [1,2]. In these experiments spin waves are excited in a ferromagnetic stripe using microwave excitation via antenna structures. Spin wave propagation in the stripes is well described by the Landau Lifshitz Gilbert equation with the appropriate boundary conditions. When the spin waves are subjected to an electric current flowing in the stripe the interaction between the spin waves (representing a smoothly varying non-collinear magnetic structure) and the spin polarized conduction electron current shifts the spinwave frequency according to $\omega = uk$ . Here k represents the k-vector of the spin waves and u is the so-called spin drift velocity which linearly depends on the spin polarization of the conduction electrons. Thus by determining u it is possible to extract the spin polarization as long as the k-vector of the spin waves is known exactly. In inductive experiments [1,2], typically the k-vector is inferred from the Fourier transform of the real space geometry of the antenna structure, see Figure 1. To launch a narrow k-vector distribution typically antenna structures with a number of repeats are used which mathematically leads to a rather narrow k-vector distribution. In the following we will show that this effort is counteracted by the finite Gilbert damping of the material.

 
J.-Y. Bigot et al. (eds), *Ultrafast Magnetism I*, Springer Proceedings in Physics 159
DOI 10.1007/978-3-319-07743-7_27

## Excitation geometry and Fourier transform

**Fig. 1.** Typical antenna structures used to lunch magneto-static spin waves into a ferromagnetic Permalloy race track.

In Figure 1 we show typical antenna structures used in the experiments. The antenna on the left consists of a shorted wave guide and the corresponding Fourier spectrum of the excitation field produced by this antenna is shown in black in Figure 2. In the middle a meandering antenna is shown in red corresponding to the red line in Figure 2 and finally on the right hand side a meandering antenna with five repeats is shown in green with the corresponding green line in Figure 2.

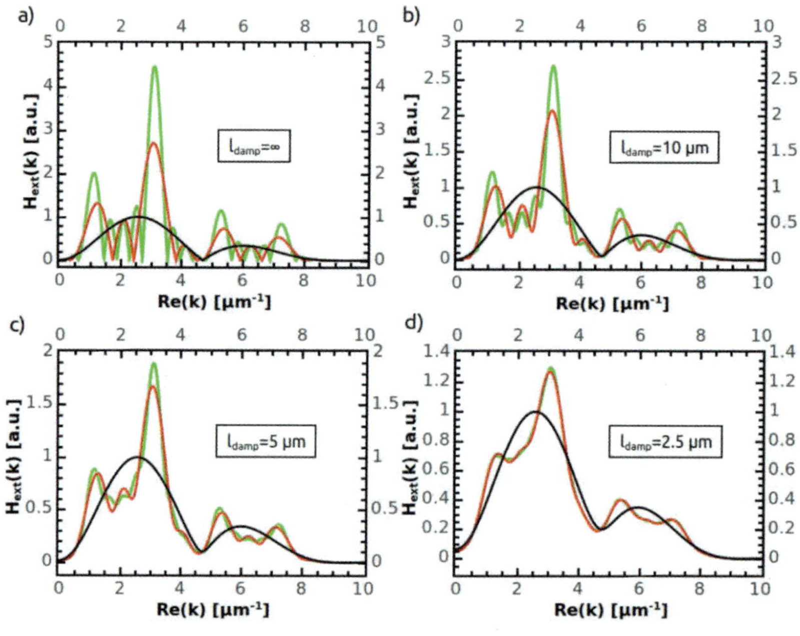


**Fig. 2.** Fourier components of the excitation field calculated form the antenna structures shown in Figure 1. The black line presents the Fourier components of the magnetic excitation field produced by the antenna shown on the left in Figure 1. The same is shown in red for the antenna with three repeats in Figure 1 and in green for the antenna with five repeats. From a) – d) the damping length of the spin waves is shortened from infinity to 2.5 micrometer.

The excitation spectrum of the antenna with five repeats is rather sharp (Figure 2a) and peaks at a k-vector corresponding to the size of the wave guide structure. Reducing the number of repeats considerably broadens the k-vector spectrum, see black line in Figure 2a). However, as soon as a Gilbert like damping is turned on (damping length < infinity) also the k-vector distribution of the excited k-vectors in the magnetic stripe broadens considerably thus counteracting the effort to reach a rather sharp k-vector distribution, see Figure 2b)-d). In fact when reaching a realistic damping length of 2.5 – 5 micrometers (Figure 2c) and d)), the k-vector distribution of all three antenna structures becomes almost indistinguishable.

This means that a reliable extraction of the current spin polarization can only be realized when the true k-vector distribution is measured independently.

**Acknowledgments** The authors thank the DFG under SFB 689 for financial support. J-Y. C. would like to acknowledge the A.vH. foundation for financial support.

## References

[1] V. Vlaminck and M. Bailleul, “Current induced spin wave Doppler shift” Science **322**, 410 (2008).

[2] M. Zhu, C.L. Dennis and R.D. McMichael, “Temperature dependence of magnetization drift velocity and current polarization in $Ni_{80}Fe_{20}$ by spin-wave Doppler measurements” Phys. Rev. B **81**, 140407 (2010).

# Spin-wave and spin-current dynamics in ultrafast demagnetization experiments

**M. Münzenberg**

I. Physikalisches Institut, Georg-August-Universität Göttingen, Friedrich Hund Platz 1, 37077, Göttingen, Germany

mmuenze@gwdg.de

**Abstract** Many facets of femtosecond excitation of ferromagnets have been investigated in the history of ultrafast demagnetization dynamics. Already the first model developed for nickel [1] in 1996 pointed out the importance of the three subsystems – electrons, spins and phonons – and the interaction within them. Here I want to review some possibilities to understand and manipulate the different knobs on ultrafast dynamics derived from that.

## Introduction

In the first discovery of ultrafast behavior of the spin system under laser excitation, the beforehand unprecedented speed of spin-dynamics on the 50-150 fs time scale was explained by the strong coupling of the electron and spin system leading to a fast heat transfer. Meanwhile different concepts have been developed to describe the nature of the ultrafast heating of the spin system. Similar to the local moment and the delocalized band picture that we have in static magnetism, also in spin dynamics, both pictures lead to different insights. In a picture of a local moment, the spins get heated and thus reduced in the averaged moment by thermal statistics. In a band picture, collapsing exchange energy reduces the moment and lateral inhomogeneous excitation can lead to heat currents (or super diffusion). In all cases, specific experiments disentangle the rich number of effects making ultrafast-spin dynamics to a vibrant field.

## From hot spins to spin waves

Spin-flip scattering is central importance in spintronics: giant-magneto resistance or spin accumulation is a non-local phenomena connected with transport properties. In spin transport, the Valet-Fert model describes the flow of the spins by adapting Mott's two current model with different propagation properties for both



J.-Y. Bigot et al. (eds), *Ultrafast Magnetism I*, Springer Proceedings in Physics 159
DOI 10.1007/978-3-319-07743-7_28

spin channels with a spin-flip scattering intermixing both channels for multilayer structures. A scattering event of an electron wave to another state, at a defect, phonon or any other source of scattering, has a certain probability of ending up in a different spin-state. Thus spin scattering in a metal involves its electronic transport properties. The probability is described by the spin-mixing parameter of the wave functions which is given by $V_{SO}/V_{exch}$. At least three possibilities to increase the heat transfer between electrons and spin system in a metal are indeed obvious: increasing the electron scattering rate, increasing the spin-orbit coupling ($V_{SO}$) or decreasing the exchange coupling ($V_{exch}$).

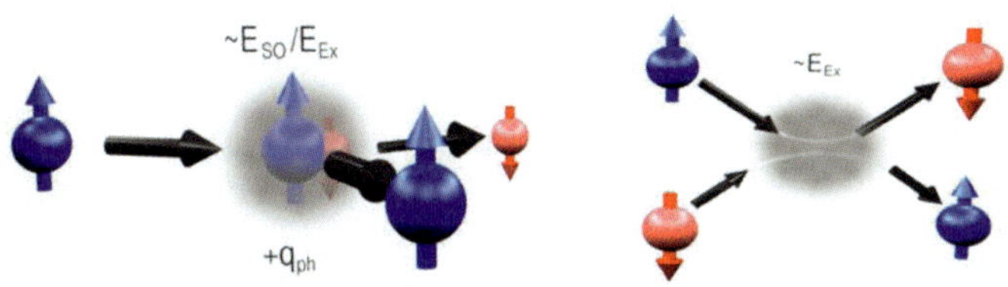


**Fig 1**. Spin-mixing versus exchange scattering (drawing courtesy by J. Walowski) [2].

Because of the fact that spin scattering is a transport property, it is an obvious idea that also spin-transport influences the spin dynamics. However this was modeled only recently by Battiato et al. [3] and in fact different experiments allowed to detect these currents directly, for example by THz emission of radiation [4]. Modeling of the spin currents involves the non-equilibrium distributions. In case of Fe, the minority spins have a very low Fermi velocity compared to the majority spins and are therefore localized, while in Ni the minority spins have a very fast relaxation time to the Fermi level compared to the majority spins that remain exited and mobile and for both in an ultrafast magnetization reduction.

Exchange scattering is known from inelastic electron scattering and allows an electron of opposite spin and high energy, impinging on a local electron, to flip its spin. This energy equilibration process in the spin-system is an important mechanism to relax local spin flips, leading to high frequency spin-wave dynamics: via the exchange energy the local spin flipped by the process relaxes into spin-waves. Subsequently, a local spin excitation will be distributed over a spin chain as a consequence of exchange energy on femtosecond time scales. This has been also observed in micromagnetic simulations on picosecond time scales and is leading to the thermal picture of demagnetization dynamics (Fig. 2). The electron system heats up the spins by spin flips. The spin system can be described by a thermal average of spin excitations and the degree coupling to the electron system can be varied by changing the spin-mixing, or more general, the coupling is described by transitions given by a Fermi golden rule model [6]. As a consequence the behavior is found to be intrinsically nonlinear. The demagnetization dynamics becomes slow – for a small increase in temperature a large demagnetization has to be achieved. It is more difficult to demagnetize the sample further.

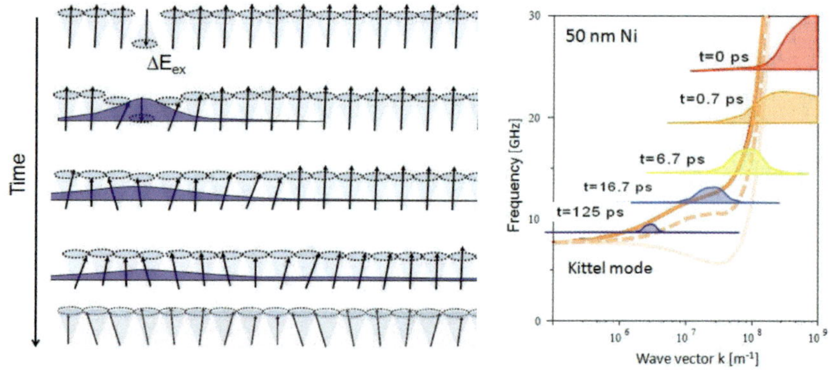


**Fig 2**. Artist impression of a Stoner excitation (local spin-flip) decaying into a spin-wave excitation (delocalized spin excitation). We can see in micromagnetic coarse grained models the same effect of a subsequent spin-wave population of lower lying wave numbers [5].

The thermal model allows a description of the ultrafast dynamics for many systems, as it was shown for example for Ni [7]. In a recent work we could pinpoint the maximum electron temperatures reached as decisive for the spin dynamics [8]. This opens up a new future direction to design the spin dynamics.

## References

[1] E. Beaurepaire, J. C. Merle, A. Daunois, and J. Y. Bigot, "Ultrafast spin dynamics in ferromagnetic Nickel", Phys. Rev. Lett. **76**, 4250 (1996).

[2] J. Walowski, G. Müller, M. Djordjevic, M. Münzenberg, M. Kläui, C. A. F. Vaz and J. A. C. Bland, "Energy equilibration processes of electrons, magnons, and phonons at the femtosecond time scale", Phys. Rev. Lett. **100**, 237401 (2008).

[3] M. Battiato, K. Carva, and P. Oppeneer, "Superdiffusive spin transport as a mechanism of Ultrafast Demagnetization", Phys. Rev. Lett. **105**, 027203 (2010).

[4] T. Kampfrath, M. Battiato, P. Maldonado, G. Eilers, J. Nötzold, S. Mährlein, V. Zbarskyy, F. Freimuth, Y. Mokrousov, S. Blügel, M. Wolf, I. Radu, P. M. Oppeneer, M. Münzenberg, "Terahertz spin current pulses controlled by magnetic heterostructures", Nature Nanotech. **8**, 256 (2013).

[5] M. Djordjevic, M. Münzenberg, "Connecting the timescales in picosecond remagnetization experiments", Phys. Rev. B **75**, 012404 (2007).

[6] A. Mann, J. Walowski, M. Münzenberg, S. Maat, M. J. Carey, J. R. Childress, C. Mewes, D. Ebke, V. Drewello, G. Reiss, A. Thomas, "Insights into ultrafast demagnetization in pseudogap half metals", Phys. Rev. X **2**, 041008 (2012).

[7] U. Atxitia, O. Chubykalo-Fesenko, J. Walowski, A. Mann, and M. Münzenberg, "Evidence for thermal mechanisms in laser-induced femtosecond spin dynamics", Phys. Rev. B **81** 174401 (2010).

[8] J. Mendil, P. C. Nieves, O. Chubykalo-Fesenko, J. Walowski, M. Münzenberg, T. Santos, S. Pisana,"Speed limit of FePt spin dynamics on femtosecond timescales", arXiv:1306.3112.

# Novel optical properties of spin-wave excitations in non-centrosymmetric oxides: the case of $Ba_2CoGe_2O_7$

**Sándor Bordács[1], Yoshinori Tokura[1,2]**

[1]Quantum Phase-Electronics Center, Department of Applied Physics, The University of Tokyo, 7-3-1 Hongo, Bunkyo-ku,Tokyo 113-8656, Japan

[2]RIKEN Center for Emergent Matter Science (CEMS), 2-12-1, Hirosawa, Wako, Saitaama 351-0198, Japan

bordacs@ap.t.u-tokyo.ac.jp

**Abstract** We have recently demonstrated that multiferroics host exotic optical phenomena, such as non-reciprocal directional dichroism [1] and natural optical activity [2], for spin-wave excitations owing to the strong dynamic magnetoelectric coupling. The observed large optical anisotropies can pave the way for novel THz applications e.g. unidirectional waveguides or polarization rotators.

## Introduction

Magnetoelectric (ME) coupling in non-centrosymmetric magnets is manifested not only in their static properties but also their spin and charge excitations are entangled. The most striking example for such a dynamic ME effect is the non-reciprocal directional dichroism that is the absorption difference between counter propagating light beams [3]. In the case of multiferroic compounds with coexisting ferroelectric and magnetic orders this phenomenon, termed as optical magnetoelectric effect (OME), can be observed when the light beam propagates along the cross product of the magnetization and the ferroelectric polarization, while in a chiral magnet such directional anisotropy arises parallel to the magnetization and called magnetochiral dichroism (MChD). Despite the early prediction of this effect [3] it has been found minimal in conventional materials and hence of little technological potential.

Here we summarize our THz studies on the multiferroic Ba2CoGe2O7 showing that the OME effect is resonantly enhanced, exceeding 40% change of the absorption coefficient, for the both electric and magnetic dipole active spin-waves, i.e. electromagnons [1]. Furthermore, the unique non-centrosymmetric crystal structure of the same compound allowed us to magnetically induce and control its chi-

 
J.-Y. Bigot et al. (eds), *Ultrafast Magnetism I*, Springer Proceedings in Physics 159
DOI 10.1007/978-3-319-07743-7_29

rality. In this chiral phase strong natural optical activity and MChD close to unity have been observed for the magnon modes [2].

## Dynamical magnetoelectric effects in $Ba_2CoGe_2O_7$

$Ba_2CoGe_2O_7$ is crystallized in a tetragonal structure (P-$42_1$m) which is neither polar nor chiral. Below $T_N$=6.7 K the S=3/2 spins of the magnetic $Co^{2+}$ ions are ordered in an easy-plane antiferromagnetic structure. The ordered magnetic moments tend to align along the {110} directions, however, a small magnetic field, B>1 T is enough to freely rotate them within the (001) easy-plane [4]. When the magnetization points along the [110] direction the magnetic point symmetry becomes m'm2' and ferroelectric polarization appears parallel to [001] [4].

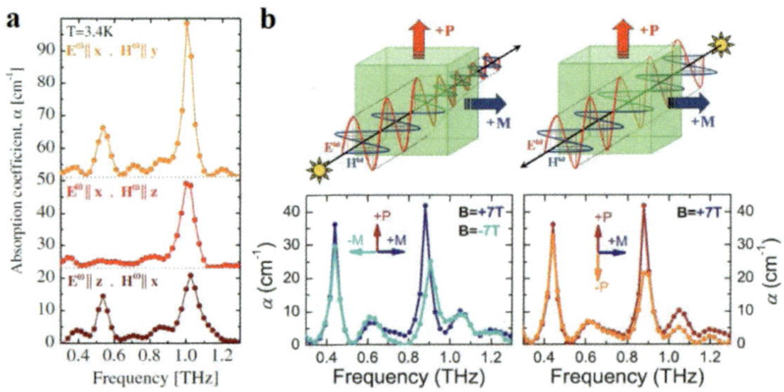

**Fig. 1.** a, Selection rules for the spin-wave excitation in $Ba_2CoGe_2O_7$. b, The large optical magnetoelectric effect (OME), i.e. the difference in the absorption coefficient for counter-propagating beams can be controlled by either the polarization or the magnetization direction [1].

Below the Néel temperature two sharp peaks are observed in the THz absorption spectrum of the $Ba_2CoGe_2O_7$ when the electric and the magnetic component of the light lie within the (001) tetragonal plane (see Fig. 1.a). The absorption peak at f~1 THz corresponds to an electromagnon excitation since it is sensitive to the direction of both the oscillating electric and magnetic fields, while the mode around f~0.5 THz is assigned to be a conventional magnon as its strength changes only when the magnetic component of the light is rotated out of the plane. According to the results of exact diagonalization studies [5] and analytical calculations [6] the electromagnon mode can be understood as an excitation through the anisotropy gap allowed by the lack of inversion symmetry.

When the light beam propagates along the [1-10] direction – parallel to the cross product of the magnetization, **M**||[110] and the ferroelectric polarization, **P**||[001],– a large OME effect is observed for the electromagnon excitation as shown in Fig.1.b. The ~40% anisotropy of the absorption coefficient can be controlled by the reversal of either the magnetization or the polarization.

In the non-centrosymmetric crystal structure of $Ba_2CoGe_2O_7$ a chiral phase corresponding to the magnetic point group 22'2' or 2'22'emerges when the magnetization points toward the [100] or [010] directions, respectively [2]. It means that the switching between the left and right handed enantiomers can be realized by 90° rotation of the magnetization.

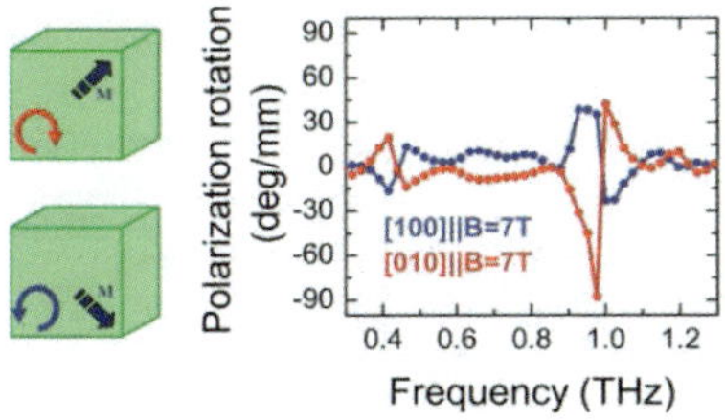


**Fig. 2.** Field induced switching between the chiral enantiomers is demonstrated by the reversal of the natural optical rotation [2].

In the magnetically induced chiral state natural optical rotation and MChD of spin-wave excitations were detected for the first time [2]. In agreement with the symmetry considerations the optical activity changes sign when the chirality is reversed as shown in Fig. 2.

Recently, optical magnetoelectric effect has also been demonstrated in manganites [7], thus, we believe that a large variety of multiferroic compounds possesses these unique optical functionalities [8] which may find applications in photonics.

**Acknowledgments** This work was supported by the Funding Program for World-Leading Innovative R&D on Science and Technology (FIRST Program), Japan.

## References

[1] I. Kézsmárki, et al., "Enhanced Directional Dichroism of Terahertz Light in Resonance with Magnetic Excitations of the Multiferroic $Ba_2CoGe_2O_7$ Oxide Compound" Phys. Rev. Lett. **106**, 057403 (2011).

[2] S. Bordács, et al., "Chirality of matter shows up via spin excitations" Nat. Phys. **8**, 734-738 (2012).

[3] L. Barron, "Molecular Light Scattering and Optical Activity", (Cambridge, 2004), Chap. 1, pp. 22-23.

[4] H. Murakawa et al., "Ferroelectricity Induced by Spin-Dependent Metal-Ligand Hybridization in $Ba_2CoGe_2O_7$" Phys. Rev. Lett. **105**, 137202 (2010).

[5] S. Miyahara, N. Furukawa, "Theory of Magnetoelectric Resonance in Two-Dimensional S=3/2 Antiferromagnet $Ba_2CoGe_2O_7$" J. Phys. Soc. Jpn. **80**, 073708 (2011).

[6] K. Penc et al., "Spin-Stretching Modes in Anisotropic Magnets: Spin-Wave Excitations in the Multiferroic $Ba_2CoGe_2O_7$" Phys. Rev. Lett. **108**, 257203 (2012).

[7] Y. Takahashi et al., "Magnetoelectric Resonance with Electromagnons in a Perovskite Helimagnet" Nat. Phys. 8, 121-125 (2011).

[8] D. Szaller, S. Bordács, I. Kézsmárki, "Symmetry Conditions for Nonreciprocal Light Propagation in Magnetic Crystals" Phys. Rev. B **87**, 014421 (2013).

# Nano-orbitronics in silicon

**B.N. Murdin[1], K. Litvinenko[1], Juerong Li[1], E. Bowyer[1], M. Pang[1], P.T. Greenland[2], B. Villis[2], G. Aeppli[2], A.F.G. van der Meer[3], B. Redlich[3], H. Engelkamp[3], and C.R. Pidgeon[4]**

[1]Advanced Technology Institute, University of Surrey, Guildford GU2 7XH, UK

[2]London Centre for Nanotechnology, University College London, London WC1H 0AH, UK

[3] Institute for Molecules and Materials, Radboud University Nijmegen, Toernooiveld 7, NL-6525 ED Nijmegen, The Netherlands

[4]Department of Physics, Heriot-Watt University, Edinburgh EH14 4AS, UK

b.murdin@surrey.ac.uk

Shallow donor impurities in silicon, once frozen out at low temperature, share many properties in common with free hydrogen atoms [1]. They have long been the subject of spectroscopic investigation, but it is only very recently [2,3] that it has been possible to investigate the time-domain dynamics of orbital excitations such as the 1s to 2p, due to the difficulty of obtaining short, intense pulses in the relevant wavelength range. These new techniques make shallow donors (and also acceptors [4]) attractive for studying atomic physics effects, and for applications in quantum information. We have measured the population dynamics of electrons orbiting around phosphorus impurities in commercially-available silicon, and shown that the lattice relaxation lifetime is about 200ps, only 1 order of magnitude shorter than the radiative lifetime of free hydrogen.

Coherent oscillation, where many particles cycle in phase, is responsible for classical phenomena like the emission of strong radio waves by many individual electrons in an antenna. At the quantum scale, coherent superposition of spin polarisations are central to Magnetic Resonance Imaging and its analogues, in which coherence is excited and then reappears later producing a delayed radio pulse (the spin "echo"). Quantum computer logic will also rely on coherence, and spins are often chosen as "qubits" because they are only weakly connected to, and disturbed by, the environment. Paradoxically, connection with the outside world is crucial for control, making charge (i.e. orbital) oscillations in semiconductors attractive. We have shown that silicon donor electrons can be put into a coherent superposition of orbital states that lasts for nearly as long as the lattice relaxation time. Our results pave the way for new devices where information is stored in single electron orbits ("coherent orbitronics") in silicon, the material that has dominated the classical computing industry for half a century.

 
J.-Y. Bigot et al. (eds), *Ultrafast Magnetism I*, Springer Proceedings in Physics 159
DOI 10.1007/978-3-319-07743-7_30

# References

[1] B.N. Murdin, Juerong Li, M.L.Y. Pang, E.T. Bowyer, K.L. Litvinenko, S.K. Clowes, H. Engelkamp, C.R. Pidgeon, I. Galbraith, N.V. Abrosimov, H. Riemann, S.G. Pavlov, H.W. Hübers & P.G. Murdin "Si:P as a laboratory analogue for hydrogen on high magnetic field white dwarf stars" Nature Communications **4**, 1469 (2013).

[2] N. Q. Vinh, P. T. Greenland, K. Litvinenko, B. Redlich, A. F. G. van der Meer, S. A. Lynch,† M. Warner, A. M. Stoneham, G. Aeppli, D. J. Paul, C. R. Pidgeon, and B. N. Murdin, " Silicon as a model ion trap: Time domain measurements of donor Rydberg states" Proc Nat Acad Sci USA **105**, 10649 (2008).

[3] P. T. Greenland, S. A. Lynch, A. F. G. van der Meer, B. N. Murdin, C. R. Pidgeon, B. Redlich, N. Q. Vinh & G. Aeppli, "Coherent control of Rydberg states in silicon" Nature **465**, 1057 (2010).

[4] N. Q. Vinh, B. Redlich, A. F. G. van der Meer, C. R. Pidgeon, P. T. Greenland, S. A. Lynch, G. Aeppli, and B. N. Murdin "Time-Resolved Dynamics of Shallow Acceptor Transitions in Silicon" Phys Rev *X* **3**, 011019 (2013).

# Evanescent exchange magnons in a 1D magnonic crystal

**M. Pereiro, C. Etz, L. Bergqvist, A. Bergman, and O. Eriksson**

Department of Physics and Astronomy, Uppsala University, P.O. Box 516, 751 20 Uppsala, Sweden

manuel.pereiro@physics.uu.se

**Abstract** We investigate the transport properties of propagating and non-propagating exchange modes in a 1D magnonic crystal composed of stacked alternating layers of cobalt (Co) and permalloy ($Fe_{20}Ni_{80}$). We observe that the exchange magnons excited with energies lying in the energy gap die as soon as the Co layer gets wider.

## Introduction

Among the metamaterials currently engineered by material's scientists, the magnonic crystals deserve a special attention because they are excellent candidates for the fabrication of nanoscale microwave devices, which are typically excited by either the application of time-varying external magnetic fields produced by microsize antennas [1] or by using the spin-transfer torque [2]. The magnonic crystal can be considered as a periodic composite, comprised of different magnetic materials so that the wavelengths of propagating modes, spin-waves or magnons, are orders of magnitude shorter than those of photons, of the same frequency. This feature has an immediate advantage over the electron-based devices for the overall demand of ever shrinking circuit units. For this reason, magnonics has become an emerging field which aims to control the generation and propagation of information carrying magnons in a way analoguous to the control of photons in photonic crystals. We devise applications of magnonics magneto-electronic devices, e.g. carrying out logical operations, and also as magnonic waveguides.

In this article, we study the transport properties of propagating and non-propagating modes in a 1D magnonic crystal composed of stacked alternating layers of cobalt (Co) and permalloy ($Fe_{20}Ni_{80}$).

 
J.-Y. Bigot et al. (eds), *Ultrafast Magnetism I*, Springer Proceedings in Physics 159
DOI 10.1007/978-3-319-07743-7_31

## Method and Results

Our aim is to combine first-principles calculations with atomistic spin dynamics simulations to study the evolution in time of the excited magnons of the 1D magnonic crystal composed of alternating layers of Co and Py in a simulation box of 20x20x100 unit cells. The method is described in Ref. [3] in more detail. We map an itinerant electron system onto an effective hamiltonian with classical spins,

$$H = -\sum_{i \neq j} J_{ij} \boldsymbol{m}_i \boldsymbol{m}_j + K \sum_i (\boldsymbol{m}_i \boldsymbol{e}_k)^2 \quad 1$$

where $i$ and $j$ are atomic indices, $\boldsymbol{m}_i$ are the classical atomic moments, and $J_{ij}$ are the exchange constants. $K$ represents the strength uniaxial anisotropy field along the direction $\boldsymbol{e}_k$. We used the SPR-KKR package [4] for calculations of the parameters to be used within the generalized hamiltonian of Eq. (1). The damping parameter was taken from experimental data ($\alpha$=0.008) [5].

We pumped the $Py_4Co_n$ (n=4,6,8) multilayer system with a time-dependent magnetic pulse. The pulse was applied just only to the two first rows of the atomic layers during 1 ps, which represents 12.5 % of the total simulation time (8 ps). The amplitude of the magnetic field was chosen to be 100 T in order to have clear results for the oscillation of the magnetic moments but it has no net effects in the main core results. The system was excited with energies in the permitted and forbidden region, i.e. 8 meV and 26 meV for the $Py_4Co_4$, 5 meV and 18 meV for the $Py_4Co_6$, and 25 meV and 44 meV for the $Py_4Co_8$, all of them in the THz regime.

In Fig. 1, we have plotted the dynamical structure factor for the three studied multilayer systems. We observe that the exchange magnon relation dispersion is quite similar to the one of dipolar magnons for a magnonic crystal. The increase of the Co layer width results in a reduction of the first band gap and in an enlargement of the second and higher order band gaps. This is in clear contradiction with the experiments reported for dipolar magnons [6] because the nature of the exchange interaction is different than the dipolar one. This fact has a clear advantage over the dipolar magnons, and it is that we can excite higher order modes with less energy consumption.

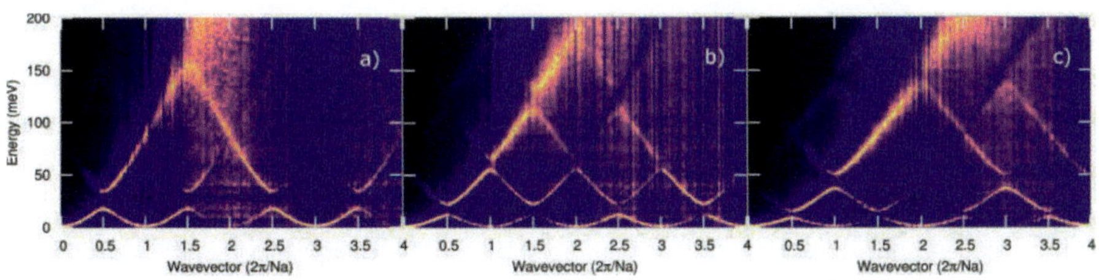


**Fig. 1.** Dynamic structure factor S(q,ω) for the multilayer a) $Py_4Co_4$, b) $Py_4Co_6$, and c) $Py_4Co_8$. In the figure "a" stands for the lattice constant and N=100 is the edge length of the simulation cell.

We plot in Fig. 2 a) the time evolution of the average magnetic moment. The results show that during the first picosecond, in which we applied the magnetic pulse (yellow area), the system is rapidly demagnetized and after that, it recovers very slowly the original magnetization. It is important to notice that as we increase the Co layer width, the slope of the curves are less steep and the gap between the permitted and forbidden modes reduces so that for the $Py_4Co_8$ system the permitted modes are below the forbidden ones.

In Figs. 2 a)-c), we show the time evolution of the x-component of some selected atomic moments. We used the x-component of the magnetic moment because the geometry of PyCo multilayers allows the x-component to be considered as a good quantum number in the language of quantum mechanics. We observe that in the forbidden region of the magnetic spectrum the evanescent exchange magnons die faster with the enlargement of the Co layer width, while the modes in the permitted region travel along the whole PyCo multilayer system.

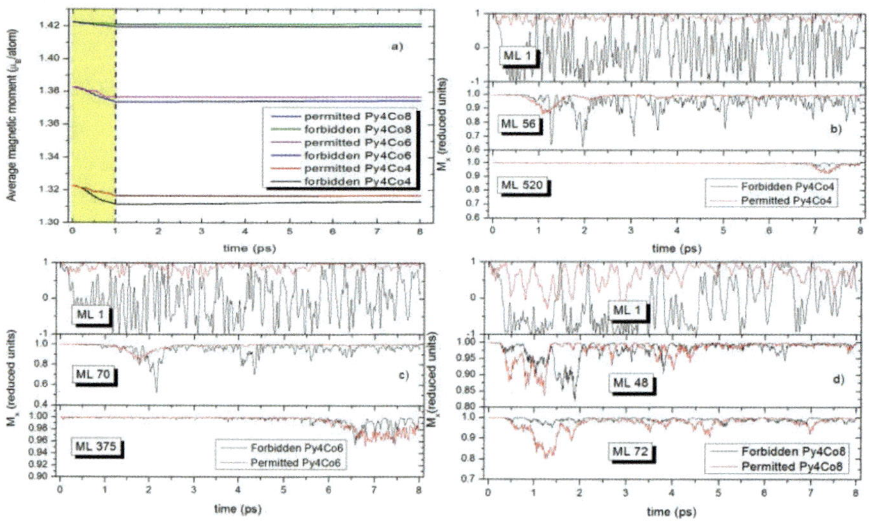


**Fig. 2.** Time evolution of the average and x-component of the magnetic moments in panel a) and b), c), and d), respectively. The PyCo system is excited with energies lying in the forbidden region (black line) and in the permitted one (red line). ML stands for monoatomic layer and every subpanel represents the number of the ML in which we have measured the magnetic moment.

**Acknowledgments** The authors thank the European Research Council (ERC Project No. 247062-ASD), the Swedish Research Council (VR), the Knut and Alice Wallenberg Foundation, the Carl Tryggers Foundation, and the Göran Gustafsson Foundation for financial support. L.B. and A.B. also acknowledge SeRC and eSSENCE, respectively. The computer simulations wereperformed on resources provided by the Swedish National Infrastructure for Computing (SNIC) at the National Supercomputer Centre (NSC), UppMAX, and High Performance Computing Center North (HPC2N).

# References

[1] S. O. Demokritov, A. N. Slavin, "Magnonics: From fundamentals to applications", (Springer, 2013).
[2] Z. Zeng et al., "Ultralow-current-density and bias-field-free spin-transfer nano-oscillator" Sci. Rep. **3**, 1426 (2013).
[3] B. Skubic et al., "A method for atomistic spin dynamics simulations: implementation and examples" J. Phys. Cond. Matter. **20**, 315203 (2008).
[4] H. Ebert, D. Ködderitzsch, and J. Minár, "Calculating condensed matter properties using the KKR-Green's function method—recent developments and applications" Rep. Prog. Phys. **74**, 096501 (2011).
[5] S. Serrano-Guisan et al., "Inductive and magneto-resistive measurements of Gilbert damping in Ni81Fe19 thin films and microstructures" J. Phys. D: Appl. Phys. **41**, 164015 (2008).
[6] Z. K. Wang et al., "Nanostructured magnonic crystals with size-tunable bandgaps" ACS nano **4**, 643 (2010).

# Magneto-optic Study of Picosecond Magnetization Dynamics in Garnet Films

**M.V. Logunov*[1], S,A. Nikitov[2], M.V. Gerasimov[1], A.V. Spirin[1], A.V. Balyasov[1]**

[1]National Research Ogarev Mordovia State University, Saransk 430005, Russia

[2]Kotel'nikov Institute of Radio Engineering and Electronics of RAS, Mokhovaya 11-7, Moscow 125009, Russia

* logunovmv@bk.ru

**Abstract** Magneto-optical system for the study of nano- and picosecond magnetization dynamics in garnet films by using Faraday or Kerr effects is presented. Time resolution of the system is determined by the duration of the laser pulses (10 ps) and the response time of the photo-detectors (30 ps).

Magnetization dynamics in the garnet films is interest for the study of magneto-optical phenomena, domain structures, photomagnetic effects [1-3]. In the talk, the system designed to study the nano- and picosecond dynamics in the garnets will be present.

The system works by using Faraday or Kerr effects. It has been designed for study dynamic domains and integral magnetization reversal signals simultaneously. Mode-locked Nd:YLF laser ($\lambda$ = 1053, 527, 351, 263 nm) with pulse duration of 10 ps was used as the light source for pump-and-probe method and for high speed photography of dynamic domain structures. Photo-detectors with response time of 30 ps were used for registration of integral magnetization reversal signals. Time resolution of the system is determined by the duration of the laser pulses and the response time of the photo-detectors. The study of garnet films can be realized in the temperature range 6 – 350 K due to optical cryostat for microscopy.

Experiments on the registration of the induction response from the effects of linearly polarized picosecond laser pulses onto the single-domain state garnet film (fig.1) and circularly polarized laser pulses onto the film with labyrinthine domain structure were produced. The initial phase of the response signal is determined either orientation of the initial magnetization of the film, or the direction of rotation of the electromagnetic field of circularly polarized laser radiation.

We used method of spatial filtering of optical signals in the diffraction pattern of the domain structure [4] to increase the sensitivity when the study of domain wall dynamics in small magnetic fields was necessary (fig.2).

Thus, several modes of operations of magneto-optical system for the study of nano- and picosecond magnetization dynamics were realized.



J.-Y. Bigot et al. (eds), *Ultrafast Magnetism I*, Springer Proceedings in Physics 159
DOI 10.1007/978-3-319-07743-7_32

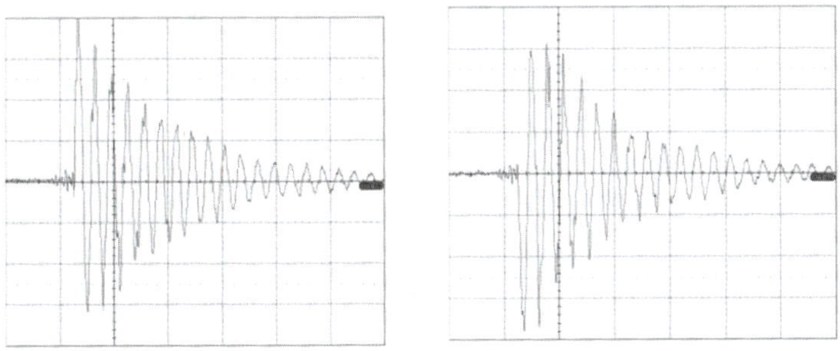

**Fig. 1.** Oscilloscope traces of e.m.f. signal obtained with the induction method for garnet film with opposite magnetizations (left, right). A source of linear polarized radiation: 10 ps pulse. Y: arb. units, X: 50 ns/div.

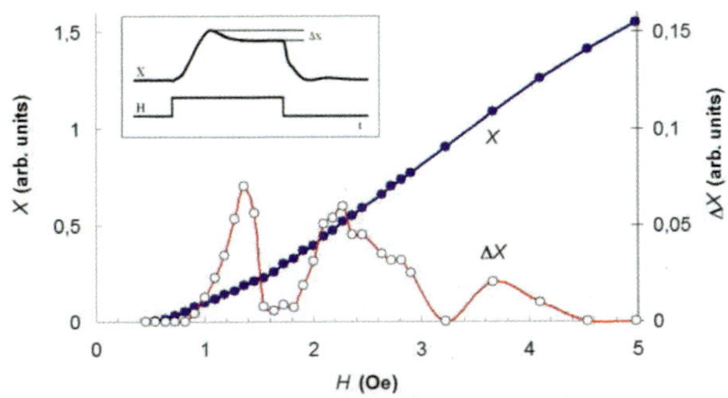


**Fig. 2.** Changes of maximal domain wall displacement $X$ with pulse magnetic field $H$ and corresponding value of the “ballistic” effect $\Delta X$.

**Acknowledgments** This work was supported by Ministry of Science and Education of Russia under grant 11.519.11.3023.

# References

[1] A. Kirilyuk, A. V. Kimel, T. Rasing, “Ultrafast optical manipulation of magnetic order” Rev. Mod. Phys. **82**, 2731 (2010).

[2] M. Logunov, S. Nikitov, M. Gerasimov, D. Kashkin, N. Loginov, A. Spirin, “Magneto-Optic Fraunhofer Diffraction on 2D Magnetic Domain Patterns” Solid State Phenomena **190**, 737-741 (2012).

[3] G. M. Genkin, Yu. N. Nozdrin, A. V. Okomel’kov, and I. D. Tokman, “Magnetization of a garnet film through a change in its multidomain structure under circularly polarized light” Phys. Rev. B **86**, 024405 (2012).

[4] A. K. Zvezdin, V. A. Kotov, “Modern Magnetooptics and Magnetooptical Materials”, (CRC Press, 2010), p. 404.

# Spin-polarized electron scattering in permalloy films: a spin-wave study

**Mohammad Haidar, Matthieu Bailleul**

Institut de Physique et Chimie des Matériaux de Strasbourg, UMR7504 CNRS / Université de Strasbourg, Strasbourg, France

bailleul@ipcms.unistra.fr

**Abstract** The current-induced spin-wave Doppler shift technique was used to measure the degree of spin-polarization of the electrical current in permalloy films of thickness 6-20nm. We find that the spin-polarization decreases as the film thickness decreases. This is interpreted within the two current model of spin-polarized transport.

The understanding of spin-polarized electron scattering in metal ferromagnets is of major importance to optimize spintronic devices. Although spin-polarized scattering by impurities and phonons in the bulk has been described in details in the 1970's [1], little is known about spin-polarized scattering by surfaces in simple ferromagnetic metal films. To address this issue, we follow the degree of spin-polarization of the electrical current $P=(J_{\uparrow}-J_{\downarrow})/(J_{\uparrow}+J_{\downarrow})$ for permalloy ($Ni_{80}Fe_{20}$) films as a function of their thickness (t=6, 10, 20nm) [2], resorting to the newly developed technique of current induced spin wave Doppler shift [3,4] which gives direct access to the parameter P.

Films of composition $Al_2O_3$ 21nm / Py t / $Al_2O_3$ 5nm were grown on intrinsic silicon substrates by magnetron sputtering. For each film thickness, we fabricated a device such as the one shown in figure 1(a). It consists of a permalloy strip and a pair of spin-wave antennae allowing one to measure the propagation of spin-waves with wave-vector k along this strip. Figure 1(b) shows the propagating spin-wave signals measured when a DC current I flows along the strip. One observes clear frequency shifts between the signals measured for the two polarities of I and k. With the help of a suitable symmetry analysis, we can separate the current-induced spin wave Doppler shift $\Delta\omega=P\mu_B Jk/-eM_s$ from other contributions (namely a current-independent frequency non-reciprocity, and a wave-vector independent residual Oersted field).

The degree of spin-polarization P extracted from this analysis is shown in Figure 2(a). One recognizes a clear decrease as the thickness decreases, from 0.62 at 20nm to 0.43 at 6nm. Qualitatively, this effect is related to a spin-depolarization induced by the surfaces, whose role gets more important as the thickness decreases. For a more quantitative understanding, we propose the two-current model

 
J.-Y. Bigot et al. (eds), *Ultrafast Magnetism I*, Springer Proceedings in Physics 159
DOI 10.1007/978-3-319-07743-7_33

sketched in figure 2(b). The electrical transport is assumed to flow in two nearly independent channels corresponding to each spin-direction and the resistivities corresponding to the different sources of electron scattering are assumed to add up within each channel.

To conclude, we have shown that the current-induced Doppler shift technique combined with a suitable two-current analysis could be used to obtain a detailed picture of spin-polarized transport in ferromagnetic metal films.

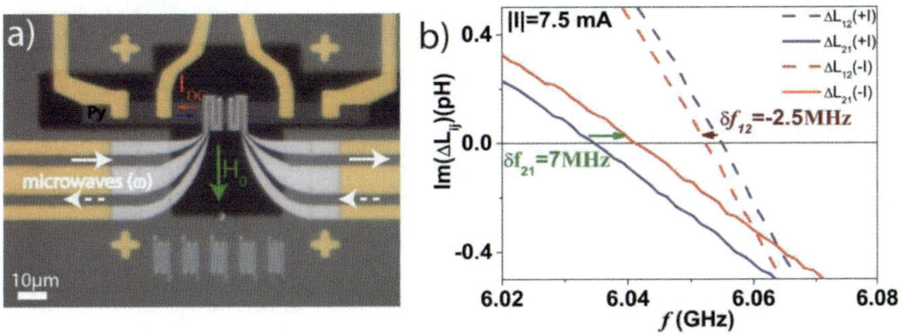


**Fig. 1.** (a) Optical microscope picture of a propagating spin-wave device. (b) Propagating spin-wave signals measured for k>0 (full) and k<0 (dashed) under I>0 (blue) and I<0 (red). Parameters are t=10nm, $k=3.8\mu m^{-1}$ and $\mu_0H_0=28mT$.

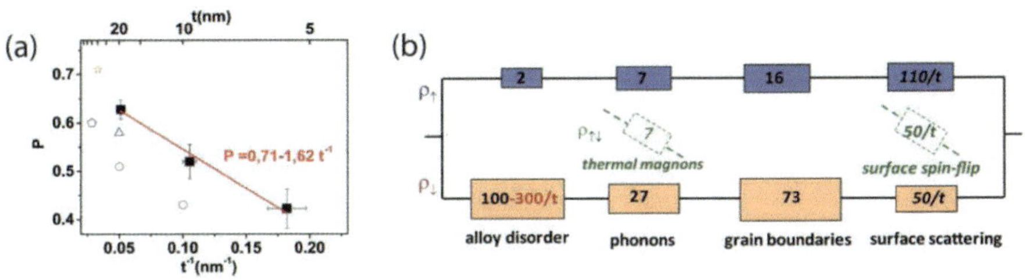


**Fig. 2.** (a) Degree of spin-polarization of the electrical current as a function of the inverse of the film thickness. Open symbols are published values. (b) Two current model accounting for the degree of spin-polarization and the resistivity measured on our films. Resistivities are in μΩ.cm and thickness in nm.

**Acknowledgments** The authors thank M. Acosta, C. Uhlaq, M. Romeo, P. Bernhardt, Y. Henry and F. Gautier for their help. This work was supported by ANR (grant ANR-11-BS-10-003).

## References

[1] I. Campbell and A. Fert, in "Ferromagnetic Materials", vol.3 (North Holland, 1982, Wohlfarth Ed.).
[2] M. Haidar, M. Bailleul, " Thickness dependence of the degree of spin polarization of the electrical current in permalloy thin films", cond-mat > arXiv:1303.1692.
[3] V. Vlaminck, M. Bailleul, "Current-induced spin wave Doppler shift" Science **322**, **410** (2008).
[4] M. Zhu, C. L. Dennis, R. D. McMichael, "Temperature dependence of magnetization drift velocity and current polarization in Ni80Fe20 by spin-wave Doppler measurements" Phys. Rev. B **81**, 140407(R) (2010).

# Spin-wave modes in a CoFeB magnonic crystal waveguide

**M. Mansurova, M. Münzenberg**

I. Physikalisches Institut, Georg-August Universitaet Goettingen, Friedrich Hund Platz 1, 37077, Goettingen, Germany

mmansur@gwdg.de, mmuenze@gwdg.de

**Abstract** Femtosecond laser pulses are used to excite and measure the magnetization dynamics on a CoFeB magnonic waveguide using the double-modulated pump-probe technique. Damon-Eshbach spin-wave modes are detected on a continuous film when exciting the magnonic crystal. This suggests that the spin waves in the waveguide are induced by the magnonic crystal.

## Introduction

The development of spin-wave applications such as logic gates which prototype is a simple spin waveguide [1] requires the understanding of propagation properties of spin waves. In particular, it is important to investigate whether spin-wave modes excited on the magnonic crystal can be detected on continuous film and how the excitation of these modes can be enhanced. Here we report a promising result for magnonic spin-wave computing.

## Sample details and experiment

In this experimental work, femtosecond laser pulses are used to excite and measure magnetization dynamics on a magnonic waveguide, consisting of a 13,01 $\mu$m wide defect line in a rectangular CoFeB antidot lattice, oriented either perpendicularly or at 45° to the external magnetic field. Samples were prepared by magnetron sputtering continuous 50 nm thick CoFeB film, where later the antidot lattice was milled using Focused Ion Beam (FIB) technique. CoFeB is a very promising material for magnonic devices because of its very low damping (Gilbert damping $\alpha$=0.006) and long spin wave-propagation of ~100 $\mu$m.

Spin waves are excited by an intense (13 mJ/cm$^2$) pump pulse that locally raises the temperature of the sample, creating an effective anisotropy field pulse and

 
J.-Y. Bigot et al. (eds), *Ultrafast Magnetism I*, Springer Proceedings in Physics 159
DOI 10.1007/978-3-319-07743-7_34

subsequently triggering the magnetization precession around an equilibrium position. Dynamic magnetization precession is detected also optically using time-resolved MOKE technique described elsewhere [2].

Propagation properties of spin waves in the waveguide for each orientation case (90° or 45° to the bias magnetic field) are investigated by systematically exciting the sample along the waveguide or from the continuous film.

Magnetization precession on the waveguide was measured during 1 ns after the arrival of the excitation pulse, and Fourier-transformed to detect fundamental frequencies of this oscillation at different magnetic fields (spin-wave spectra). Furthermore the pump beam was displaced either along the waveguide or into the continuous film, and the relative Fourier power amplitude of each mode was obtained.

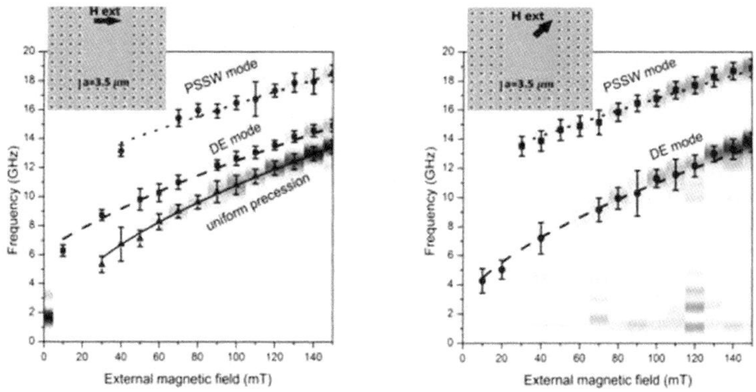


**Fig 1**. Magnetization dynamics spectra on a magnonic CoFeB waveguide a) perpendicular and b) 45° to external magnetic field. The wave vector of the Damon-Eshbach (DE) mode is determined by reciprocal lattice vector $\pi/a$.

## Results and discussion

At each orientation of the magnetic field, spin-wave spectra of the waveguide show the same modes as rectangular CoFeB antidot lattice, namely, Damon-Eshbach (DE) surface mode, perpendicular spin standing wave (PSSW) and Kittel uniform precession (90° orientation only) in agreement to the previously reported results and parameters [3] as showed in Fig1. This suggests that the spin waves in the waveguide are induced by the antidot lattice. This was confirmed by the fact that when exciting the sample on the antidot lattice only, the same modes are detected on the waveguide.

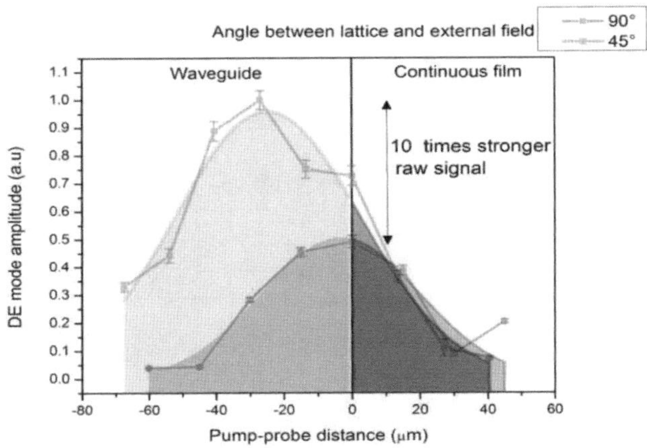


**Fig 2.** Relative FFT power amplitude as pumping from the waveguide or continuous film at 130 mT external magnetic field.

The magnonic modes propagate both into the waveguide and continuous film, although DE mode is not normally present in the continuous CoFeB thin film spectra [3]. The propagation is enhanced when propagating into the waveguide. Moreover, at 45° to the external magnetic field orientation, the DE mode propagation length into the waveguide is larger compared to 90°-orientation case. This can be understood by the specific symmetry of band structure of square lattice magnonic crystals with in-plane magnetization [3]. Thus we present the first proof of our ideas that magnonic waveguide modes can propagate and their propagation can be simply manipulated by the direction of the applied magnetic field.

**Acknowledgments** Project is funded by DFG MU 1780/ 6-1 Photo-Magnonics

## References

[1] B. Lenk, H. Ulrichs, F. Garbs, M. Muenzenberg, "The building blocks of magnonics" Physics Reports **507**, 107 (2011).
[2] J. Walowski, M. D. Kaufmann, B. Lenk, C. Hamann, J. Mc-Cord, M. Muenzenberg, "Intrinsic and non-local Gilbert damping in polycrystalline nickel examined applying Ti:Sapphire laser femtosecond spectroscopy" Journal of Physics D: Applied Physics **41**, 164016 (2008).
[3] H. Ulrichs, B. Lenk, and M. Muenzenberg, "Magnonic spin-wave modes in CoFeB antidot lattice" Appl. Phys. Lett. **97**, 092506 (2010).

# Laser-induced Giant Skyrmions and Skyrmion-compounds in a thin magnetic film of TbFeCo

**M. Savoini,[1,2] M. Finazzi,[2] A.R. Khorsand,[1] A. Tsukamoto,[3] A. Itoh,[3] L. Duò,[2] M. Ezawa,[4] A. Kirilyuk,[1] and Th. Rasing[1]**

[1] Radboud University Nijmegen, Institute for Molecules and Materials, Heyendaalseweg 135, 6525 AJ Nijmegen, The Netherlands

[2] Dipartimento di Fisica, Politecnico di Milano, Piazza Leonardo da Vinci 32, 20133 Milano, Italy

[3] College of Science and Technology, Nihon University, 7-24-1 Funabashi, Chiba, Japan

[4] Department of Applied Physics, University of Tokyo, Hongo 7-3-1, 113-8656, Japan

m.savoini@science.ru.nl

**Abstract** Here we report about the formation of topologically protected Skyrmions through illumination of a thin metallic film with short laser pulses (about 150 fs). By tuning the laser fluence it is possible to change the topological properties of the created structure.

Skyrmions, originally introduced to explain how baryons could topologically emerge from a continuous meson field [1], have found many exciting applications in condensed-matter physics, where they describe topological states of matter in a wide range of systems [2]. Topological states in solids are attracting increasing attention because of their potential practical applications in a whole new range of electronic and magnetic applications and possibly also for quantum computation [3,4].

In this Proceeding, we report the totally unexpected discovery of the creation of a new type of magnetic Skyrmions in a ferrimagnet film generated *without any external field* and with the help of an ultrashort laser pulse. The resulting Skyrmions have much larger sizes than previously reported [5] structures and their observation is possible with scanning near-field optical microscopy (SNOM). With this new approach, the created and observed Skyrmions and Skyrmion-bound states of a Skyrmion-antiSkyrmion pair are of submicron dimension and very stable in time. Moreover, by varying the laser fluence we are able to control the size of the Skyrmions and their topological order [6].

In our experiment we have first prepared a thin $Tb_{22}Fe_{69}Co_9$ ferrimagnetic alloy film (thickness = 20 nm) in a state consisting of large stripes with out-of-plane magnetization, obtained by applying a magnetic field of about 1 T in the direction perpendicular to the film plane and simultaneously heating the sample with a laser

 
J.-Y. Bigot et al. (eds), *Ultrafast Magnetism I*, Springer Proceedings in Physics 159
DOI 10.1007/978-3-319-07743-7_35

beam (spot diameter ~50 μm, λ = 800 nm). We have then irradiated these stripes with 150 fs single laser pulses (spot diameter ~2 μm, λ = 800 nm). The sample was kept at room temperature. A fundamental difference in the experimental conditions with respect to previous works [7-9] is that no external magnetic field was applied during this step. Finally, we mapped the out-of-plane component of the local magnetization with sub-diffraction resolution by measuring the Faraday rotation of linearly polarized light (λ = 635 nm) transmitted through the film from a hollow pyramid tip (aperture diameter: ~100 nm) that was scanned over the sample [10]. The Faraday rotation, i.e. the rotation of the plane of polarization of linearly polarized light traveling through a magnetic material, is proportional to the component of the magnetization in the direction of propagation. Therefore, by scanning the tip over the sample we are able to obtain a map of the out-of-plane component of the local magnetization with sub-diffraction resolution.

As shown in Fig. 1a, magnetic "bubbles" are obtained after single laser pulse irradiation with a pulse energy density of about 5 mJ/cm2. Within experimental accuracy, these sub-μm magnetic domains display a local magnetization in their center that is completely reversed with respect to the one of the surrounding material, meaning that these structures are indeed Skyrmions. Their radius depends on the energy of the single laser pulse that has been used to create them and can be as small as 100 nm. The observation that such Skyrmion structures can be created on the same sample with opposite magnetization orientations rules out the possibility that their stabilization could be mediated by the possible presence of uncontrolled external sources of magnetic fields and/or by the Earth field. Moreover, their lifetime is very long as they have not shown any noticeable change when they were imaged several months after the first measurement.

As anticipated, the isolated Skyrmions described above are not the only structures related to topological states that can be created in TbFeCo films. At higher laser fluences with respect to those leading to Skyrmion generation (corresponding to an energy density up to 7 mJ/cm2 for each single laser pulse) we observe the formation of "doughnut"-shaped magnetic domains like the one shown in Fig. 1b. In these structures the magnetization in the center is the same as the one of the surrounding medium, and it is completely reversed in the annular region around the core. This texture can be viewed as two Skyrmions with opposite topological numbers (i.e. a Skyrmion and an antiSkyrmion) nested one into the other. This gives a state with zero Skyrmion number, corresponding to a nontopological soliton. Such an arrangement can be named Skyrmionium (Sk), in analogy with states of matter constituted by a particle bound to its antiparticle [11].

Although the current structures leave room for improvements as far as film homogeneity and control of sizes is concerned, the present results clearly demonstrate that by illuminating a thin TbFeCo film with single ultrafast laser pulses we can locally create stable magnetic configurations corresponding to spin textures characterized by well-defined topological quantum numbers. By tuning the laser fluence we can control the topology and create various bound states of Skyrmion molecules (Skyrmion "chemistry"). An important feature is that, at variance from the topological magnetic excitations observed by previous works, the ones reported here are stable even in the absence of an externally applied magnetic field. The ability of controlling the dimension and the number of Skyrmions or antiSkyrmions created during a single laser illumination event and the fact that such structures are topologically protected and thus intrinsically stable might have far reaching implications for the further development of non-volatile high-density magnetic recording and long term memories of the next generation.

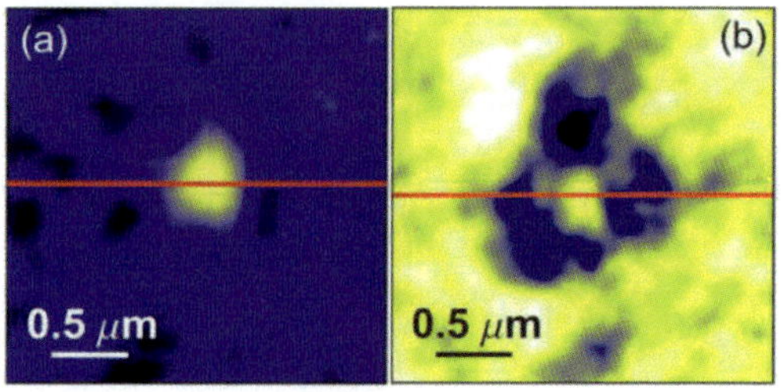


**Fig. 1.** Laser-induced (a) Skyrmion and (b) Skyrmionium mapped by the use of Scanning Near-Field Optical Microscopy [6].

**Acknowledgments** The authors thank A.V. Kimel for fruitful discussions. This research has received funding from Fondazione Cariplo through project PONDER (Rif. 2009-2726), from MIUR through PRIN Project No. 2008J858Y7, from the European Union (EU) Nano Sci-European Research Associates (ERA) project FENOMENA, from Stichting voor Fundamenteel Onderzoek der Materie (FOM), De Nederlandse Organisatie voor Wetenschappelijk Onderzoek (NWO), and EC FP7 [Grants No. NMP3-SL-2008-214469 (UltraMagnetron) and No. 214810 (FANTOMAS)]. Grant-in-Aid for Scientific Research from the Ministry of Education, Science, Sports and Culture of Japan No. 22740196 is also gratefully acknowledged.

## References

[1] T. Skyrme, "A unified field theory of mesons and baryons" Nucl. Phys. 31, 556-569 (1962).
[2] G. E. Brown, & M. Rho, "The Multifaced Skyrmions" (World Scientific, Singapore, 2010).
[3] M. Z. Hasan, & C. L. Kane, "Colloquium: Topological insulators" Rev. Mod. Phys. 82, 3045-3067 (2010).
[4] D. Hsieh et al., "Observation of unconventional quantum spin textures in topological insulators" Science 323, 919-922 (2009).
[5] M. Ezawa, "Giant Skyrmions Stabilized by Dipole-Dipole Interactions in Thin Ferromagnetic Films" Phys. Rev. Lett. 105, 197202 (2010).
[6] M. Finazzi et al., "Laser-Induced Magnetic Nanostructures with Tunable Topological Properties" Phys. Rev. Lett. 110, 177205 (2013).
[7] S. Mühlbauer, B. Binz, F. Jonietz, C. Pfleiderer, A. Rosch, A. Neubauer, R. Georgii, P. Böni, "Skyrmion Lattice in a Chiral Magnet" Science 323, 915-919 (2009).
[8] T. Ogasawara, N. Iwata, Y. Murakami, H. Okamoto, Y. Tokura, "Submicron-scale spatial feature of ultrafast photoinduced magnetization reversal in TbFeCo thin film" Appl. Phys. Lett. 94, 162507 (2009).

[9] X. Z. Yu et al., "Near room-temperature formation of a skyrmion crystal in thin-films of the helimagnet FeGe" Nat. Mater. 10. 106-109 (2011).
[10] M. Savoini, P. Biagioni, L. Duò, and M. Finazzi, "All-optical subdiffraction multilevel data encoding onto azo-polymeric thin films" Opt. Lett. 34, 761-763 (2009).
[11] D. B.Cassidy, & A. P. Jr. Mills, "The production of molecular positronium" Nature 449, 195-197 (2007).

# Part IV
# Theory of Spin Dynamics

# Theory of femtosecond laser-induced demagnetization

**Karel Carva[1,2], Marco Battiato[1], Dominik Legut[3], and Peter M. Oppeneer[1]**

[1]Department of Physics and Astronomy, Uppsala University, SE-751 20 Uppsala, Sweden

[2]Charles University, Department of Condensed Matter Physics, KeKarlovu 5, CZ-12116 Prague 2, Czech Republic

[3]Nanotechnology Centre, VSB-Techn. University of Ostrava, CZ-708 33 Ostrava, Czech Republic

peter.oppeneer@physics.uu.se

**Abstract** Using *ab initio* calculations we computed the ultrafast demagnetization that can be achieved by Elliott-Yafet electron-phonon spin-flip scatterings in laser-excited ferromagnets. Our calculations show that nonequilibrium laser-created distributions contribute mostly to the ultrafast demagnetization. Nonetheless, the total Elliott-Yafet contribution is too small to account for the fs-demagnetization.

More than a decade ago it was discovered that excitation of a metallic 3d ferromagnet with an intensive femtosecond laser-pulse causes an ultrafast demagnetization within 300 fs [1]. In spite of the huge technological potential of this discovery the mechanism underlying the femtosecond demagnetization could not yet be uncovered and remains a controversial issue [2-5]. Several theories have been proposed – mostly based on the assumption that there must exist some form of an ultrafast channel for the dissipation of spin angular momentum [6-9]. A different rationale is to consider ultrafast transport of highly mobile, laser-excited spin-polarized electrons [10]. Such spin transport in the superdiffusive regime could effectively cause demagnetization, by transporting spins rapidly out of the laser-excited region.

Clearly, further developments of theory and experiment are needed to discriminate between the various proposed models and eliminate some of them. On the experimental side experiments need to be devised which can distinguish whether an occurring spin-flip (SF) was actually due to scattering with a phonon, magnon, or other quasi-particle. On the theory side it is needed to compute *ab initio* the SF scattering rate due to electron-phonon or electron-magnon scattering, but such calculations can be regarded as a tour-de-force at the current forefront of the field. Conversely, ultrafast transport of spin angular momentum by laser-excited hot electrons can be detected by specially constructed element-sensitive experiments. Several experiments in this direction have been reported recently [11-13].

 
J.-Y. Bigot et al. (eds), *Ultrafast Magnetism I*, Springer Proceedings in Physics 159
DOI 10.1007/978-3-319-07743-7_36

One of the prominent mechanisms proposed recently to provide a very fast SF channel is the Elliott-Yafet electron-phonon SF scattering [9]. Estimations of the effectiveness of electron-phonon SF scattering were given within the Microscopic Three Temperature Model (M3TM). Several of the demagnetization ratios were large (up to 50%), and could be reproduced by the M3TM when SF probabilities $\alpha_{sf}$ of about 0.15 to 0.20 were assumed in the fits. Such magnitudes of SF probabilities were supported by calculations based on the Elliott approximation [9]. As a large demagnetization was already estimated for the thermalized electron system, it was concluded that highly excited nonequilibrium(NEQ) electron distributions can be disregarded for explaining the ultrafast magnetization dynamics.

To establish accurately the contribution ofthe Elliott-Yafet electron-phonon SF scattering to fslaser-induced demagnetization,we have developed an *ab initio* theory [14] on the basis of a generalized Eliashberg function, which describes the electron-phonon scattering. We use the relativistic bandstructure, which has been obtained from density functional theory calculations. Also the employed phonon dispersions have been *ab initio*computed. A similar investigation was recently carried out by Essert and Schneider [15].

For the description of laser-excited metals we introduce as central quantity a generalized energy- and spin- dependent Eliashberg function, $\alpha^2_{\sigma\sigma'}F(E,\Omega)$, where $\sigma$□ $\sigma'$ are the spin indices pertaining to the electron states in the scattering, $E$ the energy with respect to the Fermi energy ($E_F$), and $\Omega$ is the phonon energy (which can be negative for phonon absorption). The energy dependence is required to treat three distinct situations: 1) the 3*d*ferromagnet at low-temperature equilibrium, 2) the case of hot electrons in the thermalized electron regime, and 3) the case of laser-excited electrons in strong NEQ. In the conventional Eliashberg formulation the electron-phonon scattering occurs practically at $E_F$. After laser irradiation however there is initially a considerable redistribution of electrons, creating many holes at energies deep below $E_F$ as well as a population of hot electrons above $E_F$; this is a strongly NEQ distribution which cannot be described by any Fermi-Dirac distribution. Through electron-electron scattering these electrons thermalize in some 200-300 fs and can then be described by a Fermi-Dirac distribution with a high electron temperature $T_e$. Note that the electrons and phonon system have not yet thermalized with one another; the electron-lattice thermalization requires a few picoseconds in Ni [16].

Once the spin-dependent Eliashberg functions have been computed, we can straightforwardly compute spin-resolved transition rates $S^{\sigma\sigma'}$which, for convenience, we separate in the spin-decreasing rate $S^-$and spin-increasing rate$S^+$. It is important to note that not all SF electron-phonon scattering events lead to a reduction of the moment, they can equally well cause an increase of the moment. The total SF probability is hence given by $P_S \alpha S^- + S^+$ but for the actual demagnetization the balance of spin-decreasing and -increasing rates is essential, which is given by a normalized demagnetization ratio,$D_S \alpha S^- - S^+$. Demagnetization follows from $dM/dt = 2\mu_B(S^- - S^+)$. Note that the total SF probability $P_S$ does not enter.

The developed theory has been used to compute SF probabilities and amounts of demagnetization for bcc Fe and fcc Co and Ni [17]. The results are collected in Table 1. The calculations reveal, first, that there exists a large difference between $P_S$ and $D_S$; a factor of 10 is not uncommon, but it is less for laser-created NEQ distributions. An essential quantity is the demagnetization fraction, $\Delta M/M_0$, given in % per 250 fs. The latter number stems from the experimental fact that within about 250 fs the electrons have thermalized and the NEQ distribution has vanished. As can be recognized, the demagnetization amount for hot electrons in the electron-thermalized regime, even with a high $T_e$, is just a few percent per 250 fs. This is clearly much less than the experimentally observed demagnetizations. A larger net demagnetization is computed for NEQ distributions, but also this contribution is not sufficient to explain the measured ultrafast demagnetization.

**Table 1.** *Ab initio* calculated spin-flip probabilities $P_S$, demagnetization ratios $D_S$, and demagnetization fractions $\Delta M/M_0$, given in percent as per 250 fs, for Fe, Co, and Ni.

| | $P_S$ | $D_S$ | $\Delta M/M_0$ |
|---|---|---|---|
| Fe (low Tequil.) | 0.04 | 0 | 0 |
| Fe ($T_e$ = 3000K) | 0.09 | 0.008 | 4.5 |
| Fe (NEQ) | 0.07 | 0.030 | 11.4 |
| Co (low Tequil.) | 0.01 | 0 | 0 |
| Co ($T_e$ = 3000K) | 0.017 | 0.002 | 0.9 |
| Co (NEQ) | 0.022 | 0.010 | 2.3 |
| Ni (low Tequil.) | 0.04 | 0 | 0 |
| Ni ($T_e$ = 3000K) | 0.07 | 0.003 | 3.1 |
| Ni (NEQ) | 0.09 | 0.025 | 16.7 |

Several observations can be made. It is remarkable that the outcome of the *ab initio* calculations is precisely *opposite* to that of the M3TM simulations, which stipulated that NEQ distributions could be fully disregarded [9]. We obtain conversely a non-negligible contribution of NEQ distributions (but a negligible one from thermalized distributions), which is due to a large SF scattering amounting from energetically deep states beneath $E_F$. The surprising disparity between our *ab initio* calculations and the M3TM has been related recently to the fact that the here-employed band structure is "static" or "rigid" [18], where"rigid"is used to imply that the exchange splitting in first-order approximation remains unchanged. Note however that for metals like Fe the reduction of exchange splitting even at Curie temperature is almost negligible, and for Ni it is still quite small [19]. Therefore our model is actually much more realistic than the assumption of completely vanishing exchange splitting at the Curie temperature as used in Ref. 18. Our calculations do pertain to the first part of the demagnetization process where the occurring SFs have not yet given rise to second order changes in the magnetization. However, as we shall detail in a forthcoming publication, the origin of the large difference can be related to several other factors. One of these can already be recognized from Table 1. The thermalized SF probabilities $\alpha_{sf}$ (used in [9]) are about 0.15 for Co and 0.185 for Ni; we obtained smaller values ($P_S$=0.02 for Co and 0.07 for Ni), *however*, as explained above, the essential quantity $D_S$ is only 0.003, i.e., much smaller. In addition there is an important difference in how the energy of the electron is changed in an electron-phonon scattering process, which also contributes to the much higher demagnetization ratios estimated in the M3TM.

In conclusion, ultrafast Elliott-Yafet electron-phonon SF scattering does contribute to femtosecond laser-induced demagnetization, but it cannot explain the magnitude of the phenomenon. Other contributions need to be considered, such as superdiffusive spin transport or other channels for ultrafast spin dissipation.

**Acknowledgments** The authors gratefully acknowledge financial support from the EU (grant No. 281043, "FemtoSpin").

## References

[1] E. Beaurepaire, J.-C. Merle, A. Daunois, J.-Y. Bigot, "Ultrafast Spin Dynamics in Ferromagnetic Nickel" Phys. Rev. Lett. **76**, 4250 (1996).
[2] A. Kirilyuk, A. Kimel, Th. Rasing, "Ultrafast optical manipulation of magnetic order" Rev. Mod. Phys. **82**, 2731 (2010).
[3] P. M. Oppeneer, A. Liebsch, "Ultrafast demagnetization in Ni: theory of magneto-optics for non-equilibrium electron distributions" J. Phys.: Condens. Matter **16**, 5519 (2004).
[4] G. P. Zhang, W. Hübner, G. Lefkidis, Y. Bai, T. F. George, "Paradigm of the time-resolved magneto-optical Kerr effect for femtosecond magnetism" Nature Phys. **5**, 499 (2009).
[5] K. Carva, M. Battiatio, P.M. Oppeneer, "Reply: "Is the controversy over femtosecond magneto-optics really solved?"" Nature Phys. **7**, 665 (2011).
[6] G. P. Zhang, W. Hübner, "Laser-Induced Ultrafast Demagnetization in Ferromagnetic Metals" Phys. Rev. Lett. **85**, 3025 (2000).
[7] E. Carpene, E. Mancini, C. Dallera, M. Brenna, E. Puppin, S. De Silvestri, "Dynamics of electron-magnon interaction and ultrafast demagnetization in thin iron films" Phys. Rev. B **78**, 174422 (2008).
[8] M. Krauss, T. Roth, S. Alebrand, D. Steil, M.Cinchetti, M. Aeschlimann, H. C. Schneider, "Ultrafast demagnetization of ferromagnetic transition metals: The role of the Coulomb interaction" Phys. Rev. B **80**, 180407 (2009).
[9] B. Koopmans, G. Malinowski, F. Dalla Longa, D. Steiauf, M. Fähnle, T. Roth, M. Cinchetti, M. Aeschlimann, "Explaining the paradoxical diversity of ultrafast laser-induced demagnetization" Nature Mater. **9**, 259 (2010).
[10] M. Battiato, K. Carva, P.M. Oppeneer, "Superdiffusive Spin Transport as a Mechanism of Ultrafast Demagnetization" Phys. Rev. Lett. **105**, 027203 (2010).
[11] D. Rudolf, C. La-O-Vorakiat, M. Battiato, R. Adam, J. M. Shaw, E. Turgut, P. Maldonado, S. Mathias, P. Grychtol, H. T. Nembach, T. J. Silva, M. Aeschlimann, H. C. Kapteyn, M. M. Murnane, C. M. Schneider, P. M. Oppeneer, "Ultrafast magnetization enhancement in metallic multilayers driven by superdiffusive spin current" Nature Commun. **3**, 1037 (2012).
[12] B. Vodungbo et al., "Laser-induced ultrafast demagnetization in the presence of a nanoscale magnetic domain network." Nature Commun. **3**, 999 (2012).
[13] B. Pfau et al., "Ultrafast optical demagnetization manipulates nanoscale spin structure in domain walls" NatureCommun. **3**, 1100 (2012).

[14] K. Carva, M. Battiatio, P.M. Oppeneer, "Ab Initio Investigation of the Elliott-Yafet Electron-Phonon Mechanism in Laser-Induced Ultrafast Demagnetization" Phys. Rev. Lett. **107**, 207201 (2011).
[15] S. Essert, H. C. Schneider, "Electron-phonon scattering dynamics in ferromagnetic metals and their influence on ultrafast demagnetization processes" Phys. Rev. B **84**, 224405 (2011).
[16] X. Wang, S. Nie, J. Li, R. Clinite, J. E. Clark, J. Cao, "Temperature dependence of electron-phonon thermalization and its correlation to ultrafast magnetism" Phys. Rev. B **81**, 220301 (2010).
[17] K. Carva, M. Battiatio, D. Legut, P.M. Oppeneer, "Ab initio theory of electron-phonon mediated ultrafast spin relaxation of laser-excited hot electrons in transition-metal ferromagnets" Phys. Rev. B **87**, 184425 (2013)
[18] A.J. Schellekens, B. Koopmans, "Comparing Ultrafast Demagnetization Rates Between Competing Models for Finite Temperature Magnetism" Phys. Rev. Lett. **110**, 217204 (2013)
[19] A. L. Wysocki, R. F. Sabirianov, M. van Schilfgaarde, K. D. Belashchenko, "First-principles analysis of spin-disorder resistivity of Fe and Ni" Phys. Rev. B **80**, 224423 (2009).

# Relaxation dynamics of majority and minority electrons after ultrashort laser excitation

**B.Y. Mueller, M. Cinchetti, M. Aeschlimann, H.C. Schneider, and B. Rethfeld**

Department of Physics and Research Center OPTIMAS, University of Kaiserslautern, Erwin-Schrödinger-Str. 46, 67663 Kaiserslautern, Germany

bmueller@physik.uni-kl.de

**Abstract** We analyze the ultrafast demagnetization after ultrashort laser excitation by a kinetic model based on Elliott-Yafet scattering processes. By applying complete Boltzmann scattering integrals we trace the majority and minority electrons as well as phonons in the sample. Additionally, we allow for dynamical changes in the exchange splitting between majority and minority electrons. We find that our model has the potential to describe the magnetization dynamics and provides insights in the relaxation dynamics of the non-equilibrium electron system.

It is experimentally well established that irradiating ferromagnetic films with an ultrashort laser pulse leads to a quenching of the magnetization on a sub-picosecond timescale [1–4]. Since the discovery of the ultrafast demagnetization in a nickel film in 1996, there are intense discussions about the fundamental microscopic mechanism of this ultrafast process [5– 14] and a lot of experiments on permalloys [15], Heusler alloys [16] and multilayers [17] have been performed to investigate this effect in more detail. There are several proposals for the underlying microscopic mechanism such as, for instance, spin-flip scattering due to electron-electron [11, 16, 18], electron-phonon [2, 18–20] as well as electron-magnon [6, 21] interactions, coherent processes [5, 22] and superdiffusive transport [7, 10, 17, 23, 24]. Recently, "superdiffusive transport" has been introduced in 2010 by Battiato et al. [10] and seems to be a promising way to describe the magnetization dynamics for bulk or multilayer systems. However, we show that also the Elliott-Yafet scattering can describe the demagnetization in thin films and conclude that one has to consider both effects for a consistent model of ultrafast demagnetization dynamics.

In this work we use a dynamical approach to Elliott-Yafet [25] type spin-flip scattering that includes electron-electron and electron-phonon interactions at the level of Boltzmann scattering integrals [18, 26, 27] as well as an effective spin-mixing $b^2$ due to the spin-orbit interaction [18]. We directly calculate the microscopic time, energy and spin dependent distributions functions $f^{\uparrow,\downarrow}(E, t)$ for majority and minority electrons by using energy resolved scattering integrals and mate-

 
J.-Y. Bigot et al. (eds), *Ultrafast Magnetism I*, Springer Proceedings in Physics 159
DOI 10.1007/978-3-319-07743-7_37

rial-specific spin-dependent densities of states. The corresponding Boltzmann equation has the form

$$\frac{\partial f_{\Delta}^{\eta}}{\partial t} = \sum_{\sigma,\nu,\lambda} |\langle \eta, \sigma | \nu, \lambda \rangle|^2 \Gamma_{\text{el-el}}^{\eta,\sigma,\nu,\lambda}[\Delta] + \sum_{\sigma} |\langle \eta | \sigma \rangle|^2 \Gamma_{\text{el-ph}}^{\eta,\sigma}[\Delta] + \sum_{\sigma} |\langle \eta | \sigma \rangle|^2 \Gamma_{\text{exc}}^{\eta,\sigma}[\Delta] \quad \text{1a}$$

$$\frac{\partial g}{\partial t} = \sum_{\sigma} |\langle \eta | \sigma \rangle|^2 \Gamma_{\text{ph-el}}^{\eta,\sigma}[\Delta] \quad \text{1b}$$

with $\eta, \sigma, \nu, \lambda \in \{\uparrow, \downarrow\}$denoting the spin state and $\Gamma$ indicating the electron-electron and electron-phonon Boltzmann collision integral corresponding to the respective index. The notation $f_{\Delta}^{\eta}$ and $[\Delta]$ indicates the dependence of the distribution functions and, consequently, the Boltzmann scattering rates on the exchange splitting via the shifted energy dispersions for spin-up and spin-down electrons in a Stoner model. Importantly, we include a dynamic exchange splitting $\Delta(t) = U_{eff}M(t)$ between the majority and minority density of states which is calculated by an effective Coulomb interaction $U_{eff}$ which determines the Curie temperature and the transient magnetization [28].

In the framework of our model, in which the band structure is characterized by majority and minority densities of states and the dynamical exchange splitting, we study the importance of the equilibration of temperatures and chemical potentials between majority and minority carriers in the ultrafast magnetization dynamics.

As a result, Fig. 1 illustrates several simulations for a nickel density of states [29] for different spin-mixing parameters $b^2$ and the Curie temperatures $T_C$. The Gaussian laser pulse with FWHM of $\tau = 50$ fs has a wavelength of 800 nm and a fluence of 3:5 mJ=cm$^2$.

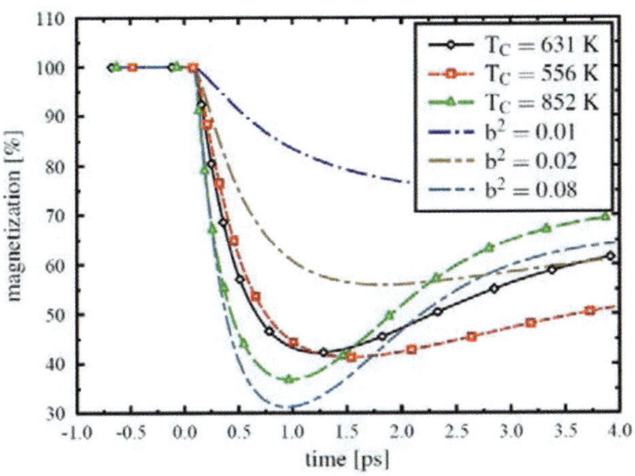


**Fig. 1.** Ultrafast magnetization dynamics for different the spin-mixing parameters $b^2$ (with constant $T_C = 631$ K) and Curie temperatures $T_C$ with constant $b^2 = 0{:}05$.

We conclude that both superdiffusive transport and Elliott-Yafet type scattering may play a considerable role in ultrafast demagnetization dynamics. With our model, we show that the dynamical exchange splitting yields a much larger quenching of magnetization than what is possible in a fixed band structure [19].

**Acknowledgments** Financial support of the Deutsche Forschungsgemeinschaft through the Emmy-Noether and Heisenberg Programm as well as the AHRP for providing the computational facilities is gratefully acknowledged.

## References

[1] E. Beaurepaire, J.-C. Merle, A. Daunois, and J.-Y. Bigot, Phys. Rev. Lett. **76**, 4250 (1996).
[2] B. Koopmans, G. Malinowski, F. Dalla Longa, D. Steiauf, M. Fähnle, T. Roth, M. Cinchetti, and M. Aeschlimann, "Explaining the paradoxical diversity of ultrafast laser-induced demagnetization" Nature Materials **9**, 259 (2010).
[3] C. Stamm, T. Kachel, N. Pontius, R. Mitzner, T. Quast, K. Holldack, S. Khan, C. Lupulescu, E. F. Aziz, M. Wietstruk, H. A. D¨urr, and W. Eberhardt, "Femtosecond modification of electron localization and transfer of angular momentum in nickel" Nature Materials **6,** 740 (2007).
[4] M. Cinchetti, M. S´anchez Albaneda, D. Hoffmann, T. Roth, J.-P. W¨ustenberg, M. Krauß, O. Andreyev, H. C. Schneider, M. Bauer, and M. Aeschlimann, "Spin-Flip Processes and Ultrafast Magnetization Dynamics in Co: Unifying the Microscopic and Macroscopic View of Femtosecond Magnetism" Phys. Rev. Lett. **97**, 177201 (2006).
[5] J.-Y. Bigot, M. Vomir, and E. Beaurepaire, "Coherent ultrafast magnetism induced by femtosecond laser pulses" Nature Physics **5**, 515 (2009).
[6] E. Carpene, E. Mancini, C. Dallera, M. Brenna, E. Puppin, and S. De Silvestri, "Dynamics of electron-magnon interaction and ultrafast demagnetization in thin iron films" Phys. Rev. B **78**, 174422 (2008).
[7] G. P. Zhang and W. Hübner, "Laser-Induced Ultrafast Demagnetization in Ferromagnetic Metals" Phys. Rev. Lett. **85**, 3025 (2000).
[8] N. Kazantseva, U. Nowak, R.W. Chantrell, J. Hohlfeld, and A. Rebei, "Slow recovery of the magnetisation after a sub-picosecond heat pulse" Europhysics Letters **81**, 27004 (2008).
[9] B. Koopmans, J. J. M. Ruigrok, F. Dalla Longa, and W. J. M. de Jonge, "Unifying Ultrafast Magnetization Dynamics" Phys. Rev. Lett. **95**, 267207 (2005).
[10] M. Battiato, K. Carva, and P. M. Oppeneer, "Superdiffusive Spin Transport as a Mechanism of Ultrafast Demagnetization" Phys. Rev. Lett. **105**, 027203 (2010).
[11] M. Krauss, T. Roth, S. Alebrand, D. Steil, M. Cinchetti, M. Aeschlimann, and H. C. Schneider, "Ultrafast demagnetization of ferromagnetic transition metals: The role of the Coulomb interaction" Physical Review B **80**, 180407 (2009).
[12] D. Steiauf and M. F¨ahnle, "Elliott-Yafet mechanism and the discussion of femtosecond magnetization dynamics" Phys. Rev. B **79**, 140401 (2009).
[13] D. Steiauf, C. Illg, and M. F¨ahnle, "Demagnetization on the fs time-scale by the Elliott-Yafet mechanism" Journal of Physics: Conference Series **200**, 042024 (2010).
[14] M. Fähnle, J. Seib, and C. Illg, "Relating Gilbert damping and ultrafast laser-induced demagnetization" Phys. Rev. B **82**, 144405 (2010).
[15] S. Mathias, C. La-o vorakiat, P. Grychtol, P. Granitzka, E. Turgut, J. M. Shaw, R. Adam, H. T. Nembach, M. E. Siemens, S. Eich, C. M. Schneider, T. J. Silva, M. Aeschlimann, M. M. Murnane, and H. C. Kapteyn, "Probing the timescale of the exchange interaction in a ferromagnetic alloy" PNAS **109**, 4792 (2012).

[16] D. Steil, S. Alebrand, T. Roth, M. Krauß, T. Kubota, M. Oogane, Y. Ando, H. C. Schneider, M. Aeschlimann, and M. Cinchetti, "Band-Structure-Dependent Demagnetization in the Heusler Alloy $Co_2Mn_{1-x}Fe_xSi$" Phys. Rev. Lett. **105**, 217202 (2010).
[17] D. Rudolf, C. La-O-Vorakiat, M. Battiato, R. Adam, J. M. Shaw, E. Turgut, P. Maldonado, S. Mathias, P. Grychtol, H. T. Nembach, T. J. Silva, M. Aeschlimann, H. C. Kapteyn, M. M. Murnane, C. M. Schneider, and P. M. Oppeneer, "Ultrafast magnetization enhancement in metallic multilayers driven by superdiffusive spin current" Nature comm. **3**, 1037 (2012).
[18] B. Y. Mueller, T. Roth, M. Cinchetti, M. Aeschlimann, and B. Rethfeld, "Driving force of ultrafast magnetization dynamics" New Journal of Physics **13**, 123010 (2011).
[19] S. Essert and H. C. Schneider, "Electron-phonon scattering dynamics in ferromagnetic metals and their influence on ultrafast demagnetization processes" Phys. Rev. B **84**, 224405 (2011).
[20] K. Carva, M. Battiato, and P. M. Oppeneer, "Ab *Initio* Investigation of the Elliott-Yafet Electron-Phonon Mechanism in Laser-Induced Ultrafast Demagnetization" Phys. Rev. Lett. **107**, 207201 (2011).
[21] A. B. Schmidt, M. Pickel, M. Donath, P. Buczek, A. Ernst, V. P. Zhukov, P. M. Echenique, L. M. Sandratskii, E. V. Chulkov, and M. Weinelt, "Ultrafast Magnon Generation in an Fe Film on Cu(100)" Phys. Rev. Lett. **105**, 197401 (2010).
[22] G. Lefkidis, G. Zhang, and W. Hübner, "Angular Momentum Conservation for Coherently Manipulated Spin Polarization in Photoexcited NiO: An Ab *Initio* Calculation" Phys. Rev. Lett. **103**, 217401 (2009).
[23] A. Eschenlohr, M. Battiato, P. Maldonado, N. Pontius, T. Kachel, K. Holldack, R. Mitzner, A. Föhlisch, P. M. Oppeneer, and C. Stamm, "Ultrafast spin transport as key to femtosecond demagnetization" Nature materials **12**, 332 (2013).
[24] M. Battiato, K. Carva, and P. M. Oppeneer, "Theory of laser-induced ultrafast superdiffusive spin transport in layered heterostructures" Phys. Rev. B **86**, 024404 (2012).
[25] I. Žutić, J. Fabian, and S. Das Sarma, "Spintronics: Fundamentals and applications" Reviews of Modern Physics **76**, 323 (2004).
[26] B. Rethfeld, A. Kaiser, M. Vicanek, and G. Simon, "Ultrafast dynamics of nonequilibrium electrons in metals under femtosecond laser irradiation" Phys. Rev. B **65**, 214303 (2002).
[27] B. Y. Mueller and B. Rethfeld, "Relaxation dynamics in laser-excited metals under nonequilibrium conditions" Phys. Rev. B **87**, 035139 (2013).
[28] W. Nolting and A. Ramakanth, Quantum Theory of Magnetism (Springer, 2009).
[29] Z. Lin, L. V. Zhigilei, and V. Celli, "Electron-phonon coupling and electron heat capacity of metals under conditions of strong electron-phonon nonequilibrium" Phys. Rev. B **77**, 075133 (2008).

# A local approach to ultrafast magnetization dynamics in ferromagnetic transition metals

**W. Töws and G.M. Pastor**

Institut für Physik, Universität Kassel, Heinrich-Plett-Straße 40, 34132 Kassel, Germany

wtoews@uni-kassel.de

**Abstract** A many-body theory of femtosecond laser-induced magnetization dynamics in itinerant ferromagnets is presented, which includes Coulomb and spin-orbit interactions, as well as interatomic hybridizations on the same footing. Applications to small systems demonstrate that ultrafast demagnetization can be understood as a purely electronic spin-orbit coupling triggered mechanism.

## Introduction

The phenomenon of laser-induced ultrafast demagnetization of ferromagnetic transition metals has been revealed by pioneering experiments almost two decades ago [1]. Since then several pump and probe experiments on thin films have shown that a breakdown of the magnetic order occurs after the absorption of an intense fs laser pulse on a time scale of about 100 fs. Despite a large amount of experimental and theoretical work aiming to explain the microscopic origin of such an ultrafast demagnetization [2,3,4,5], the underlying mechanisms are not thoroughly understood. Most of the theoretical approaches in this field are based on the Elliott-Yafet spin-flip scattering mechanism [6], which relies on a single-particle band-like or delocalized description of the electronic states. This appears to be a conceptual limitation, since magnetism in *d*-electron systems is known to be a many-body effect, which results from non-trivial electronic correlations favoring the formation of local magnetic moments.

The purpose of this work is to present a local perspective to the problem of laser-induced ultrafast demagnetization, which is expected to complement the most common *k*-space approaches. To this aim we consider a many-body Hamiltonian which takes into account hybridizations, Coulomb interactions and spin-orbit coupling (SOC) on the same footing [7]. The model is solved by including the electron correlations and the local magnetic degrees of freedom resulting from them. In this approach the transfer between spin $S$ and orbital angular momentum $L$ as well as its interplay with the electronic motion in the lattice are treated exactly within the model.

 
J.-Y. Bigot et al. (eds), *Ultrafast Magnetism I*, Springer Proceedings in Physics 159
DOI 10.1007/978-3-319-07743-7_38

## Electronic many-body model

A many-body model of the relevant valence electrons is used to describe the laser-induced magnetization dynamics. The corresponding lattice Hamiltonian

$$\hat{H} = \hat{H}_0 + \hat{H}_C + \hat{H}_{SO} + \hat{H}_{ext}(t) \quad (1)$$

takes into account various electronic contributions on the same footing: $H_0$ describes the single-particle electronic structure in the tight-binding approximation. It is responsible for electron delocalization, band formation and metallic behavior. $H_C$ models the electron-electron interactions by focusing on the dominant intra-atomic Coulomb repulsion within the $3d$ orbitals. This term introduces the correlations between the electrons and, in particular, leads to the formation and stabilization of local magnetic moments at the different atoms. $H_{SO}$ is the SOC, which is ultimately responsible for the transfer of angular momentum between spin and orbital degrees of freedom. Finally, $H_{ext}(t)$ represents the intra-atomic dipole approximation. It accounts for the initial laser excitation.

Notice that the lattice degrees of freedom (phonons) are not taken explicitly into account. This is motivated by two main reasons. First, we expect that at the fs time scale the electrons can be decoupled from the much slower lattice dynamics. Second, it is fundamentally important to understand the role of the spin relaxation on a purely electronic level and the role of SOC triggered angular momentum transfer from $S$ towards $L$ in the ultrafast demagnetization. In this context it is worth to observe that as a result of $H_0$ angular momentum is transferred from the atomic orbits to the lattice.

## Results and discussion

The many-body model (1) has been used to determine the time evolution of the spin relaxation exactly. In order to keep the numerical effort affordable, we have reduced the number of atomic orbitals that are explicitly taken into account in the dynamics: from 5 to 3 for the $3d$ and from 3 to 1 for the $4p$ orbitals. Moreover, a small cluster having $N = 3$ atoms and periodic boundary conditions is considered. The results for the initially ferromagnetic state subject to a fs laser pulse show that the spin $S_z$ decreases by a factor ½ during the first few hundreds of femtoseconds following the absorption. At the same time the orbital angular momentum $L_z$ remains almost quenched. The local magnetic moments remain stable throughout the relaxation process.

These theoretical results suggest an alternative interpretation of the microscopic demagnetization mechanism. The laser pulse pumps energy into the electronic system, creates a cloud of excited electrons and opens an important number of channels for spin-to-orbital relaxation. Since the SOC conserves the total angular momentum, angular momentum is transferred locally from $S$ towards $L$ on a time scale of several tens to hundreds of femtoseconds. However, $L$ is subsequently quenched by the crystal field due to electronic hoppings throughout the lattice. This takes place on a time scale of a few femtoseconds, i.e., almost instantaneously in the time scale of the experiment. The combination of these two effects results in a demagnetization within a few hundred femtoseconds.

One concludes that the essential microscopic mechanism can be regarded as local: the SOC acts indeed locally at atomic sites. Therefore, we expect this purely electronic mechanism to have universal validity in ferromagnetic transition metals. Finally, the stability of local magnetic moments indicates that the ultrafast demagnetization can be traced back to local fluctuations of magnetic moments in their orientation. This means that many-body correlations play an essential role for the magnetization dynamics on the ultrashort time scale.

**Acknowledgments** One of the authors (W.T.) thanks the Otto-Braun Foundation for financial support.

## References

[1] E. Beaurepaire, J.-C. Merle, A. Daunois, and J.-Y. Bigot, "Ultrafast Spin Dynamics in Ferromagnetic Nickel" Phys. Rev. Lett. **76**, 4250-4253 (1996).
[2] G. P. Zhang and W. Hübner, "Laser-Induced Ultrafast Demagnetization in Ferromagnetic Metals" Phys. Rev. Lett. **85**, 3025-3028 (2000).
[3] H.-S. Rhie, H. A. Dürr, and W. Eberhardt, "Femtosecond Electron and Spin Dynamics in Ni/W(110) Films" Phys. Rev. Lett. **90**, 247201 (2003).
[4] C. Stamm, T. Kachel, N. Pontius, R. Mitzner, T. Quast, K. Holldack, S. Khan, C. Lupulescu, E. F. Aziz, M. Wietstruk, H. A. Dürr, and W. Eberhardt, "Femtosecond modification of electron localization and transfer of angular momentum in nickel" Nature Mat. **6**, 740 (2007).
[5] B. Koopmans, G. Malinowski, F. Dalla Longa, D. Steiauf, M. Fähnle, T. Roth, M. Cinchetti, and M. Aeschlimann, "Explaining the paradoxical diversity of ultrafast laser-induced demagnetization" Nature Mat. **9**, 259 (2010).
[6] D. Steiauf and M. Fähnle, "Elliott-Yafet mechanism and the discussion of femtosecond magnetization dynamics" Phys. Rev. B **79**, 140401(R) (2009).
[7] W. Töws and G. M. Pastor, to be published (2013).

# Ultrafast quenching of the exchange interaction in a Mott-insulator

**Johan H. Mentink, Martin Eckstein**

Center for Free-Electron Laser Science (CFEL), Max Planck Institute for the Structure and Dynamics of Matter (MPSD) , Notkestraße 85, 22607 Hamburg, Germany

johan.mentink@mpsd.cfel.de

**Abstract** The exchange interaction determines the ordering of microscopic spins in magnetic materials. While often considered static, here we report calculations of time-dependent nonequilibrium exchange interactions in a Mott insulator. The results suggest that free charge carriers created by (photo) excitation can cause an ultrafast quench of the exchange interaction.

The ordering of microscopic spins in magnetic materials is determined by the exchange interaction, the strongest force in magnetism. Ultrafast magnetism is associated with the dynamics of magnetism on a timescale comparable, or smaller, than the timescale set by the period of spin motion in the exchange field. The exchange interaction itself is rooted in the Coulomb interaction between the electrons and hence can possibly be controlled directly by a laser-excitation. Nevertheless, many experimental results, including laser-induced ultrafast demagnetization [1] and magnetization reversal [2], can be modeled by considering the exchange interaction as essentially unchanged by the laser-excitation [3, 4]. As a consequence, the dynamics of the exchange interactions under electronic nonequilibrum conditions has so far largely been ignored in the study of ultrafast magnetism.

In this contribution we address the problem of nonequilibrium exchange interactions by combining two recent major methodological developments. Simulating the dynamics of interacting electrons out of equilibrium has become possible owing to the nonequilibrium extension of dynamical mean field theory [5,6]. When applied to the magnetically ordered case [7,8], this framework allows for the direct evaluation of time-dependent effective exchange interactions, using the formal expressions that have recently been derived by A. Secchi and coworkers [9], by a generalization of the well-known formulas for the exchange interaction in the static case [10,11].



J.-Y. Bigot et al. (eds), *Ultrafast Magnetism I*, Springer Proceedings in Physics 159
DOI 10.1007/978-3-319-07743-7_39

## Mathematical framework

The nonequilibrium exchange interactions are defined within the framework of the Kadonoff-Baym formalism, where the fundamental object is the Greens function

$$G_{ab}^{\sigma\sigma'}(t,t') = -i\langle \mathcal{T}_{\mathcal{C}} c_{a\sigma}(t) c_{b\sigma'}^{+}(t')\rangle \quad 1$$

This Green's function describes the electronic single-particle excitations in a system that is initially in equilibrium, and whose time-evolution explicitly includes all external perturbations such as the pump laser pulse. Here $a, b$ label site and orbital indices, $\sigma = \uparrow, \downarrow$ is the spin index, $c^{+}(c)$ is a creation (annihilation) operator of electrons and $t, t'$ are times on a L-shaped contour $\mathcal{C}$ as defined by the time-evolution operator. $\mathcal{T}_{\mathcal{C}}$ ensures the appropriate time ordering of the operators along $\mathcal{C}$. Note that under nonequilibrium conditions time-translation invariance is lost and hence the Greens function depends on the two times t and t' separately. To calculate the actual dependence of the Greens function on the two times we need to solve the Dyson equation:

$$\left[(\underline{G_0}^{-1} - \underline{\Sigma}) \circ \underline{G}\right](t,t') = \underline{\delta}(t,t') \quad 2$$

Here all quantities are matrices in the site, spin and orbital indices, $\circ$ denotes convolution over the contour and $\underline{G_0}^{-1}$ is a differential operator on the contour with site and orbital matrix elements defined as

$$G_{0,ab}^{-1}(t,t') = \delta_{ab}\delta_{\mathcal{C}}(t,t')[i\partial_t + \mu] - \delta_{\mathcal{C}}(t,t')\, h_{ab}(t) \quad 3$$

Here $h_{ab}(t)$ are the hopping matrix elements that can be time-dependent due to an externally applied field and $\mu$ is the chemical potential. Further $\underline{\delta}(t,t')_{ab} = \delta_{ab}\delta_{\mathcal{C}}(t,t')$ and we have defined the self-energy Sigma $\underline{\Sigma}(t,t')$, which stems from the interactions between the electrons. In the DMFT approximation the self-energy is local in space, but it keeps time-nonlocal correlations. Within this Kadonoff-Baym formalism, exchange interactions are defined as the objects that appear after rotating the local quantization axes of the spins by a small angle from a given magnetically ordered reference state. Subsequent averaging over electronic states yields a symmetric spin-spin interaction that can be written as [9]:

$$\begin{aligned}\mathcal{J}_{ab}(t,t') = -\frac{1}{4}\mathrm{Re}\Big[ & I_{ab}^{\downarrow}(t,t') I_{ba}^{\uparrow}(t',t) + J_{ab}^{\downarrow}(t,t') J_{ba}^{\uparrow}(t',t) \\ & - \left[\Sigma_{ab}^{\downarrow}(t,t') + \left[\Sigma^{\downarrow} \circ J^{\downarrow}\right]_{ab}(t,t')\right] G_{ba}^{\uparrow}(t',t) \\ & - G_{ab}^{\downarrow}(t,t') \left[\Sigma_{ba}^{\uparrow}(t',t) + \left[I^{\uparrow} \circ \Sigma^{\uparrow}\right]_{ba}(t',t)\right]\Big]\end{aligned} \quad 4$$

Here the convolutions involve the quantities $I_{ab}^{\sigma}(t,t') = [\Sigma^{\sigma} \circ G^{\sigma\sigma}]_{ab}(t,t')$ and $J_{ab}^{\sigma}(t,t') = [G^{\sigma\sigma} \circ \Sigma^{\sigma}]_{ab}(t,t')$.

Comparison of the nonequilibrium exchange interaction with the well-known static equilibrium exchange interaction is possible when the system evolves to a quasi-stationary nonequilibrium state. In this case the dependence on relative time $t - t'$ is much faster than the dependence on average time $T = (t + t')/2$. Integration over relative time then gives effective time-dependent nonequilibrium exchange interactions that depending on average time $T$ only.

## Results

To investigate nonequilibrium exchange interactions we choose the prototype Mott-Hubbard insulator, which displays antiferromagnetic order for sufficiently low temperatures. In this system, (photo) excited electron and hole-like charge carriers can have a relatively slow recombination rate, i.e., they live long enough such that a quasi-stationary nonequilibrium magnetic state can be established [8]. Our calculations of the time-dependent nonequilibrium exchange interactions suggest that such slowly decaying charge carriers cause an ultrafast quench of the effective exchange interaction.

**Acknowledgments** J.H.M. acknowledges funding from the Nederlandse Organisatie voor Wetenschappelijk onderzoek (NWO) by a Rubicon Grant.

## References

[1] E. Beaurepaire, J.-C. Merle, A. Daunois, J.-Y. Bigot, "Ultrafast spin dynamics in ferromagnetic Nickel" Phys. Rev. Lett. **76**, 4250-4553 (1996).
[2] I. Radu et al., "Transient ferromagnetic-like state mediating ultrafast reversal of antiferromagnetically coupled spins" Nature **472**, 205–208 (2011).
[3] A. Kirilyuk, A. Kimel, T. Rasing., "Ultrafast optical manipulation of magnetic order" Rev. Mod. Phys. **82**, 2731–2784 (2010).
[4] J.H. Mentink et al., "Ultrafast Spin Dynamics in Multisublattice Magnets" Phys. Rev. Lett. **108**, 057202 (2012).
[5] J.K. Freericks et al., "Nonequilibrium Dynamical Mean-Field Theory" Phys. Rev. Lett. **97**, 266408 (2006).
[6] A. Georges et al., "Dynamical mean-field theory of strongly correlated fermion systems and the limit of infinite dimensions" Rev. Mod. Phys. **68**, 13–125 (1996).
[7] N. Tsuji et al., "Nonthermal antiferromagnetic order and nonequilibrium criticality in the Hubbard model", Phys. Rev. Lett. **110**, 136404 (2013).
[8] Werner et al., "Nonthermal symmetry-broken states in the strongly interacting Hubbard model", Phys. Rev. B **86**, 205101 (2012).
[9] A. Secchi et al., "Nonequilibrium magnetic interactions in strongly correlated systems" Ann. Phys. **333**, 221-271 (2013).
[10] A.I. Liechtenstein et al., "Exchange interactions and spin-wave stiffness in ferromagnetic metals" J. Phys. F: Met. Phys. **14**, L125-L128 (1984).
[11] M.I. Katsnelson and A.I. Lichtenstein, "First-principles calculations of magnetic interactions in correlated systems" Phys. Rev. B **61**, 8906-8912 (2000).

# Spin dynamics and exchange interactions from the first- and second-principles calculations

**Mikhail I. Katsnelson**

Theory of Condensed Matter, Institute for Molecules and Materials, Radboud University Nijmegen, Heyendaalseweg 135, 6525 AJ, Nijmegen, The Netherlands

M.Katsnelson@science.ru.nl

Magnetic ordering and related phenomena are of essentially quantum and essentially many-body origin and require strong enough electron-electron interactions [1]. Also, they are very sensitive to the details of electronic structure of specific materials [2]. This makes a truly microscopic description of exchange interactions a challenging task. Usually, one cannot use any natural perturbation parameters related with the strength of interactions. Long ago we suggested a general scheme of calculations of exchange interactions responsible for magnetism based on the "magnetic force theorem", when one considers a response of a system on small rotations of spins with respect to a (collinear) ground state magnetic configuration [3,4]. It was formulated originally as a method to map the spin-density functional to effective classical Heisenberg model, the exchange parameters turned out to be, in general, essentially dependent on initial magnetic configuration and not universal [5]. However, they are directly related to the spin-wave spectrum and, thus, can be verified experimentally [6]. Now it is the standard scheme used for many different classes of magnetic materials, from dilute magnetic semiconductors [7] to molecular magnets [8]. This approach also lies in the base of "ab initio spin dynamics" within the density functional approach [9].

It is well known now that this scheme is, in general, insufficient for strongly correlated systems and should be combined with the mapping to the multiband Hubbard model and use of, say, dynamical mean-field theory to treat the latter [10,11]. Our original approach can be reformulated for this "second-principle" method, the results are expressed in terms of Green's functions and (local) electron self-energy [12,13]. It can be also generalized to the case of relativistic magnetic interactions, such as Dzyaloshinskii-Moriya interactions [14]. Very recently, we have extended this scheme to the case of time-dependent Hamiltonians which opens a way to a consequent microscopic theory of laser-induced spin dynamics in strongly correlated systems [15].

**Acknowledgments** This work is supported by the European Union ERC Grant Agreement No. 281043 (FEMTOSPIN) and by the Stichting voor Fundamenteel Onderzoek der Materie (FOM), The |Netherlands.



J.-Y. Bigot et al. (eds), *Ultrafast Magnetism I*, Springer Proceedings in Physics 159
DOI 10.1007/978-3-319-07743-7_40

# References


[1] S. V. Vonsovsky, "Magnetism", (Wiley, 1974).

[2] V. A. Gubanov, A. I. Liechtenstein, A. V. Postnikov, "Magnetism and the Electronic Structure of Crystals", (Springer, 1992).

[3] A. I. Liechtenstein, M. I. Katsnelson, V. A. Gubanov, "Exchange interactions and spin-wave stiffness in ferromagnetic metals", J. Phys. F **14**, L125 (1984).

[4] A. I. Liechtenstein, M. I. Katsnelson, V. P. Antropov, V. A. Gubanov, "Local spin density functional approach to the theory of exchange interactions in ferromagnetic metals and alloys", J. Magn. Magn. Mater. **67**, 65 (1987).

[5] S. A. Turzhevskii, A. I. Lichtenstein, M. I. Katsnelson, "Degree of localization of magnetic moments and the non-Heisenberg character of exchange interactions in metals and alloys", Sov. Phys. Solid State **32**, 1138 (1990).

[6] M. I. Katsnelson and A. I. Lichtenstein, "Magnetic susceptibility, exchange interactions and spin-wave spectra in the local spin density approximation", J. Phys.: Cond. Mat. **16**, 7439 (2004).

[7] O. Eriksson, L. Bergqvist, B. Sanyal, J. Kudrnovský, V. Drchal, P Korzhavyi, I. Turek, "Electronic structure and magnetism of diluted magnetic semiconductors", J. Phys.: Cond. Mat. **16**, S5481 (2004).

[8] D. W. Boukhvalov, V. V. Dobrovitski, M. I. Katsnelson, A. I. Lichtenstein, B. N. Harmon, P. Kogerler, "Electronic structure and exchange interactions in $V_{15}$ magnetic molecules: LDA+U results", Phys. Rev. B **70**, 054417 (2004).

[9]. V. P. Antropov, M. I. Katsnelson, B. N. Harmon, M. van Schilfgaarde, D. Kusnezov, "Spin dynamics of magnets: equation of motion and finite temperature effects", Phys. Rev. B **54**, 1019 (1996).

[10]. A. I. Lichtenstein, M. I. Katsnelson, "Ab initio calculations of quasiparticle band structure in correlated systems: LDA++ approach", Phys. Rev. B **57**, 6884 (1998).

[11]. G. Kotliar, S. Y. Savrasovm K. Haule, V. S. Oudovenko, O. Parcollet, C. A. Marianetti, "Electronic structure calculations with dynamical mean-field theory", Rev. Mod. Phys. **78**, 865 (2006).

[12] M. I. Katsnelson and A.I. Lichtenstein, First-principles calculations of magnetic interactions in correlated systems, Phys. Rev. B **61**, 8906 (2000).

[13] M. I. Katsnelson, A. I. Lichtenstein, "Electronic structure and magnetic properties of correlated metals. A local self-consistent perturbation scheme", Eur. Phys. J B **30**, 9 (2002).

[14] M. I. Katsnelson, Y. O. Kvashnin, V. V. Mazurenko, and A. I. Lichtenstein, "Correlated band theory of spin and orbital contributions to Dzyaloshinskii-Moriya interactions", Phys. Rev. B **82**, 100403(R) (2010).

[15] A. Secchi, S. Brener, A. I. Lichtenstein, and M. I. Katsnelson, "Non-equilibrium magnetic interactions in strongly correlated systems". Ann. Phys. (NY) **333**, 221 (2013).

# Λ-Processes Induced By Chirped Lasers

**G. Lefkidis and W. Hübner**

Dept. of Physics and Research Center OPTIMAS, University of Kaiserslautern, Germany

lefkidis@physik.uni-kl.de

**Abstract** We analytically treat model Λ-processes driven by chirped lasers. The obtained results (consistent with previous numericalfindings) bring insight into the microscopic origin of the ERASE functionality via frequency detuning in the presence of chirp, and explain the pertinent differences in the magnetic-switching behavior at Fe, Co and Ni atoms.

## Introduction

The discovery of the laser-induced spin manipulation in magnetically ordered materials [1] resulted in a huge amount of theoretical [2,3] and experimental [4] literature in an effort to understand the exact microscopic mechanisms. Lately the scientific community seems to agree that, at least at short times, the photon-electron interaction is mainly responsible for this phenomenon. A mechanism proposed to functionalize magnetic multicentered structures to logic elements [5] necessitates the ERASE functionality in order to prepare logic states. Although it was shown that chirped pulses can exhibit this effect,since they break the time-inversion symmetry, [6] the deeper mathematical understanding is still lacking.

## Results and discussion

Following Ref. [7] we solve the Λ system in the interaction picture. The chirp $\alpha$ enters in the varying laser frequency $\omega_{\text{laser}} = \omega_0 + \alpha t$, which becomes resonant with the transition frequencies $\omega_{\text{laser}} = \omega_0 = \hbar(E_c - E_a) = \hbar(E_c - E_b)$ at time $t_{\text{res}}$. Without loss of generality we set$t_{\text{res}} = 0$. Once the equations of motion are found we can change to any arbitrary $t_{\text{res}}$ via a translation of the time axis. Within the rotating wave approximation the system of the coupled differential equations for the wave functions of the initial $|a\rangle$, final $|b\rangle$, and intermediate $|c\rangle$, states become

$$\dot{a}(t) = -i\mu c(t)e^{-i\alpha t^2},\ \dot{b}(t) = i\mu c(t)e^{-i\alpha t^2},\ \dot{c}(t) = -i\mu[a(t) - b(t)]e^{i\alpha t^2},$$



J.-Y. Bigot et al. (eds), *Ultrafast Magnetism I*, Springer Proceedings in Physics 159
DOI 10.1007/978-3-319-07743-7_41

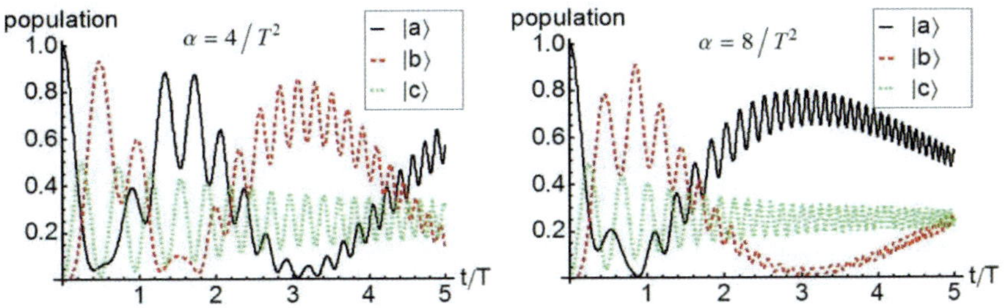


**Fig. 1.** Time evolution of the populations of the initial $|a\rangle$, final $|b\rangle$, and intermediate $|c\rangle$ states for two different chirps. Left panel for $\alpha = 4/T^2$ and right panel for $\alpha = 8/T^2$. Time is normalized over the period of the unchirped Λ scenario $T = \frac{\pi}{\sqrt{2}\mu}$. For both cases we assume $t_{\text{res}} = 0$.

with the analytic solutions

$$a(t) = \frac{1}{2\mu\Gamma(A)}\left\{2\mu K_3\Gamma(A) - (-1)^{\frac{3}{4}}\sqrt{\alpha}\left[\mathrm{M}\left(-\frac{A}{2},\frac{1}{2},-i\alpha t^2\right) - 1\right]K_1\Gamma\left(\frac{A+1}{2}\right)\right.$$
$$\left. - 2\alpha t\mathrm{M}\left(\frac{1-A}{2},\frac{3}{2},-i\alpha t^2\right)\left[K_1\Gamma\left(1+\frac{A}{2}\right) + K_2\Gamma(1+A)\right]\right\},$$

$$b(t) = \frac{1}{2\mu\Gamma(A)}\left\{2\mu K_4\Gamma(A) + (-1)^{\frac{3}{4}}\sqrt{\alpha}\left[\mathrm{M}\left(-\frac{A}{2},\frac{1}{2},-i\alpha t^2\right) - 1\right]K_1\Gamma\left(\frac{A+1}{2}\right)\right.$$
$$\left. + 2\alpha t\mathrm{M}\left(\frac{1-A}{2},\frac{3}{2},-i\alpha t^2\right)\left[K_1\Gamma\left(1+\frac{A}{2}\right) + K_2\Gamma(1+A)\right]\right\},$$

$$c(t) = K_1\mathrm{H}_{-A}\left(\sqrt[4]{-1}\sqrt{\alpha}t\right) + K_2\mathrm{M}\left(\frac{A}{2},\frac{1}{2},i\alpha t^2\right),$$

where $A = i\mu^2/2$, $\mu$ expresses the transition matrix elements $\langle a|\,\hat{x}\,|c\rangle = -\langle b|\,\hat{x}\,|c\rangle = -i\langle a|\,\hat{y}\,|c\rangle = i\langle b|\,\hat{y}\,|c\rangle = 2\mu$ (for details and derivation see Ref. [7]), $\mathrm{H}_n(z)$ are the Hermite polynomials, $\mathrm{M}(a,b,z)$ are the Kummer's confluent hypergeometric functions, and the four constants $K_i$ depend on the initial conditions.

As can be seen in Fig. 1 the intensity of chirp influences the fidelity of the Λ-process (for $\alpha = 0$ the population of $|b\rangle$ gets maximized at time $t_{\max} = k\pi/\sqrt{2}\mu$, where $k$ an odd integer [7]). It is also interesting that after the initial transfer (the quality which depends on $\alpha$) the rest of the population transfers are incomplete and carry the signature of quantum interference patterns. This explains the beatings after about 3 periods, when the laser pulse is no longer resonant.

Another factor is the time of resonance $t_{\text{res}}$ with respect to the period $T$ of the process. Clearly, if $t_{\text{res}} = t_{\max}/2 = T/4$, the laser pulse has overall the less amount of detuning during the process, and hence the fidelity gets optimized (Fig. 2). Therefore the sign of the chirp determines the best detuning policy: For $\alpha > 0$ ($\alpha < 0$) it is better to start at slightly lower (higher) frequencies. This explains why a combination of detuning and chirp controllably breaks the time-inversion symmetry of the Λ-process, i.e. selectively driving the magnetization to a desired orientation. It also explains the spin-switching differences between Fe, Co and Ni

interference), while Ni usually needs many cycles and therefore the process collapses with any small perturbation.

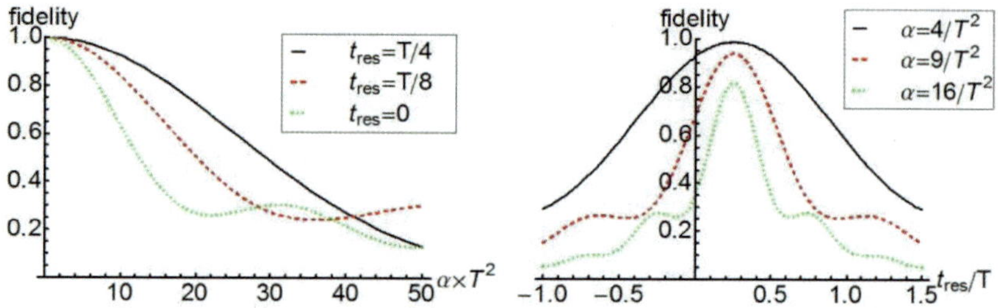


**Fig. 2.** Left: Fidelities vs. $t_{res}$for various chirps$\alpha$. The best results are always achieved when the resonance is reached at the center of the period$T$. Right: Fidelity as a function of chirp for three different $t_{res}$. The times are normalized over the period of the unchirpedcase $T = \frac{\pi}{\sqrt{2}\mu}$.

Concluding, we analytically solve the Λ-process induced by chirped laser pulses and findthat (i) the laser pulse should be resonantin the middle of the transfer, and (ii)chirp enables unidirectional control of the flipping process.

**Acknowledgments** The authors acknowledge funding fromthe German Research Foundation (DFG) via the Collaborative Research Center SFB/TRR 88 "3MET".

## References

[1] E. Beaurepaire, J.-C.Merle, A. Daunois, and J.-Y.Bigot, "Ultrafast Spin Dynamics in Ferromagnetic Nickel" Phys. Rev. Lett.**76**, 4250 (1996).
[2] B. Koopmans, J. J. M. Ruigrok, F. Dalla Longa, and W. J. M. de Jonge, "Unifying Ultrafast Magnetization Dynamics" Phys. Rev. Lett. **95**, 267207 (2005).
[3] J. Barker, R.F.L. Evans, R.W. Chantrell, D. Hinzke, and U. Nowak, "Atomistic spin model simulation of magnetic reversal modes near the Curie point", Appl. Phys. Lett. **97**, 192504 (2010).
[4] J.A. de Jong, I. Razdolski, A.M. Kalashnikova, R.V. Pisarev, A.M. Balbashov, A. Kirilyuk, Th. Rasing, and A.V. Kimel, "Coherent Control of the Route of an Ultrafast Magnetic Phase Transition via Low-Amplitude Spin Precession" Phys. Rev. Lett. **108**, 157601 (2013).
[5] W. Hübner, S. Kersten, and G. Lefkidis, "Optical spin manipulation for minimal magnetic logic in metallic three-center magnetic clusters", Phys. Rev. B **79**, 184431(5) (2009).
[6] G.P. Zhang, G. Lefkidis, W. Hübner, and Y. Bai, "Manipulating femtosecond magnetization in ferromagnets and molecular magnets through laser chirp" J. Appl. Phys. **111**, 07C508 (2012).
[7] G. Lefkidis, W. Hübner, "Analytical treatment of ultrafast laser-induced spin-flipping Λ processes on magnetic nanostructures" Phys. Rev. B **87**, 014404 (2013).
[8] C. Li, W. Jin, H.P. Xiang, G. Lefkidis, and W. Hübner, "Theory of laser-induced ultrafast magneto-optic spin flip and transfer in charged two-magnetic-center molecular ions: Role of bridging atoms" Phys. Rev. B **84**, 054415 (2011).

# Ultrafast demagnetization after laser pulse irradiation in Ni: Ab-initio electron-phonon scattering and phase space calculations

**Christian Illg, Michael Haag, Manfred Fähnle**

Max Planck Institute for Intelligent Systems, Heisenbergstr. 3, 70569 Stuttgart, Germany

faehnle@is.mpg.de

**Abstract** Electron-phonon scattering is one possible candidate to explain ultrafast demagnetization in Ni. We calculate the demagnetization time and the demagnetization rate with ab-initio density-functional theory and estimate the phase space for scattering which is related to the maximum possible demagnetization. We find that both demagnetization rate and phase space are too small for reasonable excitations to explain the experimental demagnetization.

It is well known that a thin ferromagnetic Ni film can demagnetize within some hundreds of fs after irradiation with a fs laser pulse. Since the first experiments by Beaurepaire et al. there has been a lot of effort to investigate the ultrafast demagnetization. Until now the elementary underlying processes are still not understood. Single-electron spin-flip scattering is one possible candidate. In this work we concentrate on the electron-phonon spin-flip scattering.

The transition rate $W^{\lambda}_{j\mathbf{k}\uparrow,j'\mathbf{k}'\downarrow}$ for the electron-phonon spin-flip scattering from an initial crystal wave function $\psi_{j\mathbf{k}\uparrow}$ to a final crystal wave function $\psi_{j'\mathbf{k}'\downarrow}$ is given by Fermi's golden rule where $j$ is the band index, $\mathbf{k}$ is the wave vector and $\uparrow,\downarrow$ denote the spin-up and spin-down character. For the absorption and emission of a phonon with wave vector $\mathbf{q}$ and -$\mathbf{q}$, polarization $\lambda$ and frequency $\omega_{\mathbf{q}\lambda}$ it reads [1]

$$W^{\lambda}_{j\mathbf{k}\uparrow,j'\mathbf{k}'\downarrow} = \frac{2\pi}{\hbar}\left|M^{\lambda}_{j\mathbf{k}\uparrow,j'\mathbf{k}'\downarrow}\right|^2 \frac{\hbar}{2NM\omega_{\mathbf{q}\lambda}}\,[b_{\mathbf{q}\lambda}\,\delta(E_{j'\mathbf{k}'\downarrow}-E_{j\mathbf{k}\uparrow}-\hbar\omega_{\mathbf{q}\lambda}) + (b_{-\mathbf{q}\lambda}+1)\delta(E_{j'\mathbf{k}'\downarrow}-E_{j\mathbf{k}\uparrow}+\hbar\omega_{-\mathbf{q}\lambda})]\,. \qquad 1$$

N is the number of atoms, M is the atomic mass and $b_{\mathbf{q}\lambda}$ denotes the Bose distribution function. The energies $E_{j\mathbf{k}\uparrow}$ and $E_{j'\mathbf{k}'\downarrow}$ correspond to the states $\psi_{j\mathbf{k}\uparrow}$ and $\psi_{j'\mathbf{k}'\downarrow}$. The matrix element $M^{\lambda}_{j\mathbf{k}\uparrow,j'\mathbf{k}'\downarrow}$ consists of two parts: the phonon perturbation of the potential and the phonon perturbation of the spin-orbit coupling. The transition rate $W_{\uparrow,\downarrow}$ from spin-up to spin-down is given by an integral over all states in the first Brillouin zone (BZ) with volume $\Omega$ [1]



J.-Y. Bigot et al. (eds), *Ultrafast Magnetism I*, Springer Proceedings in Physics 159
DOI 10.1007/978-3-319-07743-7_42

$$W_{\uparrow,\downarrow} = \Omega^{-2} \sum_{jj'\lambda} \iint_{BZ} d^3k\ d^3k'\ f(E_{j\mathbf{k}\uparrow}) \left(1 - f(E_{j'\mathbf{k}'\downarrow})\right) W^{\lambda}_{j\mathbf{k}\uparrow,j'\mathbf{k}'\downarrow} \tag{2}$$

where $f(E_{j\mathbf{k}\uparrow}) = (\exp((E_{j\mathbf{k}\uparrow} - E_F^{\uparrow})/kT_e) + 1)^{-1}$ and $f(E_{j'\mathbf{k}'\downarrow})$ are the Fermi distribution functions with spin-up and spin-down Fermi energies $E_F^{\uparrow,\downarrow}$, respectively, and electron temperature $T_e$. Thereby, it is our main assumption that thermalized electrons make the main contribution to the demagnetization. The demagnetization rate (more precisely: the rate for the change of the magnetic moment per atom) is $dM/dt = 2\mu_B(W_{\uparrow,\downarrow} - W_{\downarrow,\uparrow})$. The equation is exact if the states are pure spin-up or pure spin-down states. With spin-orbit coupling the states are spin-mixed and one has to take into account that a spin-flip changes the spin magnetic moment per atom by less than $2\mu_B$.

It can be shown [1] that a time-independent demagnetization time $\tau_M$ can be defined if the following approximations hold: First, the excitation is only small. Second, the demagnetization behavior is exponential, and third, the spin-mixing is not large for most of the states. Indeed, all approximations hold for Ni if the laser pulse irradiation has a small fluence. The equation for $\tau_M$ reads [1]

$$\begin{aligned}\tau_M{}^{-1} = {} & \Omega^{-2} \sum_{jj'\lambda} \iint_{BZ} d^3k\ d^3k' \times \\ & \{ W^{\lambda}_{j\mathbf{k}\uparrow,j'\mathbf{k}'\downarrow} \left[\frac{1}{\tilde{Z}^{\downarrow}(E_F)} f_0(E_{j\mathbf{k}\uparrow})\, \eta(E_{j'\mathbf{k}'\downarrow}) + \frac{1}{\tilde{Z}^{\uparrow}(E_F)} \left(1 - f_0(E_{j'\mathbf{k}'\downarrow})\right) \eta(E_{j\mathbf{k}\uparrow})\right] \\ & + W^{\lambda}_{j'\mathbf{k}'\downarrow,j\mathbf{k}\uparrow} \left[\frac{1}{\tilde{Z}^{\uparrow}(E_F)} f_0(E_{j'\mathbf{k}'\downarrow})\, \eta(E_{j\mathbf{k}\uparrow}) + \frac{1}{\tilde{Z}^{\downarrow}(E_F)} \left(1 - f_0(E_{j\mathbf{k}\uparrow})\right) \eta(E_{j'\mathbf{k}'\downarrow})\right] \}. \end{aligned} \tag{3}$$

$\eta(E)$ is the derivative of the equilibrium Fermi distribution function $f_0(E) = (\exp((E - E_F^0)/kT_0) + 1)^{-1}$ with respect to *E*. $E_F^0$ and $T_0 = 300$ K denote the Fermi energy and electron temperature in equilibrium, respectively. $\tilde{Z}^{\uparrow,\downarrow}(E_F) = \int dE\ Z^{\uparrow,\downarrow}(E)\, \eta(E)$ is an integral over the density of states $Z^{\uparrow,\downarrow}(E)$.

We calculate the electronic states with ab-initio density-functional electron theory using the linear-muffin-tin-orbital method in local spin-density approximation. The phononic states are also calculated ab-initio using the pseudopotential method (PWscf code) [2]. The matrix elements in Fermi's golden rule contain the scattering potential which is represented in the rigid-ion approximation.

The main difference between eq. 2 and eq. 3 is that only equilibrium quantities enter eq. 3 whereas non-equilibrium quantities enter eq. 2, namely the Fermi energies $E_F^{\uparrow,\downarrow}$ and the electron temperature$T_e$. Therefore, the demagnetization time in eq. 3 can be calculated with the ab-initio electronic and phononic states at hand. In order to calculate the transition rate in eq. 2 the Fermi energies $E_F^{\uparrow,\downarrow}$ and electron temperature $T_e$ have to be determined in addition. Reasonable values for $T_e$ are between 400 K and 1000 K depending on the laser fluence. In order to determine $E_F^{\uparrow,\downarrow}$ we assume that directly after laser excitation the electrons scatter dominantly via electron-electron scattering without demagnetization and thermalize very quickly. The time where the electronic system can be considered as thermalized is denoted by $t_S$. At time $t_S$ the magnetization is still the equilibrium magnetization.

This condition enables the determination of the non-equilibrium Fermi energies $E_F^{\uparrow,\downarrow}$.

We calculate the demagnetization time and the demagnetization rate for different k-point grids (up to $50^3$ k-points in the first BZ) and check for convergence. Our main result is that the demagnetization time is about $\tau_M$= 16 fs and the demagnetization rate at $t_s$ is about $dM/dt = 0.02\,\mu_B/100$ fs per atom for $T_e = 700$ K. The value is small (Ni has an equilibrium magnetic moment of $0.64\,\mu_B$ per atom). This cannot explain the experimental demagnetization rate of about $0.2\,\mu_B/100$ fs per atom. Thus, we draw a similar conclusion as in ref. 3 and 4.

Finally, we estimate the maximum possible demagnetization $\Delta M$ (more precisely: the maximum possible decrease of magnetic moment per atom) achievable in our rigid-band model by calculating the number of excited majority electrons at time $t_s$ and by calculating the number of excited minority holes at time $t_s$. The maximum possible demagnetization is achieved if all minority electrons do not flip their spin while all excited majority electrons flip their spin and all excited minority holes are filled by majority electrons via spin-flip. For electron temperatures up to 1000 K we find a maximum demagnetization $\Delta M < 0.15\,\mu_B$ per atom which is too small to explain the experimental demagnetization up to 90%. It was suggested [4] that a dynamic exchange splitting would yield greater demagnetization rates. Therefore, we also estimate $\Delta M$ for a band structure with reduced magnetic moment of $0.5\,\mu_B$ per atom and electron temperatures up to 1000 K. Indeed, the maximum demagnetization is greater, $\Delta M < 0.18\,\mu_B$ per atom, however it is also too small to explain the experimental demagnetization up to 90%. We conclude that the available phase space is too small for electron temperatures up to 1000 K and that additional processes, e.g. electron-magnon scattering, are important to explain the experiments. This is in line with the statements in ref. 5.

**Acknowledgments** The authors thank K. Carva, B. Koopmans, P. M. Oppeneer, A. J. Schellekens, and H. C. Schneider for helpful discussions.

## References

[1] D. Steiauf, C. Illg, M. Fähnle, "Extension of Yafet's theory of spin relaxation to ferromagnets" J. Magn. Magn. Mater. **322**, L5-L7 (2010).

[2] C. Illg, B. Meyer, M. Fähnle, "Frequencies and polarization vectors of phonons: Results from force constants which are fitted to experimental data or calculated ab initio" Phys. Rev. B **86**, 174309 (2012).

[3] K. Carva, M. Battiato, P. M. Oppeneer, "Ab initio investigation of the Elliott-Yafet electron-phonon mechanism in laser-induced ultrafast demagnetization" Phys. Rev. Lett. **107**, 207201 (2011).

[4] S. Essert, H. C. Schneider, "Electron-phonon scattering dynamics in ferromagnetic metals and their influence on ultrafast demagnetization processes" Phys. Rev. B **84**, 224405 (2011).

[5] A. J. Schellekens, B. Koopmans, "Comparing ultrafast demagnetization rates between competing models for finite temperature magnetism" Phys. Rev. Lett. **110**, 217204 (2013).

# Ultrafast spin flip on homodinuclear clusters

**W. Jin[1], C. Li[2], G. Lefkidis[1], W. Hübner[1]**

[1]Department of Physics and Research Center OPTIMAS, University of Kaiserslautern, PO Box 3049, 67653 Kaiserslautern, Germany

[2]School of Mechanics, Civil Engineering and Architecture, Northwestern Polytechnical University, Xi'an 710072, China

jinwei@physik.uni-kl.de

**Abstract** Using fully *ab initio* calculations we investigate the laser-induced ultrafast spin flip on the magnetic homodinuclear clusters FeOFe and NiONi. The oxygen atoms lower the total symmetry of the clusters inducing spin-density localization. Moreover, the proposed spin-flip scenarios on the two clusters exhibit different patterns matching their respective electronic structures.

## Introduction

Exploiting the spin degree of freedom for technological applications becomes more and more appealing due to its great potential to enhance the data storage density and to accelerate the read-write speed in computational devices. In the past two decades a substantial amount of experimental and theoretical literature demonstrated [1-4] that magnetic response can be achieved in the subpicosecond regime. Among various mechanisms a $\Lambda$ process [5] was proposed, via which spin can be coherently manipulated under the cooperative effects of a laser field and spin-orbit coupling (SOC). Based on this process, laser-induced ultrafast spin dynamics on different heteronuclear multicenter magnetic structures with a high degree of spin-density localization have been successfully predicted [6, 7]. In homodinuclear structures, however, it is impossible to distinguish the magnetic centers if they are equivalent by symmetry. Thus to locally manipulate the spin, the symmetry of the structures needs to be reduced. Here, two small homodinuclear magnetic clusters bridged by oxygen atoms, i.e., FeOFe and NiONi, are investigated.

## Ultrafast laser-induced spin-flip scenarios

We start from the optimized structures at the Hartree-Fock level. Fig. 1 gives the geometries of FeOFe and NiONi and their lowest 121 electronic-state energies as

 
J.-Y. Bigot et al. (eds), *Ultrafast Magnetism I*, Springer Proceedings in Physics 159
DOI 10.1007/978-3-319-07743-7_43

obtained with the SAC-CI method [8] without the inclusion of SOC. Clearly, the two identical magnetic atoms in each structure are inequivalent, which leads to spin-density localization thus fulfilling the first prerequisite for spin dynamics. The distributions of the electronic levels of the two structures behave differently due to the different numbers of the electrons (e. g., the energy levels of NiONi are broader and more disperse than those of FeOFe).

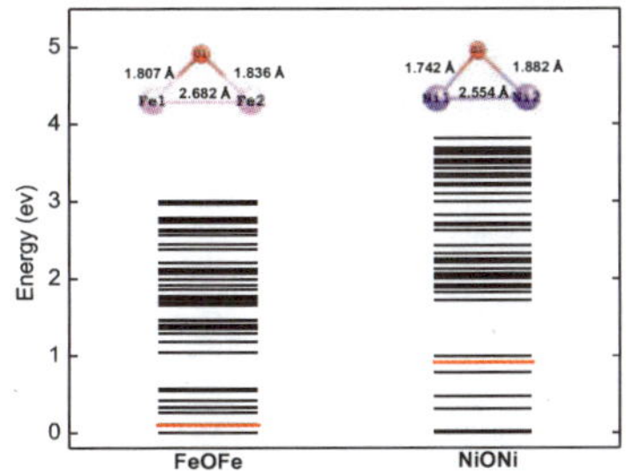


**Fig. 1.** The optimized geometries of FeOFe and NiONi and their respective energy levels from SAC-CI results without SOC. The red lines indicate the triplet states from which the initial and final states (with opposite spin directions) of the spin-flip scenarios originate.

As a next step the dynamics of the light-induced spin manipulation is calculated using time-dependent perturbation theory. This also requires the inclusion of SOC (to create excited spin-mixed states), an external magnetic field (for the Zeeman splitting), and a time-dependent laser pulse (see Ref. [7]). Two possible spin-flip scenarios at Fe1 in FeOFe and at Ni1 in NiONi are shown in Fig. 2, where the initial and final states (which stem from the same triplet states before SOC and have the opposite spin directions of each dynamics) are also marked. The energy positions of these states can be found in Fig. 1. A magnetic field with a strength of $|\mathbf{B}| = 10^{-5}$ at. un. is chosen for both scenarios. Note that in the absence of SOC the direct transition from spin-up to spin-down is forbidden within the electric-dipole approximation. The population of the initial state in both scenarios is successfully transferred to the target state (with fidelities of 97% and 86%, respectively) and so the spin gets flipped within around 800 fs (Fig. 2). In total there are 21 intermediate states involved in the scenario shown in Fig. 2(b), leading to a quite complex dynamics behavior in NiONi. In contrast, the scenario in FeOFe [Fig. 2(a)] involves only 6 intermediate states and therefore is simpler.

The parameters of the optimized laser pulses for scenarios Fig. 2(a) and (b) are $\theta = 180.2°$, $\phi = 207.6°$, $\gamma = 313.9°$, FWHM=168 fs, $I = 27.8$ J/(sm$^2$), $E_l = 1.06$ eV and $\theta = 48.9°$, $\phi = 67.2°$, $\gamma = 29.2°$, FWHM=274 fs, $I = 0.5$ J/(sm$^2$), $E_l = 1.66$ eV, respectively, where $\theta$ and $\phi$ denote the angles of incidence in spherical coordinates, respectively, $\gamma$ is the angle between the polarization of the light and the optical plane, FWHM is the full width at half maximum of the laser pulse, and $I$ and $E_l$ are the intensity and the energy of the laser pulse. Obviously, the FWHM of the scenario Fig. 2(a) is shorter than that of Fig. 2(b), however at the price of a much higher intensity. Additionally, due to the more disperse energy levels of NiONi and the higher energies of the intermediate states involved [mostly in the energy region of 2.4-2.9 eV for the scenario Fig.2 (b)], the laser energy in the second scenario is higher than in the first one.

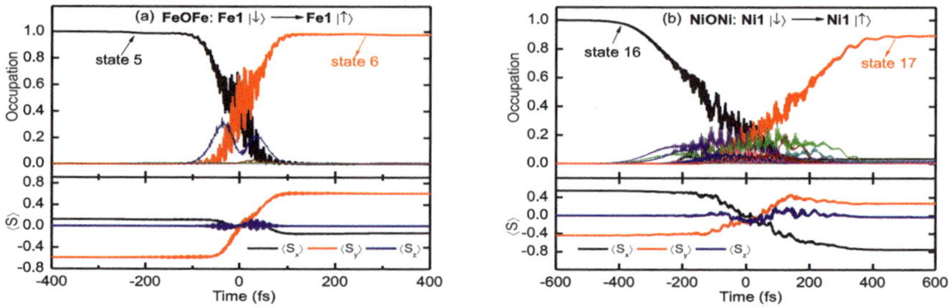


**Fig. 2**. Local spin-flip scenarios on (a) FeOFe and (b) NiONi. For each scenario, the upper panel denotes the time-resolved occupations of the involved states and the lower one shows the expectation values of the spin.

In summary, with the use of *ab initio* calculations, laser-induced ultrafast local spin-flip scenarios on FeOFe and NiONi are proposed. By introducing a bridge atom, which induces spin-density localization, the electronic structures of the two clusters acquire different patterns and thus exhibit different dynamics behavior.

**Acknowledgments** We acknowledge financial support from the DFG and the collaborative Research Center TRR/SFB 88 "3MET" as well as the Carl-Zeiss Foundation.

## References

[1] E. Beaurepaire, J.-C. Mele, A. Daunois, J.-Y. Bigot., "Ultrafast spin dynamics in ferromagnetic Nickel", Phys. Rev. Lett. **76**, 4250 (1996).
[2] F. Hansteen *et al.*, "Femtosecond photomagnetic switching of spins in ferrimagnetic garnet films", Phys. Rev. Lett. **95**, 047402 (2005).
[3] J.-Y. Bigot, M. Vomir, and E. Beaurepaire, "Coherent ultrafast magnetism induced by femtosecond laser pulses", Nature phys. **5**, 515 (2009).
[4] J. Barker *et al.*, "Atomistic spin model simulation of magnetic reversal modes near the Curie point", Appl. Phys. Lett. **97**, 192504 (2010).
[5] R. Gómez-Abal *et al.*, "All-optical subpicosecond magnetic switching in NiO(001)", Phys. Rev. Lett. **92**, 227402 (2004).
[6] C. Li *et al.*, "First-principles calculation of the ultrafast spin manipulation of two-center metallic clusters with a CO molecule attached to one center as an infrared marker", Phys. Rev. B **79**, 180413 (2009).
[7] C. Li *et al.*, "Theory of laser-induced ultrafast magneto-optic spin flip and transfer in charged two-magnetic-center molecular ions: Role of bridging atoms", Phys. Rev. B **84**, 054415 (2011).
[8] H. Nakatsuji *et al.*, http://www.sbchem.kyoto-u.ac.jp/nakatsuji-lab/sacci.html, 2005.

# Switching dynamics of two sub-lattice magnets

**Sönke Wienholdt and Ulrich Nowak**

Fachbereich Physik, Universität Konstanz, D-78457 Konstanz, Germany

soenke.wienholdt@uni-konstanz.de, ulrich.nowak@uni-konstanz.de

**Abstract** After thermal excitation ferrimagnets can switch via a transient ferromagnetic-like state. It is shown by spin model simulations that this state follows from a dissipationless dynamics on picosecond time scales, while slower dissipative relaxation leads back to the ferrimagnetic state which might or might not be switched.

## Introduction

It has been demonstrated recently that linearly polarized light can trigger a thermally driven switching in ferrimagnetic GdFeCo compounds [1, 2] via a so called "ferromagnetic-like state", where the rare-earth (RE) and transition metal (TM) sublattice magnetisations are aligned parallel on a picosecond time scale. The ultra-short time scale of the laser pulse in connection with the high electron temperatures following the excitation lead to non-equilibrium processes where longitudinal magnetisation dynamics becomes pronounced [2-6].

In a prior work [7] we have shown that the thermally driven spin-switching of RE-based ferrimagnets can be well described on the basis of an orbital-resolved spin model, distinguishing electrons in d and f orbitals of the RE. We have shown that this distinct treatment of the spins contributes significantly to the different demagnetization times of the strongly coupled TM and RE sublattice. The pronounced thermal excitation of d electrons drives the system into a non-equilibrium state which by relaxation processes that even exist in the dissipationless limit, where energy and angular momentum are distributed between the sublattices, relaxes into a transient ferromagnetic-like state on times scales of just a picosecond. However, in the following we will show that even after having reached the transient ferromagnetic-like state, the system does not necessarily switch. Recently this has also been observed experimentally with element specific measurements on TbFeCo [8].

 
J.-Y. Bigot et al. (eds), *Ultrafast Magnetism I*, Springer Proceedings in Physics 159
DOI 10.1007/978-3-319-07743-7_44

## Methods

We simulate a ferrimagnetic model of 64000 spins arranged on two intertwined sublattices on a cubic lattice with initial conditions where spins on the one sublattice (RE) are perfectly ordered while those on the other one (TM) are randomly oriented. We set the magnetic moments to $\mu_{\mathrm{RE}} = 2\mu_{\mathrm{TM}}$ and the exchange constants to $J_{\mathrm{TM}} = 3J_{\mathrm{RE}} = -3J_{\mathrm{RE-TM}}$, which results in a Curie temperature of $T_C = 10\ J_{RE}/k_B$ and a magnetisation compensation temperature of $T_M = 7\ J_{RE}/k_B$. This serves as toy model for a ferrimagnetic TM-RE sample directly after the initial excitation where the TM spins are highly excited while the RE spins are still rather ordered due to their slower dynamics. We consider Langevin dynamics, i. e. we numerically solve the stochastic LLG equation of motion for each spin (see e.g. [4] for the numerical methods).

## Results and Discussion

Fig. 1 shows the relaxation dynamics of the two sub-lattice magnetizations for two different temperatures, below (left) and above (right) the magnetic compensation temperature of the systems. In both cases the initial dynamics leads first to a transient ferromagnetic-like state. For $T$=5.5 $J_{RE}/k_B$ first a transient switching of the RE sublattice occurs but later precessional dynamics switches the system back towards its initial state. Interestingly, this happens exactly when both sublattice magnetisations have the same amplitude and the dynamics is via an antiferromagnetic switching mode [9]. For $T$=8.5 $J_{RE}/k_B$ the system relaxes linearly (without any precession) into a ferrimagnetic ground state with the sublattice magnetisations switched as compared to the initial state.

Our results suggest that switching of the RE sublattice magnetisation after an excitation with an ultrashort laser pulse can only be observed when the temperature after the excitation – when electrons and phonons are in equilibrium again – is above or only just below $T_M$.

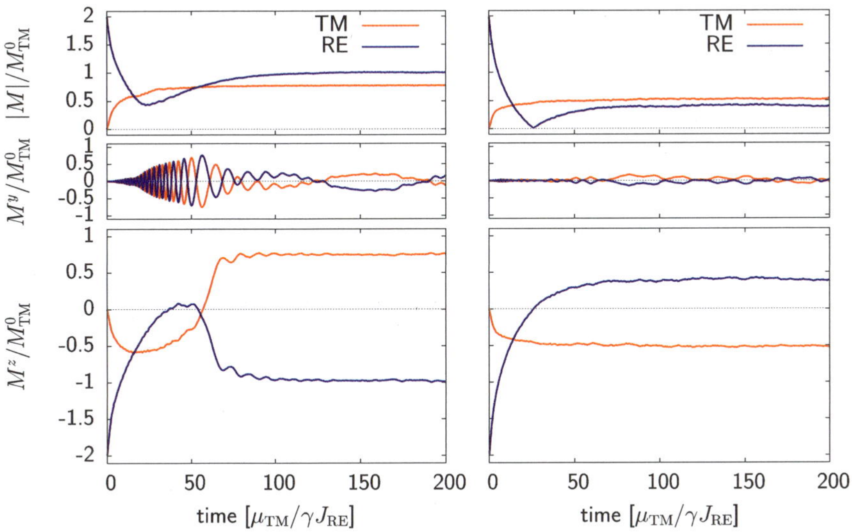


**Fig. 1.** Relaxation dynamics for two different temperatures, $T$=5.5 $J_{RE}/k_B$ (left) and $T$=8.5 $J_{RE}/k_B$ (right). Shown are (from top) the magnitude of the sublattice magnetizations, one perpendicular component, and the easy-axis component.

**Acknowledgments :** This work was supported by the European Research Council via the collaborative research network FEMTOSPIN.

# References


[1] I. Radu et al., "Transient ferromagnetic-like state mediating ultrafast reversal of antiferromagnetically coupled spins" Nature **472**, 205 (2011).

[2] T. A. Ostler et al., "Ultrafast heating as a sufficient stimulus for magnetization reversal in a ferrimagnet" Nat. Commun. **3**, 666 (2012).

[3] E. Beaurepaire, J.-C. Merle, A. Daunois, and J.-Y. Bigot, "Ultrafast Spin Dynamics in Ferromagnetic Nickel" Phys. Rev. Lett. **76**, 4250 (1996).

[4] N. Kazantseva, U. Nowak, R. W. Chantrell, J. Hohlfeld and A. Rebei, "Slow recovery of the magnetisation after a sub-picosecond heat pulse" Europhys. Lett. **81**, 27004 (2008).

[5] J. H. Mentink, J. Hellsvik, D. V. Afanasiev, B. A. Ivanov, A. Kirilyuk, A. V. Kimel, O. Eriksson, M. I. Katsnelson, and Th. Rasing, "Ultrafast Spin Dynamics in Multisublattice Magnets" Phys. Rev. Lett. **108**, 057202 (2012).

[6] U. Atxitia, P. Nieves, and O. Chubykalo-Fesenko, "Landau-Lifshitz-Bloch equation for ferrimagnetic materials" Phys.Rev. B **86**, 104414 (2012).

[7] S. Wienholdt et al., Phys. Rev. B, in press (2013)

[8] A. R. Khorsand, M. Savoini, A. Kirilyuk, A. V. Kimel, A. Tsukamoto, A. Itoh, and Th. Rasing, "Element-Specific Probing of Ultrafast Spin Dynamics in Multisublattice Magnets with Visible Light" Phys. Rev. Lett. **110**, 107205 (2013)

[9] S. Wienholdt, D. Hinzke, U. Nowak, "THz Switching of Antiferromagnets and Ferrimagnets" Phys. Rev. Lett. **108**, 247207 (2012).

# The Landau-Lifshitz-Bloch equation for quantum spin

**P. Nieves [§], D. Serantes, O. Chubykalo-Fesenko**

Instituto de Ciencia de Materiales de Madrid, CSIC, Cantoblanco, c. Sor Juana Inés de la Cruz, 3, Madrid, 28440, Spain

[§]pablo.nieves@csic.es

**Abstract** The Landau-Lifshitz-Bloch equation provides a micromagnetic framework for large-scale modeling of the ultrafast magnetization dynamics in ferromagnets. Here we present a detailed discussion of the LLB equation with a quantum spin number S. As an example, we present our recently numerical results of the ultrafast demagnetization dynamics in FePt thin films.

The Landau-Lifshitz-Bloch (LLB) equation of motion has been recommended itself as a useful micromagnetic framework to analyze ultra fast spin dynamics due to the account of the longitudinal relaxation resulting from the exchange field. In the classical case the LLB equation describes quite well the dynamics of thermally averaged spin polarization [1], i.e. is equivalent to the classical Heisenberg model with spin number $S \to \infty$ and atomistic spin dynamics. An alternative statistical description of purely quantum systems consists of a two-level system with spin up and spin down [2] and is equivalent to the LLB equation with S=1/2 [3]. Here we discuss the LLB equation for arbitrarily spin and some of its generalizations, e.g. for ferrimagnets.

The derivation of the quantum version of the LLB equation [4] starts from the density operator equation of motion (in terms of the complete set of the Hubbard operators) for a magnetic ion weakly interacting with a phonon bath through a linear spin-phonon coupling. Several approximations are made, such as a short-memory (Markoff) approximation and the secular approximation for the elements of the Hubbard operators. The obtained equation for the spin operators $\hat{\mathbf{S}}$ is rewritten for the magnetization $\mathbf{m} = \langle \hat{\mathbf{S}} \rangle / S$. The final equation represents a generalized LLB equation for a paramagnetic ion. After that the interactions are taken into account in the mean-field approximation resulting in the LLB equation in the following form:

$$\frac{d\mathbf{m}}{dt} = -\gamma \mathbf{m} \times \boldsymbol{H}_{eff} + \frac{\gamma\,\widetilde{\alpha}_{\parallel}}{\mathrm{m}^2}(\mathbf{m} \cdot \boldsymbol{H}_{eff})\mathbf{m} - \frac{\gamma\widetilde{\alpha}_{\perp}}{\mathrm{m}^2}(\mathbf{m} \times (\mathbf{m} \times \boldsymbol{H}_{eff}))$$

 
J.-Y. Bigot et al. (eds), *Ultrafast Magnetism I*, Springer Proceedings in Physics 159
DOI 10.1007/978-3-319-07743-7_45

where

$$\boldsymbol{H_{eff}} = \boldsymbol{H} + \boldsymbol{H_A} + \begin{Bmatrix} \dfrac{1}{2\tilde{\chi}_{\parallel}}\left(1 - \dfrac{m^2}{m_e^2}\right)\mathbf{m} & T \leq T_c \\ -\dfrac{1}{\tilde{\chi}_{\parallel}}\left(1 + \dfrac{3T_c m^2}{5(T - T_c)}\right)\mathbf{m} & T \geq T_c \end{Bmatrix}$$

$$\tilde{\alpha}_{\parallel} = \frac{4\lambda T q}{3T_c \sin(2q)} \; ; \quad \tilde{\alpha}_{\perp} = \lambda\left[\frac{\tanh(q)}{q} - \frac{T}{3T_c}\right] + \frac{\hbar(K_1 - K_2)(S+1)}{3T_c}$$

Here the effective field includes a sum of the applied $\boldsymbol{H}$ and anisotropy $\boldsymbol{H_A}$ fields and an additional internal exchange field represented through the longitudinal susceptibility $\tilde{\chi}_{\parallel} = \mu_0 \beta B'/(1 - \beta S^2 J_0 B')$, where $B'$ is the derivative of the Brillouin function, $\beta = 1/k_B T$, $k_B$ is the Boltzmann constant, $J_0$ is the exchange constant, $\mu_0$ is the atomic magnetic moment. Additionally, $q = 3T_c m_e/2(S + 1)T$ , $T_c$ is the Curie temperature, $\hbar$ is the reduced Planck constant, $\gamma$ is the gyromagnetic ratio, $m_e$ is the equilibrium magnetization and

$$\lambda = \frac{\hbar(S+1)K_2}{k_B T} \; ; \; K_1 = \sum_{p,q} |V_{p,q}|^2 n_p(n_q + 1)\pi\delta(w_q - w_p)$$

$$K_2 = \sum_q |V_q|^2 (n_q + 1)\pi\delta(w_q - \gamma H) + \sum_{p,q} |V_{p,q}|^2 n_p(n_q + 1)\pi\delta(w_q - w_p - \gamma H)$$

where $V_q$ and $V_{p,q}$ are the amplitudes of the direct scattering processes and the Raman processes, $n_p = 1/(\exp(\hbar w_p/k_B T) - 1)$ is the boson distribution function. It easy to see that if $S \to \infty$ we recover a classical LLB equation [1], where $\lambda$ represents an atomistic coupling to the bath parameter.

At high temperatures ($\gamma H \hbar \ll k_B \theta_D \ll k_B T$ ; $\gamma H \ll |w_q - w_p|$; $\theta_D$ is the Debye temperature) the two-phonon emission and absorption processes (Raman scattering) are dominant and $K_1 \cong K_2 \sim T^2$ [5], this yields $\lambda \sim T$. On the other hand, at low temperatures ( $k_B T \ll \gamma H \hbar$ ) the one-phonon emission and absorption processes are dominant thus $K_1 \ll K_2 = const$ [5], then in this case we have $\lambda \sim T^{-1}$. Within the LLB model the characteristic longitudinal timescale, defining the demagnetization speed, is given by $\tau_{\parallel} = \tilde{\chi}_{\parallel}/\gamma\, \tilde{\alpha}_{\parallel}$ [6]. This value is constant at low temperatures and diverges approaching $T_C$.

Recently we have done the generalization of the quantum LLB equation for a two-sublattice case [7], similar to the classical case [8]. Although strictly speaking the derivation of the quantum LLB equation was done for the model containing the coupling to the phonon bath only, similar to the LLG equation it can be phenomenologically used for the electronic bath, considering that the scattering processes are hidden in the phenomenological coupling to the bath parameter $\lambda$. Currently we are also working on its generalization for the fermionic (electron) bath [9].

As an example, recently [10] we have simulated ultra fast demagnetization in a thin film of FePt using the quantum version of the LLB equation (S=3/2) coupled to the two temperature model (which describes the dynamics of the phonon and electron temperatures). In Fig.1 we show these results, the data are shown with increased laser fluence. In agreement with the experimental observations [10], a transition from type I (fast) ultrafast dynamics to type II (slow) is obtained.

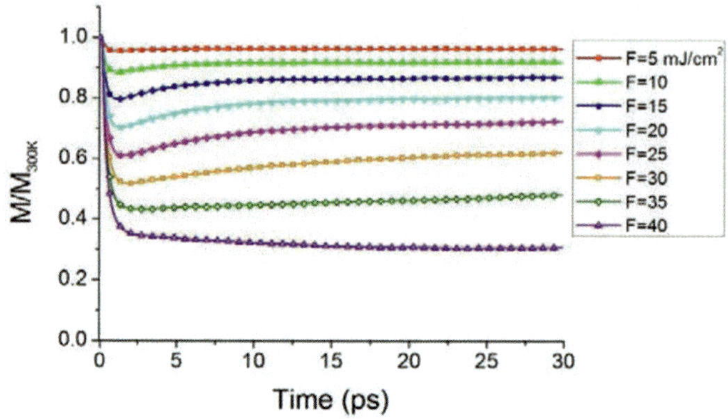


**Fig. 1.** Ultrafast demagnetization dynamics obtained by integration of the quantum LLB equation (S=3/2) coupled to the 2T model.

**Acknowledgments** The authors would like to acknowledge the Femtospin project from European Community and the Spanish grant FIS2010-20979-C02-02 from Ministry of Science and Innovation.

# References

[1] D. A. Garanin, "Fokker-Planck and Landau-Lifshitz-Bloch equations for ferromagnets" Phys. Rev. B. **55**, 3050-3057 (1997).

[2] B. Koopmans et al, " Explaining the paradoxical diversity of ultrafast laser-induced demagnetization" Nature Mat. **9**, 259-265 (2010).

[3] U. Atxitia and O.Chubykalo-Fesenko, "Ultrafast magnetization dynamics rates within the Landau-Lifshitz-Bloch model" Phys. Rev. B. **84**, 144414 (2011).

[4] D. A. Garanin, "Generalized equation of motion for a ferromagnet." Physica A **172**, 470-491 (1991).

[5] D. A. Garanin, "Integral relaxation time of single-domain ferromagnetic particles" Phys. Rev. E, **54**, 3250-3256 (1996).

[6] O. Chubykalo-Fesenko et al, "Dynamic approach for micromagnetics close to the Curie temperature" Phys. Rev. B. **74**, 94436-(5) (2006).

[7] P. Nieves et al, to be published.

[8] U. Atxitia et al, "Landau-Lifshitz-Bloch equation for ferrimagnetic materials" Phys. Rev. B. **86**, 104414 (2012).

[9] P. Nieves et al, to be published.

[10] J. Mendil et al, to be published.

# Inertial Regime of the Magnetization: Nutation resonance Beyond Precession Resonance

**J.-E. Wegrowe[1], M. Meyer[1], M. Hayoun[1], E. Olive[2]**

[1] Ecole Polytechnique, LSI, CNRS and CEA/DSM/IRAMIS, Palaiseau F-91128.

[2] GREMAN, UMR 7347, Universitée François Rabelais-CNRS, Tours F-37200

jean-eric.wegrowe@polytechnique.fr

**Abstract** An inertial regime of the magnetization dynamics has been predicted. We study numerically the corresponding equation in the frequency domain. It is shown that inertial terms generate a second resonance peak of the magnetic susceptibility, which corresponds to the nutation frequency. The interplay between precession and nutation is investigated.

The equation that describes the dynamics of the magnetization – the Landau-Lifshitz-Gilbert (LLG) equation – is a kinetic equation, and does not contain inertial terms. This is somehow an unphysical limitation. The goal of this study is to present an overview of the problem of magnetic inertia in relation with experimental studies. For that purpose, we focus here on the generalized LLG equation with inertial terms, studied in the frequency domain, i.e. in the experimental context of absorption spectroscopy in the range 1011 to 1016 rad/sec.

The dynamics of a uniformly magnetized body of magnetization $\mathbf{M} = M_s\, \mathbf{e_r}$ (where $\mathbf{e_r}$ is the radial unit vector) in an effective magnet field $H^{eff}$ is well described by the Landau-Lifshitz-Gilbert (LLG) equation. The Gilbert form of the equation reads $d\mathbf{M}/dt = \gamma\, \mathbf{M} \wedge (\mathbf{H}^{eff} - \eta\, d\mathbf{M}/dt)$, where $\eta$ is the Gilbert damping coefficient and $\gamma$ is the gyromagnetic ratio.

The same equation can be re-written in the equivalent Landau-Lifshitz form: $d\mathbf{M}/dt = \gamma /(1+\alpha^2)\, \{\mathbf{M} \wedge \mathbf{H}^{eff} - (\alpha/M_\sigma)\, \mathbf{M} \wedge (\mathbf{M} \wedge \mathbf{H}^{eff})\}$, where $\alpha = \gamma\eta M_s$ is the dimensionless damping. The first term in the right hand side accounts for the precession of the magnetization, i.e. the rotation of the extremity of the vector of constant modulus $M_s$ with the Larmor angular velocity $\omega_L = \gamma H^{eff}$. The second term accounts for the relaxation of the magnetization toward its equilibrium position along the effective magnetic field $\mathbf{H}^{eff}$.



J.-Y. Bigot et al. (eds), *Ultrafast Magnetism I*, Springer Proceedings in Physics 159
DOI 10.1007/978-3-319-07743-7_46

If an AC magnetic field $h_\perp(\omega)$ is superimposed perpendicularly to a static field, the relaxation process is compensated provided that the frequency $\omega$ is of the same order as the Larmor frequency (usually some tens of GHz). The precession is then maintained in a stationary state, and the response of the magnetization (i.e. the susceptibility $dM_\perp/dh_\perp = \chi_\perp$) as a function of the frequency is characterized by a resonance peak at $\omega_L$; this is the ferromagnetic resonance (FMR).

However, the limit of validity of the LLG equation forbids in principle the predictions at high enough frequency, or at very short time scales. Brown [1] gave in 1963 the limit of validity of the LLG equation in terms of diffusion process in a field of force with the derivation of the associated stochastic equations. The typical time scale below which the LLG equation breaks down is of the order of the picosecond. Below, the time scale separation between the magnetization and the other degrees of freedom is no longer valid, and the diffusion process should be described by considering explicitly the supplementary variables. In particular, the momentum variables are no longer at equilibrium, and an inertial regime of the uniform magnetization is expected [2]. This regime has been described recently, and the generalization of the LLG equation to inertial regime has been proposed [2-7].

In all cases, the equation takes the simple form, with an additional inertial term in the Gilbert equation:

$$\frac{d\boldsymbol{M}}{dt} = \gamma \boldsymbol{M} \wedge \left[ \boldsymbol{H}^{\text{eff}} - \eta \left( \frac{d\boldsymbol{M}}{dt} + \tau \frac{d^2\boldsymbol{M}}{dt^2} \right) \right]$$

where $\tau$ is the relaxation time of the angular momentum that can be expressed in terms of the first atomic moment of inertia $I_1$: $\tau = I_1/(\eta M_s^2)$ [2]. The equation converges to the usual LLG equation for small values of $\tau$.

Some results about the numerical study are presented in Fig. 1. The usual FMR peak for the precession resonance is located at the Larmor frequency $\gamma H^{\text{eff}}$, as expected (see inset). For one set of parameters $\alpha$ and $\tau$ presented in Fig. 1, the FMR precession peak is plotted together with that obtained from the usual LLG equation (i.e. without the inertial term), it is slightly shifted. On the other side, the nutation resonance peak is located at the frequency $(\alpha\tau)^{-1}$ ($10^{13}$ and $10^{14}$ rad/sec). The separation of two to three decades between the two resonance peaks leads to a complete decoupling of the two effects. It is however no longer the case with a choice of parameters such that the two peaks are not so much separated. A noticeable modification of the FMR resonance peak could then be expected, due to the perturbation produced by the nutation. This is the aim of the discussion.

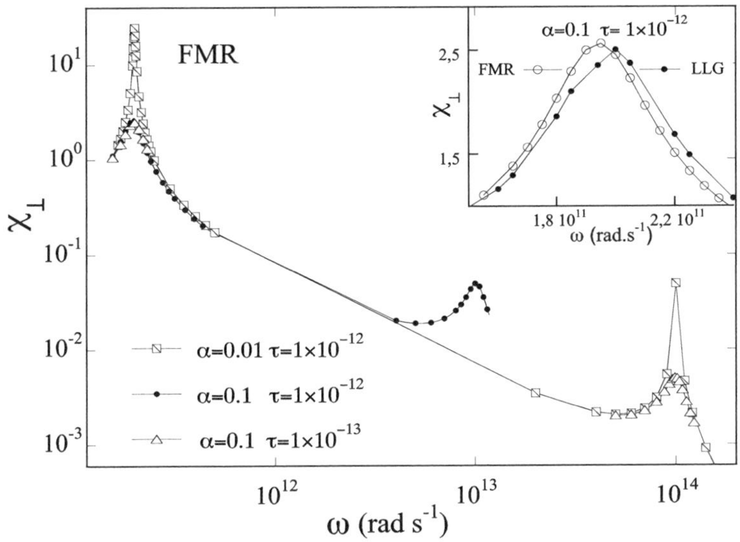


**Fig. 1.** Transverse magnetic susceptibility as a function of the oscillating pulsation $\omega$ of the perpendicular AC field. The FMR peak is fixed at the Larmor frequency, while the nutation peak is located at the pulsation $\omega_{nu}= 1/\alpha\tau$. The susceptibility is plotted for ($\alpha\tau = 10^{-13}$ and $\alpha\tau = 10^{-14}$), and for ($\alpha$=0.1 and $\alpha$=0.01). Inset: for ($\alpha$=0.1 and $\alpha\tau = 10^{-13}$), the FMR peak is slightly shifted from that deduced from the usual LLG equation. We took $\gamma$ = 1011 rad/(sT), $h_\perp$=0.1, 0.01, and 0.001 T, with a static field of 2T.

# References


[1] W. F. Brown, "Thermal fluctuations of a single-domain particle" Phys. Rev. **130**, 1677-1686 (1963).

[2] M.-C. Ciornei, J. M. Rubi, and J.-E. Wegrowe, "Magnetization dynamics in the inertial regime: Nutation predicted at short time scales" Phys. Rev. B **83**, 020410(R) (2011).

[3] J.-E. Wegrowe, M.-C. Ciornei, "Magnetization dynamics, gyromagnetic relation, and inertial effects" Am J. Phys. **80**, 607 (2012).

[4] E. Olive, Y. Lansac, and J.-E. Wegrowe, "Beyond ferromagnetic resonance: The inertial regime of the magnetization" Appl. Phys. Lett. **100**, 192407 (2012).

[5] M. Fähnle, D. Steiauf, and Ch. Illg, "Generalized Gilbert equation including inertial damping: Derivation from an extended breathing Fermi surface model" Phys. Rev. B **84**, 172403 (2011).

[6] S. Bhattacharjee, L. Nordström, and J. Fransson, "Atomistic Spin Dynamic Method with both Damping and Moment of Inertia Effects Included from First Principles" Phys. Rev. Lett. **108**, 057204 (2012).

[7] D. Böttcher and J. Henk, "Significance of nutation in magnetization dynamics of nanostructures" Phys. Rev. B **86**, 020404(R) (2012).

# Multiscale Modeling of Ultrafast Magnetization Dynamics

**T.A. Ostler[1], J. Barker[1], R.F.L. Evans[1], U. Atxitia[1], R.W. Chantrell[1], O. Hovorka[1] and O. Chubykalo-Fesenko[2]**

[1]Department of Physics, the University of York, Heslington, York, North Yorkshire, YO10 5DD, UK.

[2]Instituto de Ciencia de Materiales de Madrid, CSIC, Cantoblanco, 28049 Madrid, Spain.

thomas.ostler@york.ac.uk

**Abstract** Atomistic and mesoscopic-scale spin dynamics have proven to be a powerful tool for studying magnetization dynamics on the sub-picosecond time-scale. These approaches are ideal for combining different length and time-scales. In this paper we focus on atomistic spin dynamics in the ultrafast regime, in particular on recent results of heat-induced switching in the transition-metal rare-earth ferrimagnets.

## Introduction

Control of magnetization dynamics on the sub-picosecond time-scale is receiving wide interest recently due to the potential implications for next generation data storage and processing. To gain a full understanding of the collective dynamics of complex magnetic materials under the influence of ultrafast external stimuli, such as pulsed magnetic field pulses[1], acousto-magneto-plasmonics [2] and femtosecond laser pulses[3,4] usually requires input from theoretical and/or numerical model calculations. It was recently demonstrated that magnetization reversal, under the influence of femtosecond laser heating of the ferrimagnet, GdFeCo, that deterministic magnetization switching was possible[5], and occurred via a transient ferromagnetic-like state[6]. This thermally induced magnetic switching (TIMS) was predicted using numerical atomistic calculations and confirmed by experimental observations. The mechanism of switching in this material has remained elusive. However, by combining analytic treatment with atomic and mesoscopic magnetization dynamics[7,8] the nature of the switching mechanism is now clear.

 
J.-Y. Bigot et al. (eds), *Ultrafast Magnetism I*, Springer Proceedings in Physics 159
DOI 10.1007/978-3-319-07743-7_47

## Modeling of Laser Induced Switching in GdFeCo

The modeling of the spin dynamics of GdFeCo is based on the use of the Landau-Lifshitz-Gilbert equation of motion for individual atomic moments:

$$ {}_2\frac{\partial S_i}{\partial t} = -\frac{|\gamma_i|}{(1+\alpha_i)\mu_i}\left[S_i \times \boldsymbol{H}_i^{\mathbf{eff}} + \boldsymbol{\alpha}_i\, S_i \times S_i \times \boldsymbol{H}_i^{\mathbf{eff}}\right] \quad 1 $$

where $S_i$, is the reduced spin moment of unit length, $\boldsymbol{\gamma}_i$, $\boldsymbol{\alpha}_i$ and $\boldsymbol{\mu}_i$ are the gyromagnetic ratio, damping constant and magnetic moment respectively. $\boldsymbol{H}_i^{\mathbf{eff}}$ is the effective field acting on spin, *i,* which is made up of contributions from the Hamiltonian and the stochastic term that mimics thermal fluctuations. We solve this coupled differential equation numerically giving us the dynamic behavior of magnetic moments with atomic resolution. To determine the form of the Hamiltonian, the two common methods are: mapping *ab-initio* data[9,10] to a Heisenberg Hamiltonian, and from experimental data. The model of GdFeCo presented can be found in Ref.[11] and was based on experimental observations of static magnetic properties.

To incorporate the effects of laser heating within our approach we employ the *two-temperature model*[12]. This model defines a temperature associated hot electrons and the phonon bath. We couple the electron temperature to the spin system, which allows us to observe the time-resolved magnetization dynamics. Figure 1 (figure reproduced from Ref.[5] and [7]) shows the solution of the two-temperature model after the action of multiple heat pulses (temperature profile shown in the upper panel, a) with no external applied field, with the response of the individual sublattice magnetizations in GdFeCo (panel b) showing deterministic switching after each pulse.

This remarkable result was, until now, still unexplained. Using analysis from a mesoscopic scale model based on the Landau-Lifshitz-Bloch model[7] we have shown that for switching to occur, it is necessary that angular momentum is transferred from the longitudinal to the transverse magnetization components in the transition metal. Panel c of figure 1 shows the numerical integration of the switching behavior for the non-stochastic LLB with a small angle (15°) between sublattices (solid lines). Without the angle between the sublattices (dashed line), switching does not occur.

Using analysis from linear spinwave theory, mean field theory and atomistic spin dynamics we have shown that this switching is caused by a two spinwave mode bound state[8] in the spinwave spectrum and results in this angular momentum transfer predicted from the LLB analysis.

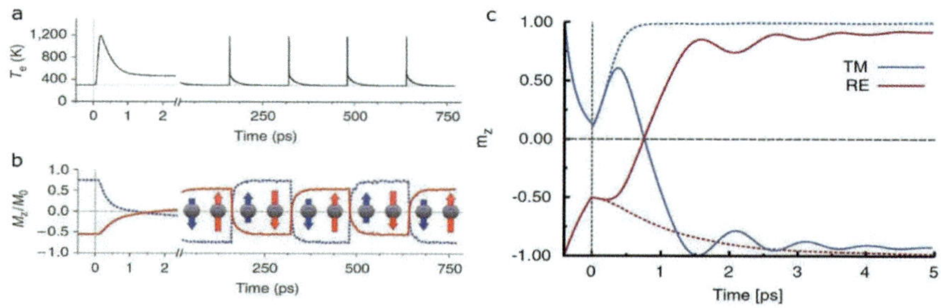


**Fig. 1.** (a) Evolution of the temperature of the electronic thermal bath during a sequence of 5 Gaussian pulses. (b) Computed time-resolved dynamics of the z-component of the magnetizations of Fe and Gd sublattices from the atomistic model. (c) Numerical integration of the switching behavior for the LLB with a small angle (15°) between sublattices (solid lines). Without angle (dashed lines) switching does not occur. The time t=0 corresponds to the end of the square shaped laser pulse with a peak temperature of 1500K. Figures reproduced from Refs.[5] and [7].

## Conclusion

We have shown that the atomistic model is a useful tool in predicting the behavior of multi-component magnetic materials under the influence of ultrafast laser heating and using such a model have predicted a novel switching mechanism using only heat. We have utilized a multiscale approach in the form of the LLB model to demonstrate the importance of the transfer of angular momentum from longitudinal magnetization components to the transverse. The use of these models across different length-scales shows the importance of multiscale modeling in the field of ultrafast magnetism.

**Acknowledgments** The authors thank the European Community's Seventh Framework Programme (FP7/2007-2013) under grant agreements NMP3-SL-2008-214469 (Ultramagnetron) N214810 (FANTOMAS), and NMP3-SL-2012-281043 (FEMTOSPIN) and the Spanish Ministry of Science and Innovation under grant FIS2010-209790-C02-02.

## References

[1] I. Tudosa, C. Stamm, A. B. Kashuba, F. King, H. C. Siegmann, J. Stöhr, G. Ju, B. Lu & D. Weller, "The ultimate speed of magnetic switching in granular recording media" Nature **428**, 831-833 (2004).

[2] V. V. Temnov, "Ultrafast acousto-magneto-plasmonics" Nature Photonics **6**, 728-736 (2012).

[3] J.-Y. Bigot, M. Vomir, E. Beaurepaire, "Coherent ultrafast magnetism induced by femtosecond laser pulses" Nature Physics **5**, 515-520 (2009).

[4] A. Kirilyuk, A. V. Kimel, and T. Rasing, "Ultrafast optical manipulation of magnetic order" Rev. Mod. Phys. **82**, 2731-2784 (2010).

[5] T. A. Ostler et al., "Ultrafast heating as a sufficient stimulus for magnetization reversal in a ferrimagnet" Nat. Commun. **3**, 666 (2012).
[6] I. Radu, K. Vahaplar, C. Stamm, T. Kachel, N. Pontius, H. A. Dürr, T. A. Ostler, J. Barker, R. F. L. Evans, R. W. Chantrell, A. Tsukamoto, A. Itoh, A. Kirilyuk, Th. Rasing & A. V. Kimel, "Transient ferromagnetic-like state mediating ultrafast reversal of antiferromagnetically coupled spins" Nature **472**, 205-208 (2011).
[7] U. Atxitia, T. Ostler, J. Barker, R. F. L. Evans, R. W. Chantrell, and O. Chubykalo-Fesenko, "Ultrafast dynamical path for the switching of a ferrimagnet after femtosecond heating" Phys. Rev. B **87**, 224417 (2013).
[8] J. Barker et al., Submitted.
[9] O. N. Mryasov, U. Nowak, K. Y. Guslienko and R. W. Chantrell, "Temperature-dependent magnetic properties of FePt: Effective spin Hamiltonian model" Europhys. Lett, **69**, 805-811 (2005).
[10] L. Szunyogh, L. Udvardi, J. Jackson, U. Nowak, and R. Chantrell, "Atomistic spin model based on a spin-cluster expansion technique: Application to the IrMn3/Co interface" Phys. Rev. B. **83**, 024401 (2011).
[11] Thomas A. Ostler, Richard F. L. Evans, Roy W. Chantrell, Unai Atxitia, Oksana Chubykalo-Fesenko, Ilie Radu, Radu Abrudan, Florin Radu, Arata Tsukamoto, A. Itoh, Andrei Kirilyuk, Theo Rasing, and Alexey Kimel, "Crystallographically amorphous ferrimagnetic alloys: Comparing a localized atomistic spin model with experiments" Phys. Rev. B **84**, 024407 (2011).
[12] J.K. Chen, D.Y. Tzou, J.E. Beraun, "A semiclassical two-temperature model for ultrafast laser heating" International Journal of Heat and Mass Transfer, **49**, 307-316 (2006).

# What can we learn about magnetization dynamics from first-principles calculations?

**Paul J. Kelly**

Faculty of Science and Technology and MESA+ Institute for Nanotechnology,

University of Twente, P.O. Box 217, 7500 AE Enschede, The Netherlands

P.J.Kelly@utwente.nl

**Abstract** A first-principles formulation of scattering theory that includes disorder, spin-orbit coupling and non-collinearity is used to study the resistance of realistic domain wall (DW) profiles, to calculate spin-flip diffusion lengths, to calculate Gilbert damping in the adiabatic and non-adiabatic limits and to determine the non-adiabatic out-of-plane torque parameter β.

## Introduction

The design of magnetoelectronic devices is based upon frequently incomplete models formulated in terms of parameters that are difficult to measure experimentally. We have developed a computational framework that allows us to calculate material-dependent parameters with a predictive capability for complex materials such as CoFeB or Permalloy (Py: $Ni_{80}Fe_{20}$) that form the backbone of existing magnetoelectronic devices such as spin valves and magnetic tunnel junctions. They play a central role in spin-transfer torque (STT) devices such as magnetic random-access memories, spin-torque oscillators, and so-called ''racetrack memories'' based upon the STT effect whereby a spin-polarized current exerts a torque on a magnetization forcing it to precess. A realistic description of electrical transport in itinerant ferromagnets is made difficult by the degeneracy of the partially filled d bands that are responsible for the magnetism and result in complicated Fermi surfaces for ordered materials. The concepts of Bloch states and Fermi surfaces that enable the development of transport theories for crystals are lost in disordered alloys. Even though the spin-orbit coupling (SOC) is in energy terms small, it has a large effect on the transport properties of magnetic alloys and must be included in any realistic description. Finally, the STT effect results when electrical currents flow between materials whose magnetizations are not collinear; in the case of DWs, the length scale over which the magnetization changes is of or-

 
J.-Y. Bigot et al. (eds), *Ultrafast Magnetism I*, Springer Proceedings in Physics 159
DOI 10.1007/978-3-319-07743-7_48

der 10–100 nm. These difficulties all stand in the way of a satisfactory description of transport properties of materials such as CoFeB or Py, currently important candidates for applications.

To be able to address the above issues, we extended an efficient scattering formalism of spin transport [1] based upon tight binding muffin tin orbitals (TB-MTOs) and the atomic spheres approximation (ASA) to include SOC. This successfully describes the transport properties of alloys such as Py [2-4]. We extended this to treat noncollinearity and reported on a first application to the resistance of Py DWs in [5].

**Acknowledgments** This work was carried out in collaboration with Anton Starikov, Zhe Yuan, Yi Liu, Rien Wesselink, Arne Brataas, Gerrit Bauer and Yaroslav Tserkovnyak. It was supported by EU FP7 Contract No. NMP3-SL-2009-233513 MONAMI and ICT Grant No. 251759 MACALO. It also forms part of the research programme of ''Stichting voor Fundamenteel Onderzoek der Materie'' (FOM) and the use of supercomputer facilities was sponsored by the ''Stichting Nationale Computer Faciliteiten'' (NCF), both of which are financially supported by the ''Nederlandse Organisatie voor Wetenschappelijk Onderzoek'' (NWO).

## References

[1] K. Xia, M. Zwierzycki, M. Talanana, P.J. Kelly, and G.E.W. Bauer, "First-principles scattering matrices for spin-transport", Phys. Rev. B**73**, 064420 (2006).

[2] Anton A. Starikov, Paul J. Kelly, Arne Brataas, Yaroslav Tserkovnyak, and Gerrit E.W. Bauer, "Unified First-Principles Study of Gilbert Damping, Spin-Flip Diffusion, and Resistivity in Transition Metal Alloys", Phys. Rev. Lett. **105**, 236601 (2010).

[3] Yi Liu, Anton A. Starikov, Zhe Yuan and Paul J. Kelly, "First-principles calculations of magnetization relaxation in pure Fe, Co, and Ni with frozen thermal lattice disorder", Phys. Rev. B**84**, 014412 (2011).

[4] H. Ebert, S. Mankovsky, and D. Ködderitzsch and Paul J. Kelly, "Ab Initio Calculation of the Gilbert Damping Parameter via the Linear Response Formalism", Phys. Rev. Lett. **107**, 066601 (2011).

[5] Zhe Yuan, Yi Liu, Anton A. Starikov, Paul J. Kelly, and Arne Brataas, "Spin-orbit-coupling-induced domain-wall resistance in diffusive ferromagnets" Phys. Rev. Lett. **109**, 267201 (2012).

# Theoretical modeling of coherent ultrafast spin-light interactions: from one to many-electron systems

**P.-A. Hervieux, G. Manfredi, O. Morandi, J. Zemanian, Y. Hinschberger, and A. Dixit**

Institut de Physique et Chimie des Matériaux de Strasbourg, CNRS and University of Strasbourg, 23 rue du Loess, BP 43, 67034 Strasbourg Cedex 2, France

Hervieux@unistra.fr

**Abstract** In this contribution we review some theoretical efforts performed in our group and devoted to the description, from first-principles, of the coherent ultrafast spin-light interactions. Based on the semi-relativistic expansion of the Dirac equation, we started by investigating the cases of one, two and many electrons (in the mean-field approximation) systems knowing that our main purpose is to realistically describe the nonlinear electron and spin dynamics of a quantum-relativistic system of many interacting electrons excited by an intense and ultrashort electromagnetic field. In addition to that, a classical and macroscopic model based on the an-harmonic Drude-Voigt model and the Maxwell's equations was also developed for modeling ultrafast nonlinear coherent magneto-optical experiments performed on ferromagnetic thin films.

## Introduction

Recently, relativistic couplings between spins and photons have been evoked for explainingthe process of coherent ultrafast demagnetization induced by a femtosecond laser pulse focused ona ferromagnetic Ni film [1].So far, this hypothesis has not been confirmed or refuted and is the subject of intense research efforts in the field of ultrafast magneto-optical spectroscopy.

The electromagnetic field associated to a femtosecond laser pulse can be strong enough to significantly perturb the electronic charges and spins in condensed matter systems, so that relativistic effectsbecome important. Given the intensity of the fields involved, nonlinear effects are also expected to play animportant role. This scenario represents an ambitious theoretical challenge, as it requires the *modeling of the nonlinear dynamics of a quantum-relativistic system of many interacting electrons excited by an intense and ultrashortelectromagnetic field.*

 
J.-Y. Bigot et al. (eds), *Ultrafast Magnetism I*, Springer Proceedings in Physics 159
DOI 10.1007/978-3-319-07743-7_49

In this contribution, I will review recent theoretical developments on that issue which are performed in our group in Strasbourg. I will start by describing quantum microscopic approaches. Then, I will conclude by describing a classical macroscopic model whose predictions have been compared successfully with available experimental measurements of [1].

## Microscopic approaches

The well-known Foldy–Wouthuysen transformation (FWT) [2] is a unitary transformation that decouples the small (that represent antiparticles) and large components (related to particles) of the Dirac bispinor. It therefore allows to separate the relativistic solutions into electronic (which are investigated in condensed matter physicsand correspond to the low-energy limit of the Dirac equation) and positronic states. It can also be systematically used to get the non-relativistic approximation of the Dirac equation by expanding the relativistic Hamiltonian in power series of *1/m*. In [3], thanks to this transformation the non-relativistic limit at fifth order in powers of *1/m* of the one-electron external-electromagnetic-field Dirac equation is performed. It allows us to postulate a general analytical expression of the Hamiltonian related to the direct electron spin-field interaction valid at any order in powers of *1/m.The latter can be viewed as a generalization of the usual Zeeman Hamiltonian to the time-dependent regimes.*

In [4], by extending the previous methodology to the case of two interacting electrons (which is by far a more complex problem) and within the framework of the Dirac-Breit model, we have shown how an*external electric field could modify and control the electron-electron interaction* at third-order in *1/m*.

Very recently [5], and still using the FWT, the mean-spin angular momentum operator introduced for the first time by Foldy and Wouthuysen[2] for the case of a free electron has been extended to the case of an electron interacting with a time-dependent electromagnetic field.The equation of motion of the latterleads to the *Landau-Lifshitz-Gilbert equation revealing thus its microscopic origin.*We have therefore argued that the mean-spin operator must be used instead of the usual one to properly describe the dynamics of the spin magnetization.

In order to model more realistic systems in the fields of condensed matter and plasma physics we have developed several mean-field models which incorporate the interaction between the electrons at the lowest-order approximation of the many-body problem.In this context, the full Dirac-Maxwell equations must normally be solved. However, most of the time their resolution being unfeasible, one must resort to simpler models.

In [6], we have derived a mean-field model that is based on a two-component Pauli-like equation and incorporates quantum, spin, and relativistic effects up to second-order in *1/c*. By using a Lagrangian approach, we have obtained the self-consistent charge and current densities that act as sources in the Maxwell equations. A physical interpretation has been provided for the second-order corrections to the sources. The Maxwell equations have been also expanded to the same order [7]. The resulting self-consistent model constitutes an *appropriate semi-relativistic approximation to the full Dirac-Maxwell equations*.

A high intensity X-ray laser can induce strong modifications of the magnetic order of a nanometric system [8]. Furthermore, under extreme condition on the laser power, hot relativistic plasma of electrons whose speeds approach the speed of light could be generated by irradiating a thin metallic slab [9]. Related to these processes, the nonlinear magneto-optical response of a plasma slabwith metallic densityunder strong X-ray laser excitation is investigated within the framework of a semi-relativistic quantum fluid model which is derived from the second-order Foldy-Wouhuysen Hamiltonian [10].We have focused*on the effects of the spin dependent semi-relativistic corrections on the polarization of the reflected and transmitted light*.

## Macroscopic approach

In [11], based on the classical anharmonic Drude-Voigt theory, a rather simple model has been developed for modeling ultrafast nonlinear coherent magneto-optical experiments performed on ferromagnetic thin films [1]. Theoretical estimations of the Faraday rotation angles are compared to available experimental values of nickel thin filmand give meaningful insights about the physical mechanisms underlying the observed coherent magneto-optical phenomena. *Under realistic conditions, the model successfully explains the observed trends*. The crucial role played by the spin-orbit mechanism resulting from the direct interaction between the external electric field of the laser and the electron spins of the sample is underlined.

**Acknowledgments** The author thanks Dr. J. -Y. Bigot and Pr. M. Maamache for useful discussions. This work was supported in part by the project ATOMAG supported by the European Research Council advanced grant ERC-2009-AdG-20090325-247452.

## References

[1] J.-Y. Bigot, M. Vomir and E. Beaurepaire, "Coherent ultrafast magnetism induced by femtosecond" Nature Physics **5**, 515 (2009).
[2] L. L. Foldy and S. A. Wouthuysen, "On the Dirac Theory of Spin 1/2 Particles and Its Non-Relativistic Limit" Phys. Rev. **78**, 29 (1950).
[3] Y. Hinschberger and P.-A. Hervieux "Foldy–Wouthuysen transformation applied to the interaction of an electron with ultrafast electromagnetic fields" Phys. Lett. A **376**, 813 (2012).
[4] Y. Hinschberger and P. -A. Hervieux, submitted (2013).

[5] Y. Bouguerra, M. Maamache and P. –A. Hervieux, "Effect of a new mean-spin angular momentum operator on Gilbert damping" submitted (2013).
[6] A. Dixit, Y. Hinschberger, J.Zamanian,G. Manfredi and P. –A.Hervieux, "Lagrangian approach to the semi-relativistic electron dynamics in the mean-field approximation" Phys. Rev. A**88**, 032117 (2013).
[7] G. Manfredi, "Non-relativistic limits of Maxwell's equations" Eur. J. Phys. **34**, 859 (2013).
[8] T. Wang et al., "Femtosecond Single-Shot Imaging of Nanoscale Ferromagnetic Order In Co/Pd Multilayers Using Resonant X-Ray Holography" Phys. Rev. Lett.**108**, 267403 (2012).
[9] D. Kiefer et al., "Relativistic electron mirrors from nanoscale foils for coherent frequency upshift to the extreme ultraviolet" Nature communications **4**, 1 (2013).
[10] O. Morandi, J. Zemanian, G. Manfredi and P. –A. Hervieux, "Spin dependent semi-relativistic corrections to the laser-plasma interaction" to be submitted to Phys. Rev. A (2013).
[11] Y. Hinschberger and P. -A. Hervieux, "Classical Modeling of Ultrafast Coherent Magneto-Optical Experiments" Phys. Rev. B **88**, 134413 (2013).

# Localization of magnetic normal modes on topological defects

**F.J. Buijnsters, A. Fasolino, and M.I. Katsnelson**

Radboud University Nijmegen, Institute for Molecules and Materials,
Heyendaalseweg 135, 6525 AJ Nijmegen, The Netherlands

F.Buijnsters@science.ru.nl

**Abstract** Topological defects in magnetic materials, such as domain walls or Skyrmion magnetic bubbles, support magnetic normal modes (low-energy spin-wave excitations) which are localized on the edge of the defect [1]. Using an efficient numerical scheme [7], we present the eigenmodes of systems in nontrivial equilibrium configurations with complex domain structures.

## Introduction

Topological defects in magnetic materials have attracted a great deal of interest. Owing to their stability, magnetic topological defects, including domain walls, are a key ingredient of the proposed magnetic storage devices of the future. The dynamics of domain walls, Skyrmion magnetic bubbles, and magnetic domains shows some rich physics [1,2]. Understanding these phenomena is likely to be important for technological applications such as racetrack memory [3].

The dynamics of magnetic materials can often be accurately described at the classical level with the Landau–Lifshitz equation. Small-amplitude deviations from the equilibrium magnetization state (which may be highly non-trivial, for instance as a result of topological defects) often propagate through the material as spin waves. As we shall see, these spin waves may or may not be localized on the edge of a topological defect. The study of spin-wave excitations is an important tool in the analysis of dynamical magnetic behavior. It can be used to test and improve models of magnetic microstructures, such as magnetic rings, in a very direct fashion [4]. It may be used to find magnetic resonances, which can enhance the motion of domain walls in devices [3]. It is also useful in the theoretical analysis of the motion of other topological defects such as Skyrmion magnetic bubbles [1].

Traditionally, the theoretical prediction of magnetic normal modes relies on analytical methods [1,5], which are limited to relatively simple cases; 'brute-force' dynamical micromagnetic simulation [1,4]; or explicit numerical diagonalization of large, non-symmetric matrices [6]. The latter two approaches, while in principle capable of dealing with systems of arbitrary complexity, become very computa-

 
J.-Y. Bigot et al. (eds), *Ultrafast Magnetism I*, Springer Proceedings in Physics 159
DOI 10.1007/978-3-319-07743-7_50

tionally expensive for large systems. A recently developed method [7] overcomes the disadvantages of the latter scheme and can be used to study very large systems containing many topological defects.

In this contribution, we present the normal modes of magnetic domain walls and Skyrmion bubbles using an efficient numerical scheme [7]. This method allows us to consider the dynamics not only of single domain walls or magnetic bubbles, but also of complex configurations containing many magnetic bubbles. These bubble domains may be arranged in a lattice (Skyrmion lattice [8]) or, perhaps even more interestingly, trapped in a metastable, disordered configuration. We also discuss the domain-wall excitations of stripe domains.

## Method

We linearize the Landau–Lifshitz equation near a given magnetic configuration which is a local minimum of the Hamiltonian. We are now interested primarily in those eigenmodes of the linearized Landau–Lifshitz equation which represent the low-energy, low-frequency excitations of the system, which includes the magnetic normal modes localized on topological defects. This suggests the use of iterative numerical methods. Ref. [7] demonstrates how the magnetic normal mode problem can be reduced to the *Hermitian definite generalized eigenproblem* (HDGEP),

$$Dx = \lambda Sx \qquad 1$$

where D and S are Hermitian matrices and S is positive definite. Eigenproblems of this form can be solved particularly efficiently using schemes such as the Lanczos or conjugate-gradient eigenvector algorithms.

## Magnetic modes on domain walls or complex domain structures

It is well known that in a homogeneous medium, the lowest-energy magnetic excitations are the usual planar spin waves. The presence of a domain wall modifies this picture. In 1D, a zero mode (Goldstone mode) appears, in addition to the running spin-wave modes, which corresponds to an infinitesimal translation of the domain wall [5]. In two dimensions, the magnitude of this displacement may be modulated in space. Thus, the single Goldstone mode becomes a continuum of low-energy modes, each described by a wavevector directed along the domain wall. In this direction, the modes can propagate freely; in the perpendicular direction, they are localized. Fig. 1 shows an example of such a domain-wall mode.

In the case of a magnetic bubble domain, the periodicity condition on the circular domain wall turns the continuum of domain-wall modes into a discrete set, corresponding to radial deformations of the magnetic bubble [1].

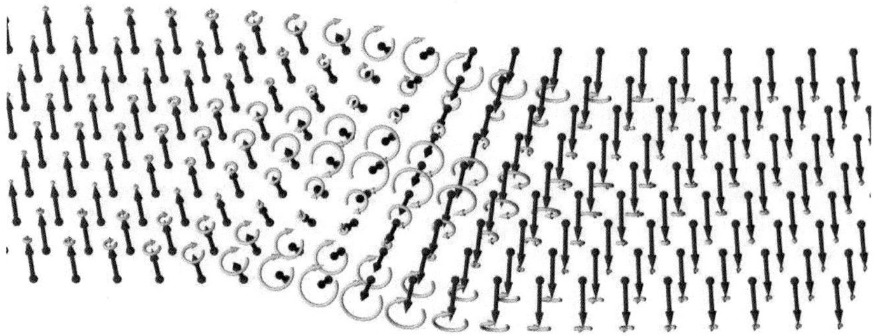

**Fig. 1.** Magnetic normal mode localized in one spatial direction on the domain wall

Ref. [8] describes the experimental observation of a regular *lattice* of Skyrmion magnetic bubbles, which are stabilized by the Dzyaloshinskii–Moriya interaction. In such a regular lattice, the coupling between the bubbles turns the individual bubble deformation modes into bands of delocalized excitations [1]. In systems where it is magnetostatic interactions that stabilize the magnetic bubbles, by contrast, magnetic bubbles tend to be a lot larger and more irregular in size and arrangement. The degree of localization of the domain-wall modes of such irregular bubble configurations, as well as of stripe domains, is an interesting question.

**Acknowledgments** This work is part of the research programme of the Foundation for Fundamental Research on Matter (FOM), which is part of the Netherlands Organisation for Scientific Research (NWO).

## References

[1] I. Makhfudz, B. Krüger, O. Tchernyshyov, "Inertia and Chiral Edge Modes of a Skyrmion Magnetic Bubble" Phys. Rev. Lett. **109**, 217201 (2012).
[2] Y. Talbi et al., "Magnetic excitation in weak stripe domains: Ferromagnetic resonance and Brillouin light scattering studies" J. Phys. Conf. Ser. **200**, 072107 (2010).
[3] S.S.P. Parkin, M. Hayashi, L. Thomas, "Magnetic Domain-Wall Racetrack Memory" Science **320**, 190-194 (2008).
[4] F. Giesen, J. Podbielski, D. Grundler, "Mode localization transition in ferromagnetic microscopic rings" Phys. Rev. B **76**, 014431 (2007).
[5] J.S. Helman, H.B. Braun, J.S. Broz, W. Baltensperger, "General solution of the Landau–Lifshitz–Gilbert equations linearized around a Bloch wall" Phys. Rev. B **43**, 5908-5914 (1991).
[6] M. Grimsditch, L. Giovannini, F. Montoncello, F. Nizzoli, Gary K. Leaf, and Hans G. Kaper "Magnetic normal modes in ferromagnetic nanoparticles: A dynamical matrix approach" Phys. Rev. B **70**, 054409 (2004).
[7] F.J. Buijnsters, A. Fasolino, M.I. Katsnelson (in preparation).
[8] S. Mühlbauer, B. Binz, F. Jonietz, C. Pfleiderer, A. Rosch, A. Neubauer, R. Georgii, P. Böni, "Skyrmion Lattice in a Chiral Magnet" Science **323**, 915-919 (2009).

# Effect of the variation of the bond length on laser-induced spin-flip scenarios at$Ni_2$

**D. Chaudhuri[1*], G. Lefkidis[1], A. Kubas[2], K. Fink[3], W. Hübner[1]**

[1]Dept. of Physics and Research Center OPTIMAS, University of Kaiserslautern, Germany

[2]Department of Physics and Astronomy, University College London, UK

[3]Institute of Nanotechnology, Karlsruhe Institute of Technology, Germany

chaudhuri@physik.uni-kl.de

**Abstract** We investigate laser-induced spin-dynamics scenarios on a nickel dimer at various interatomic distances. Successful spin-flip processes based on highly correlated *abinitio* calculations are established and analyzed. Switching is achieved through a laser-driven Λ-process and optimized with a genetic algorithm.

## Introduction

Theoretical[1,2] and experimental[3] treatment of transition metals attract growing interest since the last decade because of their complex electronic and magnetic properties. Nickel is interesting to investigate because of its open $3d$ subshells which lead to complicated electronic structures. Bonding in $Ni_2$ molecule is mainly due to 4*s* orbitals interaction with only small 3*d*-3*d* contribution due to their strongly localized character.Ultrafast dynamics[4-6] is particularly important in order to meet the industrial necessities that demand for smaller sizes and shorter time scales with respect to magnetic responses in spintronic devices.

## Results and Discussion

We describe our system with two real-space quantum chemistry methods: the equation-of-motion coupled-cluster with single and double excitations (EOM-CCSD [7] in the GAUSSIAN09 package [8]), and the$n$-electron valence-state perturbation theory (NEVPT2[9] in the ORCA package [10], Fig. 1) based on CAS (20, 12) with all 3*d* and 4*s* orbitals in the active spaceand scalar relativistic effects at the DKH2 level [11].We take four steps:(a) computation of the ground and the 30 lowest excited states,(b) perturbative inclusion of spin-orbit coupling(SOC),(c)



J.-Y. Bigot et al. (eds), *Ultrafast Magnetism I*, Springer Proceedings in Physics 159
DOI 10.1007/978-3-319-07743-7_51

spin-density inspection, and (d) the spin flip (Λ-process) optimization.The dynamics results stem from EOM-CCSD.The triplet ground state has a zero-field splitting of 60 ± 20 $cm^{-1}$. For the Λ-process four levels are required:two ground states with distinct spin-up and spin-down orientation and two excited states coupled by SOC to yield mixed-spin states (accessible from both ground states). The two (otherwise degenerate and non-distinguishable) ground states are energetically separated by an external **B**-field (Zeeman effect).Whenincluding SOC spin is not a good quantum number; yet it is possible to compute its expectation value and, in a loose manner, characterize some states as spin-up or spin-down if $\langle S_z \rangle \approx \pm 1$.

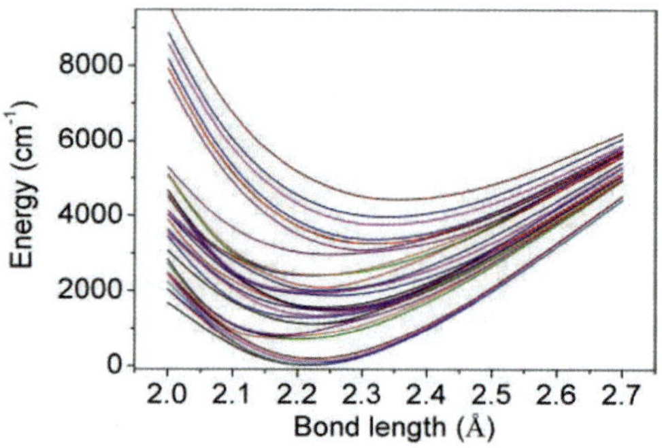


**Fig. 1.** Energies of the electronic statesof $Ni_2$as a function of the bond length, computed withthe NEVPT2 method and the Def2-QZVPP [24s18p10d3f1g/11s6p5d3f1g] basis set.

The laser pulse (optimized with a genetic algorithm [12]) is defined by six parameters: $\vartheta$ and $\varphi$ (angle of incidence on spherical coordinates), $\gamma$ (angle between the light polarization and the optical plane), $I$ (intensity), $E$ (energy) and FWHM (full width at half maximum). A typical spin-switch takes 500 fs (Fig. 2a). The effect of the laser is studied under three different aspects: (a) the laser pulse is optimized for each bond length separately (Fig. 2b, black solid line). (b) The pulse optimized for a particular bond length is also used for rest. Fig. 2b (red dashed line) shows the fidelity if the laser parameters are kept the same as for the equilibrium distance (2.5 Å), and Fig. 2c the fidelity for a pulse optimized for 2.15 Å (worst maximum efficiency). Strangely enough, the latter pulse yields a reasonable fidelity also at 2.55 Å. (c) The fidelity drop for a fixed laser could be explained simply by the loss of resonance when changing the bond length. However, using the same laser pulse while only adjusting its frequency to match the excitation energy of the relevant intermediate states proves inadequate (Fig. 2d).This disproves the simple resonance argument and underlines the importance of correlations in $Ni_2$.

Summarizing we find that (i) high-fidelity spin-switching is possible with Λ-processes, (ii) $Ni_2$ at each bond length responds mostly to a specially optimized pulse, and (iii) tuning the laser frequency to be in resonance with the intermediate excited state is not enough to improve thesehighly non-linear processes. As a conclusion the coherent control of spin with uniquely specified parameters opens an avenue for designing logic elements and future spintronic devices.

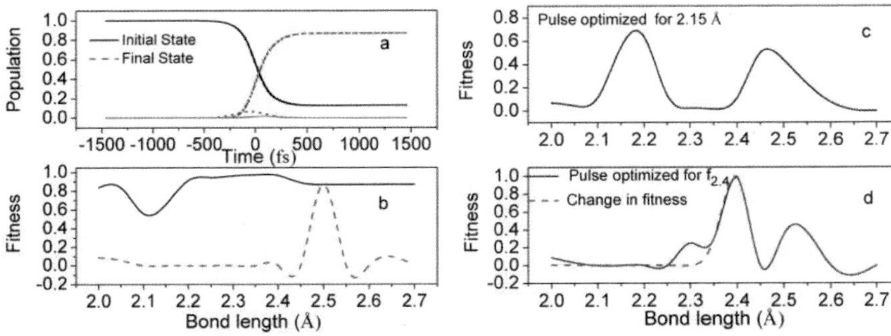


**Fig. 2.** (a) A successful spin flip scenario on $Ni_2$at 2.00Å. (b) Fitness of the spin-flip (solid black line for pulses optimized at each bond length, red dashed line for a pulse optimized at 2.50 Å). (c) Efficiency at different bond lengths witha laser pulse optimized for 2.15 Å. (d) Efficiency at different bond lengths with a laser pulse optimized for 2.4 Å. The solid line refers to exactly the original pulse($f_{2.4}$) and the dotted line to the case with all laser parameters fixed, except the frequency which is adjusted to match the energy of the relevant intermediate excited state.

**Acknowledgments** The authors acknowledge funding from the German Research Foundation (DFG) via the Trans-regional Collaborative Research Center SFB/TRR 88 "3MET".

# References

[1] R. Pou-Amerigo *et al.*, "The chemical bonds in CuH, $Cu_2$, NiH, and $Ni_2$ studied with multiconfigurational second order perturbation theory", J. Chem. Phys. **101**, 4893 (1994).
[2] J. Barker, R. F. L. Evans, R.W. Chantrell, D. Hinzke, U. Nowak, "Atomistic spin model simulation of magnetic reversal modes near Curie point", Appl. Phys. Lett. **97**, 192504 (2010).
[3] J. C. Pinegar *et al.* "$Ni_2$ revisited: Reassignment of the ground electronic state" J. Chem. Phys. **102**, 666-674 (1995).
[4] W. Hübner, S. Kersten, and G. Lefkidis, "Optical spin manipulation for minimal magnetic logic operations in metallic three center magnetic clusters", Phys. Rev. B. **79**,184431(2009).
[5] J.-Y. Bigot, M. Vomir, E. Beaurepaire, "Coherent ultrafast magnetism induced by femtosecond laser pulses", Nature Phys. **5**, 515-520 (2009).
[6] A. V. Kimel, A. Kirilyuk, A. Tsvetkov, R. V. Pisarev, and T. Rasing, "Laser-induced ultrafast spin reorientation in the antiferromagnet $TmFeO_3$", Nature **429,** 850-853 (2004).
[7] M. Kálleyand J. Gauss, "Calculation of excited-state properties using general coupled-cluster and configuration-interaction models" J. Chem. Phys. **121**, 9257-9269 (2004).
[8] Gaussian 09, Revision B.2, M. J. Frisch *et al.*, Gaussian, Inc., Wallingford CT, (2009).
[9] C. Angeli, R. Cimiraglia, and J. P. Malrieu,"Introduction of n-electron valence states for multireference perturbation theory", J. Chem. Phys. **114**, 10252-10264 (2001).
[10] ORCA V2.9.1, Developed by F. Neese. http://www.cec.mpg.de/forum.
[11] A. Wolf, M. Reiher, B. A. Hess, "The generalized Douglas–Kroll transformation", J. Chem. Phys. **117**, 9215 (2002).
[12] T. Hartenstein, C. Li, G. Lefkidis and W.Hubner, "Local light-induced spin manipulation intwo magneticcentre metallic chains", J. Phys. D: Appl Phys. **41**, 164006 (2008).

# Coarse-Graining Approach to Atomistic SpinDynamics

**T. Nystrand, J. Venemalm, J. Werpers, O. Eriksson, J. Chico and A. Bergman**

Deparment of Physics and Astronomy, Division of Materials Theory, Uppsala University, Box

516, SE-75210, Uppsala Sweeden.

jonathan.chico@physics.uu.se

**Abstract** We introduce a coarse-graining approach to study the movement of a Domain Wall (DW) under the influence of a spin polarized current, in the framework of atomistic spin dynamics. An increase in performance of up to 35% is obtained. We show the dependence of the method on both exchange range and temperature effects.

## Introduction

Magnetization dynamics is an increasingly interesting topic which has experienced much attention due to its possible technological applications. Domain wall (DW) motion has been thoroughly studied in the last decades due to their potential importance in spintronic and magnonic devices. Therefore the controlled movement of these magnetic textures under spin polarized electronic currents due to the spin transfer torque (STT) phenomenon is of extreme importance [1].

In this work we present a coarse-graining approach which allows us to study the motion of a DW in a monolayer of Fe/W(110) when it is subjected to a spin polarized current as implemented in the Uppsala Atomistic Spin Dynamics (UppASD) package [2].

## Atomistic Spin Dynamics Framework

In order to investigate the dynamics of the DW in this system, we perform atomistic spin dynamics simulations as implemented in the UppASD package. Each atomic moment is considered to be a three dimensional (3D) vector with an equation of motion given by the Landau-Lifshitz equation:

 
J.-Y. Bigot et al. (eds), *Ultrafast Magnetism I*, Springer Proceedings in Physics 159
DOI 10.1007/978-3-319-07743-7_52

$$\frac{\partial \vec{m}}{\partial t} = -\gamma \vec{m}_i \times \overrightarrow{B_i^{eff}} - \gamma \frac{\alpha}{m}\left[\vec{m}_i \times \left[\vec{m}_i \times B_i^{eff}\right]\right] \quad 1$$

where the effective field contains $\mathbf{B}_i$ a field which takes into account all the interactions considered in the Hamiltonian of the system, in this case the classic Heisenberg Hamiltonian, and an stochastic field to take into account the temperature effects by using Langevin dynamics.

In the atomistic framework, the calculation of the effective field is very expensive computationally. This high computational cost is warranted only if detailed information about the magnetization on an atomic level is needed. However, in many possible applications of the ASD method, an atomistic resolution is only needed for a small part of the studied system. A prime example of this is the movement of individual DWs in ferromagnetic materials where only the actual DW region has the need for an atomistic description while it should be possible to model the rest of the system with a coarser model.

## Coarse graining approach

The approach used in this work is based on the simple assumption that regions not exhibiting variations in magnetization on the atomic scale can be combined into larger magnetization regions. It then suffices to evaluate the Hamiltonian for one unit cell region.

The idea behind the coarse-grain approach is to divide the domain into core-, boundary and noise regions while keeping the classification on atomic level. This requires the inclusion of a similarity measure. In particular, the cosine similarity is a suitable quantity for this task. Two atomic moments are said to be similar if the similarity between their corresponding moments exceeds a user-specified threshold. A given atom is defined as a core atom if the pairwise similarity between itself and all its neighboring atoms exceeds the threshold value. This requirement must hold for all atoms located in the same unit cell.

A constellation is then defined as the union of core cells where the similarity threshold is satisfied for all participating core atoms in the constellation region. Finally atoms located on the border of a constellation are defined as boundary atoms (with respect to the neighbouring constellation), and all atoms not meeting the requirements of being core- or boundary atoms are by default classified as noise atoms.

At each time step, every atom is checked against all pre-existing constellations, and if the similarity is larger than the user defined threshold the atom is said to belong to the constellation that is being checked against. Therefore if the fraction of core atoms is large the evaluation of the Hamiltonian is simplified.

## Results

To study the speedup of this method, a monolayer of Fe/W(110) of size 100x40 repetitions is used. This system has been previously studied using atomistic spin dynamics [3-4] and is known to exhibit a very narrow DW width. A DW is established by flipping half of the spins of the system and then equilibrating the system at the temperature of interest. It is possible to see that it is possible to obtain speedups up to 35% when we consider 40 exchange shells (J40 red line).

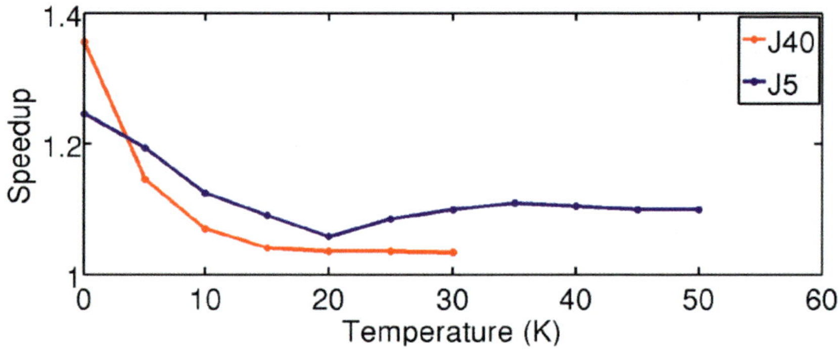


**Fig. 1.** Speedup dependence on temperature for two different exchange ranges

As can be seen as the temperature increases the speedup decreases as expected, as the stochastic field will make it harder to have regions were the similarity is high. The range of the exchange interactions also affects the speedup, as a larger exchange ranges will lead to stronger requirements for the creation of core cells. Applying an STT (Slonczewski term) over the sample allows us to move the DW towards one of the edges of the sample, the proposed coarse-graining procedure gives us speedups of up to 25%, lower than for the static situation as the number of core cells is constantly changing due to the movement of the DW.

## Conclusion

We have devised a method for coarse graining of atomistic spin dynamics simulations and observe speedups up to 35% for domain wall systems. The method can also be used for other systems such as simulating laser heated hotspots in an otherwise cooled magnetic sample.

**Acknowledgments** We gratefully acknowledge the European Research Council (ERC Project No. 247062-ASD), the Swedish Research Council (VR), the Knut and Alice Wallenberg Foundation, the Carl Tryggers Foundation, and the G\"oran Gustafsson Foundation for financial support. O.E. and A.B acknowledges eSSENCE.

# References

[1] D.C. Ralph and M.D. Stiles, "Spin transfer torques" J. Mag. Magn. Mat. **320** (7)1190-1216, (2008).
[2] B. Skubic, J. Hellsvik, L. Nordström, and O. Eriksson , "A method for atomistic spin dynamics simulations: implementation and examples" J. Phys. Cond. Matt., **20** (31):315203, (2008).
[3] A. Bergman et al. , "Magnon softening in a ferromagnetic monolayer: A first principles spin dynamics study" Phys. Rev. B **81**:144416, (2010).
[4] J. Chico et al. "Thermally driven domain wall motion in Fe on W(110)" submitted to PRB

# Coherent ultrafast spin-light interactions in one- and two-electron systems

**Y. Hinschberger and P.-A. Hervieux**

Institut de Physique et Chimie des Matériaux de Strasbourg, CNRS and University of Strasbourg, 23 rue du Loess, BP 43, 67034 Strasbourg Cedex 2, France

Yannick.Hinschberger@ipcms.u-strasbg.fr and Hervieux@unistra.fr

**Abstract** Thanks to the Foldy-Wouthuysen transformation (FWT), the non-relativistic limit at fifth order in powers of *1/m* of the one-electron external-electromagnetic-field Dirac equation is performed. It allows us to postulate a general analytical expression of the Hamiltonian related to the direct electron spin-field interaction valid at any order in powers of *1/m*. By extending the previous methodology to the case of two interacting electrons and within the framework of the Breit model, we show how an external electric field could modify and control the electron-electron interaction at the third order in *1/m*.

## Introduction

Recently, relativistic couplings between spins and photons have been evoked for explaining the process of coherent ultrafast demagnetization induced by a femtosecond laser pulse focused on a ferromagnetic Ni film [1]. So far, this hypothesis has not been confirmed or refuted and is the subject of intense research efforts in the field of ultrafast magneto-optical spectroscopy. The aim of the present study is to develop theoretical methods needed for isolating the direct (and therefore coherent) coupling between the spins and the light. In the first part, using the FWT of the one-particle Dirac's Hamiltonian, we derive an analytical expression illustrating the direct coupling between the spin degrees of freedom (*dof*) and an external time-dependent electromagnetic (*em*) field. In the second part, using the same methodology, we briefly investigate the case of two interacting electrons (which by far a more complex problem) described by the Dirac-Breit Hamiltonian.

## Ultrafast spin-light interactions from first principles

In the presence of an *em* field, the well-known FWT expands the external-electromagnetic-field Dirac Hamiltonian in powers of *1/m* and leads to the electronic Schrödinger-Pauli Hamiltonian which is usually presented up to the second order in *1/m* [2]



J.-Y. Bigot et al. (eds), *Ultrafast Magnetism I*, Springer Proceedings in Physics 159
DOI 10.1007/978-3-319-07743-7_53

$$H_e' = \frac{(\mathbf{p}-e\mathbf{A})^2}{2m} + e\Phi - \frac{e\hbar}{2m}\boldsymbol{\sigma}.\boldsymbol{B} - \frac{e\hbar^2}{8m^2c^2}\boldsymbol{\nabla}.\boldsymbol{E} - \frac{e\hbar}{4m^2c^2}\boldsymbol{\sigma}.\left(\boldsymbol{E}\wedge(\mathbf{p}-e\mathbf{A})\right)$$

$$-\frac{ie\hbar^2}{8m^2c^2}\boldsymbol{\sigma}.\boldsymbol{\nabla}\wedge\boldsymbol{E} + \mathcal{O}(m^{-3})\,. \qquad 1$$

In reference [3], an original FWT of the Dirac Hamiltonian is performed up to the third order in *1/m*. The third order Hamiltonian $H_e'^{(3)}$ exhibits a coupling between the spin *dof* and the time derivative of the external electric field and reads

$$H_e'^{(3)} = \frac{(\mathbf{p}-e\mathbf{A})^4}{8m^3c^2} + \frac{e\hbar}{2m}\left\{\frac{(\mathbf{p}-e\mathbf{A})^2}{4m^2c^2},\boldsymbol{\sigma}.\boldsymbol{B}\right\} - \left(\frac{e\hbar}{2m}\right)^2\frac{\boldsymbol{B}^2}{2mc^2} - \frac{e\hbar^2}{16m^3c^4}\left\{(\mathbf{p}-e\mathbf{A}),\frac{\partial \boldsymbol{E}}{\partial t}\right\} + \frac{e\hbar^2}{16m^3c^4}\boldsymbol{\sigma}.\left(\frac{\partial \boldsymbol{E}}{\partial t}\wedge(\mathbf{p}-e\mathbf{A}) + (\mathbf{p}-e\mathbf{A})\wedge\frac{\partial \boldsymbol{E}}{\partial t}\right). \qquad 2$$

Using the Faraday's equation, the direct spin-field terms of Eqs. **(1)** and **(2)** (*ie* the ones written in the form of a coupling between the spin and an effective magnetic field) can be expressed as follows

$$^{\sigma}H' = -\frac{e\hbar}{2m}\boldsymbol{\sigma}.\left(\boldsymbol{B} + \frac{1}{4mc^2}\left(2\boldsymbol{E}\wedge(\mathbf{p}-e\mathbf{A}) - i\hbar\frac{\partial \boldsymbol{B}}{\partial t}\right) - \frac{e\hbar^2}{8m^2c^4}\frac{\partial^2 \boldsymbol{B}}{\partial t^2}\right) + \mathcal{O}(m^{-3}). \qquad 3$$

The first two terms of Eq. **(3)** are the usual Zeeman and spin-orbit interactions, and the two others can be seen as a Zeeman interaction with the 1st and the 2nd time-derivative of the magnetic field. Let us mention that the term $m^{-3}\{(\mathbf{p}-e\mathbf{A})^2,\boldsymbol{\sigma}.\boldsymbol{B}\}$ is considered as an indirect term. In order to understand this original behavior, in [4] we have performed the FWT up to the fifth order in *1/m* and obtained the following two additional direct spin-field terms $^{\sigma}H'^{(4)} = \frac{e\hbar^3}{32m^4c^6}\boldsymbol{\sigma}.\left(2\frac{\partial^2\boldsymbol{E}}{\partial t^2}\wedge(\mathbf{p}-e\mathbf{A}) - i\hbar\frac{\partial^3\boldsymbol{B}}{\partial t^3}\right)$ and $^{\sigma}H'^{(5)} = \frac{ie\hbar^4}{64m^5c^8}\boldsymbol{\sigma}.i\hbar\frac{\partial^4\boldsymbol{B}}{\partial t^4}$. They illustrate a spin-orbit coupling with the second time-derivative of the electric field and a Zeeman interaction with the third and fourth time-derivative of the magnetic field. Hence, introducing the Compton wave-length $\omega_c = 2\pi/\lambda_c$ (with $\lambda_c$ = h/mc), all the direct spin-*em* terms up to fifth order in powers of *1/m* can be written in a compact form as

$$^{\sigma}H' = -\frac{e\hbar}{2m}\boldsymbol{\sigma}.\left(\boldsymbol{B} + \sum_{n=1,2,3,4,5}\frac{1}{(2i\omega_c)^n}\frac{\partial^n\boldsymbol{B}}{\partial t^n} + \frac{1}{2mc^2}\sum_{n=0,2}\frac{i^n}{(2\omega_c)^n}\frac{\partial^n\boldsymbol{E}}{\partial t^n}\wedge(\mathbf{p}-e\mathbf{A})\right) \qquad 4$$

where the Zeeman interaction and the spin-orbit coupling appear like the first two terms of a more general analytical expansion. Therefore we postulate that the direct coupling between the spin *dof* and an external time-dependent electromagnetic field can be generalized through the general formula of Eq. **(4)** with $n = 1, \ldots \infty$. The amplitude of the $n^{\text{th}}$ correction of Eq. **(4)** can be written as $A_n\boldsymbol{B}$ with $A_n = (\omega/2\omega_c)^n$ when working with a plane wave of pulsation $\omega$ and by assuming that$\|\mathbf{p}\|$ = mc. In the ultrafast regime, the amplitudes of the higher order terms become more and more larger, and are equal precisely at $\omega = 2\omega_c \simeq 10^{21}\ s^{-1}$. This intrinsic limit for which the series in Eq. (4) diverges represents the upper boundary above which the FWT becomes non valid and the electron-positron pair creation may occur (this process is absent in our model because the *em* field is not quantized).

## Light-induced modifications on the electron-electron interaction

Very recently we have extended the above procedure to the case of two interacting electrons. In the Dirac's formalism, the Coulomb interaction between two electrons $e^2/r_{12}$ can be described using the Breit Hamiltonian $B = e^2(\boldsymbol{\alpha}_1.\boldsymbol{\alpha}_2/2r_{12} + (\boldsymbol{\alpha}_1.\boldsymbol{r}_{12})(\boldsymbol{\alpha}_2.\boldsymbol{r}_{12})/2r_{12}^3)$ which includes retardation effects. Its non-relativistic limit leads to the well-known Breit-Pauli Hamiltonian at the 2$^{nd}$ order in powers of *1/m*. We have performed a FWT of the two-interacting electrons Hamiltonian $H = \left(\sum_{i=1}^{2} H_{Di} + e^2/r_{12} + B\right)$ up to the third order in *1/m* with $H_{Di} = c\boldsymbol{\alpha}_i.(\boldsymbol{p}_i - q\boldsymbol{A}) + mc^2\beta_i + q\Phi$. The influence of the external *em* field on the bi-electronic Hamiltonian can be summarized as follows

$$H' = \sum_{i=1}^{2}(H'_{ei} + H'^{(3)}_{ei}) - \frac{e^2}{r_{12}} + H^{(2)}_{12}(m^{-2}, \boldsymbol{A}) + H^{(3)}_{12}(m^{-3}, \boldsymbol{E}) + \mathcal{O}(m^{-4}).$$

At second order in *1/m*, the Breit-Pauli Hamiltonian is recovered, where the external field acts only on the momentum operators via the usual substitution $\boldsymbol{p}_i \to \boldsymbol{p}_i - q\boldsymbol{A}$. However, the third order Hamiltonian exhibits terms involving explicitly the external electric field, the Coulomb interaction and the two electron spins

$$H^{(3)}_{12} = \frac{\hbar e^2 q}{4m^3c^6}\left(2(\boldsymbol{\sigma}_1 - \boldsymbol{\sigma}_2)\left(\frac{i\hbar \boldsymbol{r}_{12}}{r_{12}^3}\wedge\boldsymbol{E} + \frac{1}{\boldsymbol{r}_{12}}((\boldsymbol{p}_1 - q\boldsymbol{A}) - (\boldsymbol{p}_2 - q\boldsymbol{A}))\wedge\boldsymbol{E}\right)\right.$$
$$\left. - \hbar\left(\boldsymbol{\sigma}_1\wedge\frac{\boldsymbol{r}_{12}}{r_{12}^3}\right)(\boldsymbol{\sigma}_2\wedge\boldsymbol{E}) + \hbar\left(\boldsymbol{\sigma}_2\wedge\frac{\boldsymbol{r}_{12}}{r_{12}^3}\right)(\boldsymbol{\sigma}_1\wedge\boldsymbol{E})\right)\right).$$

By noting $\boldsymbol{E} = \boldsymbol{E}_{ext}$ and $e^2/r_{12} = \boldsymbol{E}_{int}$, it's possible to see from the above equation that the electric field *can act as a third partnership* for the two interacting electrons and generates terms like $\boldsymbol{\sigma}_i.(\boldsymbol{E}_{int}\wedge\ \boldsymbol{E}_{ext})$ and $(\boldsymbol{\sigma}_i\wedge\ \boldsymbol{E}_{int}).(\boldsymbol{\sigma}_j\wedge\ \boldsymbol{E}_{ext})$. The ratio $\|H^{(3)}_{12}\|/\|H^{(2)}_{12}\|$ can be expressed as $\simeq 10^{-6}E_{ext}r_{12}$ and illustrates the competition between the internal and external electric fields.

**Acknowledgments** The authors would like to thank Dr. J. -Y. Bigot and Pr. M. Maamache for useful discussions. This work was supported in part by the project ATOMAG supported by the European Research Council advanced grant ERC-2009-AdG-20090325-247452.

## References

[1] J. -Y. Bigot, M. Vomir and E. Beaurepaire, "Coherent ultrafast magnetism induced by femtosecond laser pulses" Nature Physics **5**, 515 (2009).
[2] P. Strange, *Relativistic Quantum Mechanics*, Cambridge University Press, (2005).
[3] F. Reuse, *Electrodynamique et Optique quantiques*, Presses polytechniques romandes, (2007).
[4] Y. Hinschberger and P. -A. Hervieux, "Foldy–Wouthuysen transformation applied to the interaction of an electron with ultrafast electromagnetic fields" Phys. Lett. A **376**, 813 (2012).

# Noncollinear Ballistic and Diffusive Spin Transport: Magnetic-Field Dependence

**Steffen Kaltenborn and Hans Christian Schneider**

Physics Department and Research Center OPTIMAS, University of Kaiserslautern, 67663 Kaiserslautern, Germany

hcsch@physik.uni-kl.de

**Abstract** We study time-dependent noncollinear spin-transport dynamics in a nonmagnetic material after ultrashort excitation. In particular, we analyze its dependence on a magnetic field. The dynamics of a pure spin current in a nonmagnetic metal are described by the wave-diffusion equations, which contain both ballistic and diffusive transport as special cases.

## Introduction

Non-collinear spin transport dynamics in magnetic multilayers has been a central issue in magnetoelectronics [1]. In addition, the influence of spin transport on the *ultrafast* magnetization dynamics has gained interest [2].
We have developed a time-dependent macroscopic theory to describe spin transport in normal metals with collinear and noncollinear spin alignment [3,4] as an extension to the static Valet-Fert theory [5]. This approach shows the wave character of spin transport, which allows one to extract a finite spin-signal velocity [3,4] and is related to ballistic transport [6].

In this paper we concentrate on the influence of a magnetic field on a non-collinear spin current pumped into a nonmagnetic material as shown in Fig. 1.

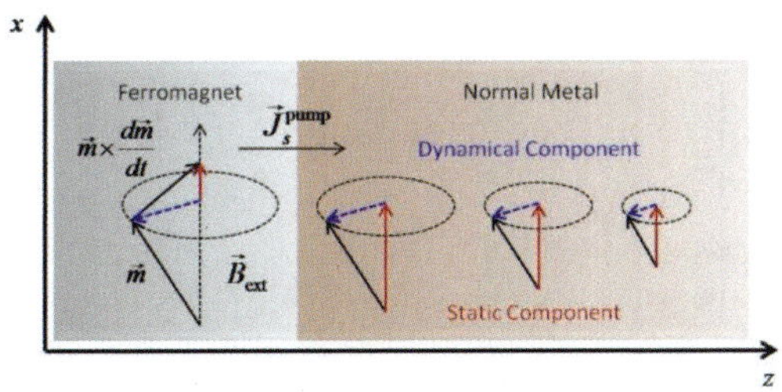


**Fig. 1.** A pure spin current $\vec{J}_s^{\text{pump}}$ is pumped from a ferromagnetic metal with a precessing magnetization $\vec{m}(t)$ into an adjacent nonmagnetic layer subject to an external magnetic field $\vec{B}_{\text{ext}}$.

 
J.-Y. Bigot et al. (eds), *Ultrafast Magnetism I*, Springer Proceedings in Physics 159
DOI 10.1007/978-3-319-07743-7_54

## Macroscopic Wave-Diffusion Theory

The time-dependent spin dynamics in $z$-direction under the influence of an external magnetic field $\vec{B}_{\text{ext}}$ can be described in terms of the spin density $\vec{n}_s(z,t)$ and spin-current density $\vec{j}_s(z,t)$ via [4]

$$\frac{\partial[\vec{n}_s(z,t)]_i}{\partial t} = -\gamma\left[\vec{n}_s(z,t) \times \vec{B}_{\text{ext}}\right]_i - \frac{[\vec{n}_s(z,t)]_i}{T_i} - \frac{\partial[\vec{j}_s(z,t)]_i}{\partial z} \qquad 1$$

and

$$\vec{j}_s(z,t) = -D\frac{\partial \vec{n}_s(z,t)}{\partial z} - \vec{n}_s(z,t) - \tau\gamma\vec{j}_s(z,t) \times \vec{B}_{\text{ext}} - \tau\frac{\partial \vec{j}_s(z,t)}{\partial t}, \qquad 2$$

where $\gamma$ denotes the gyromagnetic ratio and $D = \frac{v_F^2}{3}\tau \equiv c_S^2\tau$ is the diffusion constant. Thus, the transport dynamics close the Fermi surface are determined by the transport parameters Fermi velocity $v_F$, momentum relaxation time $\tau$ and spin relaxation time $T_i$ for each component $i$. The difference of Eq. (1) to Ref. [4] is that we keep here the possibility of different longitudinal and transverse spin relaxation time $T_1$ and $T_2$, which makes it possible to separate the decay of the static and the dynamical spin component [7] as shown in Fig. 1. The time derivative of the spin current density in Eq. (2) is responsible for the wave-diffusion duality of spin transport [3,4] and thus for the smooth transition from the ballistic to the diffusive regime [6]. The decay lengths $l_i$ corresponding to Eqs. (1) and (2) in a magnetic field pointing into the $x$-direction can be calculated analytically [4] as function of the pump excitation frequency $\omega$. If the pumped current is due to the precessing magnetic moment in the ferromagnetic layer around the same $B_{\text{ext}}$ field, the effective pump frequency is dependent on the magnetic field, $\omega_{\text{eff}} = \omega - \gamma B_{\text{ext}}$. In this case, the magnetic field dependence of the decay lengths is small and one finds $l_i = c_S\sqrt{\tau T_i}$ [4,7]. Here, we will consider a *fixed pump frequency* $\omega$ and a variable external magnetic field to investigate the influence of a magnetic field on time-dependent spin transport.

## Spin Dynamics under the Influence of a Magnetic Field

In Fig. 2 we show snapshots of $\vec{n}_s(z,t)$ and $\vec{j}_s(z,t)$ for different external magnetic fields in a permalloy/copper setup (cf. Ref. [7]). The parameters used are $f_{\text{pump}} =$ 5.5 GHz, $v_F = 1.57$ nm/fs, $\tau = 25$ fs [7] and $T_1 = T_2 = 5.96$ ps. The arrow at $z = 0$ in Fig. 2 (d), (e), (f) is the current pumped into the metal at the time of the snapshot. For small magnetic fields, the direction of the current is nearly the same as the pumped current. The spin density $\vec{n}_s(z,t)$ has a small phase shift compared to $\vec{j}_s(z,t)$, which is due to wave-diffusion characteristics of spin transport and cannot be obtained by spin-diffusion theory alone [4]. For larger magnetic fields, the spin dynamics are changed considerably. The direction of the spin current develops a wave-like envelope, which is due to the wave characteristics of spin transport. Further, the decay length decreases with increasing external field; for $B_{ext} = 2\text{T}$ it reaches 80% of the field-free value. The phase shift between the spin density and the current increases with the magnetic field. For larger magnetic fields, the wave character of spin transport is thus exhibited much more clearly.

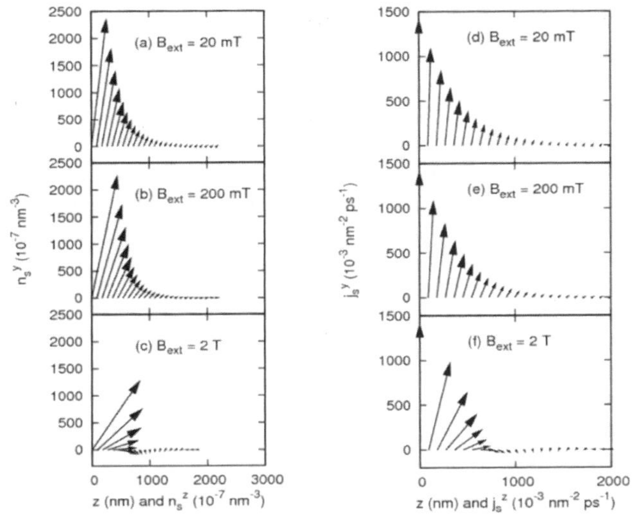


**Fig. 2.** Snapshots of $\vec{n}_s(z,t)$ ((a), (b) and (c)) and $\vec{j}_s(z,t)$ ((d), (e) and (f)) in a copper layer for different external magnetic fields and a fixed pump frequency.

## References

[1] A. Brataas, G. E. W. Bauer, and P. J. Kelly, "Non-collinear magnetoelectronics" Phys. Rep. **427**, 157-255 (2006).
[2] M. Battiato, K. Carva, and P. M. Oppeneer, "Superdiffusive Spin Transport as a Mechanism of Ultrafast Demagnetization" Phys. Rev. Lett. **105**, 027203, 1-4 (2010).
[3] Y.-H. Zhu, B. Hillebrands, and and H. C. Schneider, "Signal propagation in time-dependent spin transport" Phys. Rev. B **78**, 054429 (2008).
[4] Y.-H. Zhu, B. Hillebrands, and and H. C. Schneider, "Spin-signal propagation in time-dependent noncollinear spin transport" Phys. Rev. B **79**, 214412 (2009).
[5] T. Valet and A. Fert, "Theory of the perpendicular magnetoresistance in magnetic multilayers" Phys. Rev. B **48**, 7099 (1993).
[6] S. Kaltenborn, Y.-H. Zhu, and H. C. Schneider, "Wave-diffusion theory of spin transport in metals after ultrashort-pulse excitation" Phys. Rev. B **85**, 235101 (2012).
[7] F. Fohr, S. Kaltenborn, Y. Fukuma, L. Wang, J. Hamrle, H. Schultheiß, A. A. Serga, Y. Otani, H. C. Schneider, and B. Hillebrands, "Optical Detection of Spin Transport in Non-magnetic Metals" Phys. Rev. Lett. **106**, 226601, 1-4 (2011).

# Semi-relativistic quantum electron dynamics – a Lagrangian approach

**A. Dixit, Y. Hinschberger, J. Zamanian, G. Manfredi, P.-A. Hervieux**

Institut de Physique et Chimie des Matériaux de Strasbourg, CNRS and Université de Strasbourg, 23 rue du Loess, F-67034 Strasbourg, France

giovanni.manfredi@ipcms.unistra.fr

**Abstract** A semi-relativistic mean-field Lagrangian theory, which is based on the two-component wave-function, is developed. The Maxwell equations are coupled self-consistently to the Pauli evolution equation. The sources of the Maxwell equations (charge and current densities) are derived up to second order in $1/c$.

## Introduction

The electromagnetic field associated to a femtosecond laser pulse is strong enough to significantly perturb the electronic charges and spins in condensed matter systems, so that relativistic effects – both in the dielectric and magnetic responses – may become important [1]. Given the intensity of the fields involved, nonlinear effects are also expected to play a considerable role. Thus, this scenario represents an ambitious theoretical challenge that requires modeling the nonlinear dynamics of a quantum-relativistic system of many interacting electrons.

As a starting point, we begin with the Dirac equation for a many-electron system in a mean-field approximation, coupled to the Maxwell equations in the Lorentz gauge. We then perform a Foldy-Wouthuysen transformation [2] on the Dirac equation up to second order in $1/c$, which gives the following Hamiltonian:

$$H = mc^2 + q\phi + \frac{(p-qA)^2}{2m} - \frac{q\hbar}{2m}\sigma\cdot(\nabla\times A) - \frac{q\hbar}{8m^2c^2}\Delta\phi - \frac{q\hbar}{8m^2c^2}\nabla\cdot\partial_t A \qquad 1$$
$$- \frac{q\hbar}{4m^2c^2}\sigma\cdot[(\nabla\phi+\partial_t A)\times(p-qA)] + \frac{q\hbar}{8m^2c^2}\sigma\cdot(p\times\partial_t A),$$

where $\phi$ and $A$ are the scalar and vector potential, $\boldsymbol{\sigma}$ are the Pauli matrices, $m$ and $q$ are the mass and charge of the electron.

The first term in Eq. (1) is the rest mass of the electron and the next two terms are those in the standard Schrodinger Hamiltonian in presence of electromagnetic fields. The fourth term is the Pauli spin term (Zeeman effect). The fifth and sixth

 
J.-Y. Bigot et al. (eds), *Ultrafast Magnetism I*, Springer Proceedings in Physics 159
DOI 10.1007/978-3-319-07743-7_55

terms constitute the so-called Darwin correction and the last two terms represent the spin-orbit coupling. We neglected the relativistic correction to the kinetic energy.

## Lagrangian approach

We derived a Lagrangian density: $L(\Psi, \Psi^*, A, \phi)$ which returns the Pauli evolution equation:

$$i\hbar\partial_t \Psi = H\Psi$$

(where $H$ is the above second-order Hamiltonian) when the Euler-Lagrange equations are applied to $\Psi$ or $\Psi^*$. The same Lagrangian density returns the Maxwell equations (in the Lorentz gauge) when the Euler-Lagrange equations are applied to $\phi$ and $A$:

$$\Delta A - \frac{1}{c^2}\frac{\partial^2 A}{\partial t^2} = -\mu_0 J$$

$$\Delta\phi - \frac{1}{c^2}\frac{\partial^2 \phi}{\partial t^2} = -\frac{\rho}{\varepsilon_0}$$

The bonus of this approach is that the Maxwell equations obtained in such a fashion automatically provide the charge and current densities that are compatible with the Hamiltonian (1). At zeroth order, one finds the usual Schrödinger charge density $\rho_0 = \Psi\Psi^*$ as well as the current:

$$J_0 = \frac{i\hbar}{2m}\left[\left(\nabla\Psi^*\right)\Psi - \Psi^*\left(\nabla\Psi\right)\right] - \frac{q}{m}A\Psi^*\Psi$$

This technique also yields the so-called spin current

$$J_{spin} = \frac{\hbar}{2m}\nabla\times\left(\Psi^*\sigma\Psi\right)$$

In addition to the above zeroth order terms, the Lagrangian approach also provides second-order corrections to the sources. The actual expressions of the Lagrangian density and the second-order sources are rather involved and will be provided – along with their detailed derivation – in a forthcoming paper. These second-order corrections reflect the Darwin and spin-orbit terms in the original Hamiltonian (1).

It can also be proven that the full sources (up to second order in $1/c$) obey a continuity equation that is compatible with the conservation of charge that is implicit in the Maxwell equations.

## Conclusion

The Lagrangian formalism that we developed allowed us to obtain, in a straightforward manner, the charge and current densities that are compatible with a second-order (in $1/c$) Pauli-like Hamiltonian. These sources can be used in a mean-field model that couples self-consistently the Pauli evolution equation with the Maxwell equations. In addition, the Maxwell equations should also be expanded to the same order in $1/c$, along the lines presented for instance in Ref. [3]. This aspect will be discussed in a forthcoming study.

This type of model may be useful for applications to dense and weakly degenerate electron plasmas created via intense laser pulses [4], as well as to inertial confinement fusion plasmas and nanometric systems excited with ultrashort laser pulses [1].

**Acknowledgments** This work was partially funded by the European Research Council through the Advanced Grant "Atomag" (ERC-2009-AdG-20090325 247452).

## References

[1] J.-Y. Bigot, M. Vomir, and E. Beaurepaire, "Coherent ultrafast magnetism induced by femtosecond laser pulses" Nature Phys. **5**, 515 (2009).
[2] L. L. Foldy, "The Electromagnetic Properties of Dirac Particles" Phys. Rev. **87**, 688 (1952).
[3] G. Manfredi, "Non-relativistic limits of Maxwell's equations" Eur. J. Phys. **34**, 859 (2013).
[4] D. Kiefer, M. Yeung, T. Dzelzainis, P.S. Foster, S.G. Rykovanov, C.LS. Lewis, R.S. Marjoribanks, H. Ruhl, D. Habs, J. Schreiber, M. Zepf & B. Dromey, "Relativistic electron mirrors from nanoscale foils for coherent frequency upshift to the extreme ultraviolet" Nature Comm. **4**, 1763 (2013).

# Electron Lifetimes in a 2D Electron-Gas with Rashba SO-Coupling: Screening Properties

**S. Vollmar*, A. Ruffing, S. Jakobs*, A. Baral, S. Kaltenborn, M. Cinchetti, M. Aeschlimann, S. Mathias, and H.C. Schneider**

Department of Physics and Research Center OPTIMAS, University of Kaiserslautern, 67663 Kaiserslautern, Germany

*Graduate School MAINZ, University of Kaiserslautern, 67663 Kaiserslautern, Germany

SvenjaVollmar@web.de, hcsch@physik.uni-kl.de

**Abstract** We calculate lifetimes due to electron-electron scattering in a 2D Rashba band structure and study the influence of the substrate screening. A comparison with measurements on the quantum-well system Bi/Cu(111) is presented.

## Introduction

Giant spin splittings have been found in non-magnetic metallic QWs [1], and have been interpreted by effective Bychkov-Rashba [2] band structures. To understand the electronic dynamics in these systems, we use a 2D-electron gas model to study electron-electron scattering around the intersection of the spin-split bands. The results are compared with experimental data obtained from time- and angle-resolved 2-photon photoemission spectroscopy (tr-2PPE), carried out on the Bi/Cu(111)-QW-system, where the Rashba bands are unoccupied [1]. Also, we present our derivation of the statically screened potential in the thin Bi film and discuss the effect of the Cu-screening parameter $\kappa_{\mathrm{Cu}}$.

## Lifetime calculation and comparison with experiment

For the calculation of the lifetime due to electron-electron scattering we use a linear Rashba hamiltonian for electrons confined in the $xy$-plane with Rashba parameter $\alpha$, effective mass $m_1^*$ and energy offset $E_\Delta$.

$$H = H_0 + H_{\mathrm{Rashba}} = \frac{\hbar^2 k^2}{2m_1^*} + \alpha(k_x \sigma_y - k_y \sigma_x) + E_\Delta. \tag{1}$$



J.-Y. Bigot et al. (eds), *Ultrafast Magnetism I*, Springer Proceedings in Physics 159
DOI 10.1007/978-3-319-07743-7_56

The corresponding eigenenergies and eigenvectors are $\mathrm{E}_\sigma(k) = \frac{\hbar^2 k^2}{2m_1^*} + \sigma\alpha k + E_\Delta$ and $|\mathbf{k}, \sigma\rangle = \frac{1}{\sqrt{2}}(-\sigma i e^{-i\varphi_k}, 1)$, where $\sigma \in \{+, -\}$ is the Rashba-band index. The model is completed by adding two experimentally established facts for the Bi/Cu(111) system [1]: an additional occupied parabolic band, $\varepsilon(k) = \frac{\hbar^2 k^2}{2m_2^*} - \varepsilon_{\min}$, whose electrons act as scattering partners for the excited electrons [3], and a nearly constant interband-scattering rate $\Gamma_{\text{inter}}$ in the considered energy range. The latter contribution is taken into account to obtain a total inverse lifetime $\tau_{\sigma,\text{tot}}^{-1}(k) = \tau_\sigma^{-1}(k) + \Gamma_{\text{inter}}$. The important momentum (and, consequently, energy dependent) contribution to the inverse lifetime from electronic transitions in the (initially empty) Rashba bands due to Coulomb interaction with the occupied quantum-well band is

$$\tau_\sigma^{-1}(k) = -\frac{2}{\hbar} \sum_{\mathbf{q}} \sum_{\mu \in \{+,-\}} \left(V_q^{\mathrm{S}}\right)^2 \left|M_{k,q}^{\sigma,\mu}\right|^2 b(\hbar\omega) \chi''(\mathbf{q}, \hbar\omega). \tag{2}$$

Here we have introduced the imaginary part of the susceptibility $\chi''(\mathbf{q}, \hbar\omega)$ for the occupied band [4] in a linearized Boltzmann scattering integral. $\left|M_{k,q}^{\sigma,\mu}\right|^2$ is the matrix element for transitions $\sigma, k \to \mu, |\mathbf{k} + \mathbf{q}|$, $\hbar\omega = E_\mu(|\mathbf{k} + \mathbf{q}|) - E_\sigma(k)$ is the transferred energy, and $b(\hbar\omega)$ is the Bose-Einstein distribution function.

An important quantity in Eq. (3) is the statically screened Coulomb potential. We use here $V_q^{\mathrm{S}} = \frac{e^2}{(2\varepsilon_0)} \frac{1}{\sqrt{q^2 + \kappa_{\mathrm{Cu}}^2} + q}$ with a dependence on the 3D-screening constant of the substrate $\kappa_{\mathrm{Cu}}$, which is much bigger than $\kappa_{\mathrm{Bi}}$. For the derivation of $V_q^{\mathrm{S}}$ we solve the Poisson equation for the potential $\Phi(\vec{r})$ with an externally controlled charge at $\vec{r} = \vec{0}$ ($\rho_{\text{ext}} = e\delta(\vec{r})$ ) and with the induced charge density expressed by the density of states of bismuth $D_{\mathrm{Bi}}(E_F)$ $\left(\rho_{\text{ind}} = -e^2 D_{\mathrm{Bi}}(E_F)\Phi(\vec{r})\right)$. After a 2D-Fourier-Transformation in the plane we find a general solution for the potential in the Bi layer depending on $\kappa_{Bi}^2 = \frac{e^2}{\varepsilon_0} D_{\mathrm{Bi}}(E_F)$. Additionally we take the solutions of the potential in the vacuum and in the Cu-substrate depending on $\kappa_{\mathrm{Cu}}^2$ and require continuity. Because the screening in Bi is very small [5] and because the thickness of the layer is around 5 nm, we arrive at the expression for $V_q^{\mathrm{S}}$.

We are interested here in the influence of the substrate screening constant on the lifetime, as shown in Fig. 1. The calculated results were obtained with the parameters $\alpha = 200$ meVnm, $m_1^* = 0.34 m_e$, $E_\Delta = 2.65$ eV, $\varepsilon_{\min} = -1.24$ eV, $m_2^* = 0.24 m_e$ and $\Gamma_{\text{inter}} = (10.8\text{ fs})^{-1}$, compared to the experimental data. The longest lifetimes are obtained at the minima of the bands, since the available phase space for intraband scattering processes decreases when approaching the band minimum. The dependence on the screening parameter is rather pronounced. For $\kappa_{\mathrm{Cu}} = 25\text{ nm}^{-1}$ [6], we obtain lifetimes close to the ones observed in the experiment. The lifetimes become shorter for smaller substrate screening constants when the Coulomb interaction is screened less.

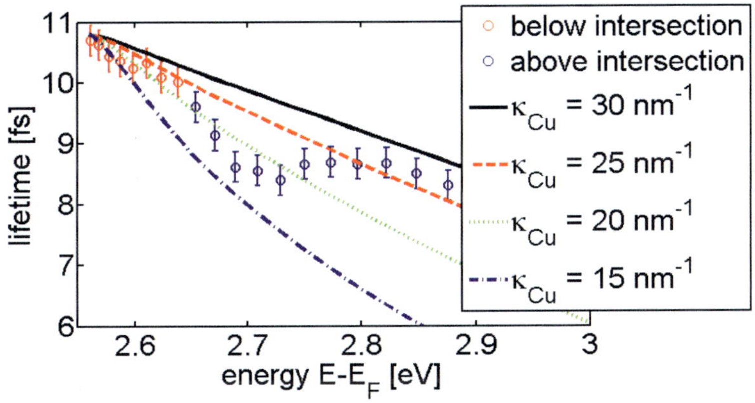


**Fig. 1.** Calculated (lines) and measured (circles; Rashba bands intersect at $E - E_F \approx 2.65$ eV) lifetimes, from time-resolved 2PPE, in the Bi/Cu(111)-system. Calculations with different screening parameters in the copper substrate $\kappa_{Cu}$ are presented.

The results of Fig. 1 show that for quantitative calculations of lifetimes in Bi quantum wells, a correct treatment of screening, including the substrate, is important. The increase of lifetimes towards the band bottom is well reproduced, but the variation of the lifetimes around the Rashba intersection point cannot be explained based on a pure Rashba band dispersion. It is likely that the dip in the measured lifetimes is due to a deviation of the band structure from the idealized Rashba model.

## Conclusion

In summary, we presented a study on the influence of screening on electronic lifetimes in empty Rashba bands, as they are assumed to be realized in Bi quantum wells. We showed the importance of the contribution of the substrate screening for the calculation of lifetimes. We ascribed characteristic differences between theory and experiment to a deviation of the band structure from a pure Rashba band structure.

**Acknowledgments** S. Mathias was supported by the European Community's FP7 under Marie Curie International Outgoing Fellowship GA 253316. S. Vollmar and S. Jakobs are recipients of a fellowship through the Excellence Initiative (DFG/GSC266).

## References

[1] S. Mathias et al., “Quantum-Well-Induced Giant Spin-Orbit Splitting” Phys. Rev. Lett. **104**, 066802 (2010).
[2] Y. Bychkov, E. Rashba, “Properties of a 2D electron gas with lifted spectral degeneracy” JETP Lett. **39**, 78 (1984).
[3] S. Mathias et al., “Quantum oscillations in coupled two-dimensional electron systems” Phys. Rev. Lett **103**, 026802 (2009).
[4] G. Giuliani, G. Vignale, “Quantum theory of the electron liquid” (Cambridge University Press, 2005), pp. 160-162.
[5] V. S. Edel’man, “Scanning tunnel microscopy and spectroscopy of an atomically clean bismuth surface” Physics-Uspekhi **48**, 1057 (2005).
[6] S. Ogawa, H. Nagano, H. Petek, “Hot-electron dynamics at Cu(100), Cu(110), and Cu(111 ) surfaces: Comparison of experiment with Fermi-liquid theory” Phys. Rev. B **55**, 10869-10877 (1997).

# Non-equilibrium spin-spin interactions in strongly correlated systems

**A. Secchi[a,]*, S. Brener[b], A.I. Lichtenstein[b], M.I. Katsnelson[a]**

[a] Institute for Molecules and Materials, Radboud University Nijmegen, Heyendaalseweg 135, 6525 AJ Nijmegen, The Netherlands.

[b] Institut für Theoretische Physik, Universitat Hamburg, Jungiusstraße 9, D-20355 Hamburg, Germany

andrea.secchi@gmail.com

**Abstract** We develop a theory for magnetism of strongly correlated systems driven out of equilibrium by an external time-dependent electric field. We provide expressions for computing the effective interaction parameters between electronic spins, including a new interaction that we name *twist exchange*. Our theory is suitable for laser-induced ultrafast magnetization dynamics.

## Introduction and Method

The theoretical study of magnetization in realistic systems is a challenging problem. In equilibrium, magnetic interactions in magnetic metals and semiconductors are non-Heisenberg [1-4], i.e., the lengths of magnetic moments and values of exchange parameters depend on the magnetic configuration for which they are computed, which calls for an accurate *ab initio* formulation. The expressions for the exchange-interaction parameters were given years ago [5, 6], and they can be used within a classical Heisenberg model to simulate spin dynamics. However, this treatment is not expected to be appropriate for ultrafast dynamics [7], since the time scale of the excitations is faster than the typical scale of exchange interactions (10÷100 fs). Exchange parameters are thus time-dependent. Our purpose is to derive their expressions in a general non-equilibrium framework, including external time-dependent fields [8]. The resulting formulas are valid for any time scale, thus they are applicable to ultrafast spin dynamics.

We consider a system described by the time-dependent multi-band Hubbard Hamiltonian,

$$\hat{H}(t) \equiv \sum_{a}\sum_{b}\sum_{\sigma=\uparrow,\downarrow}\hat{\phi}^{+}_{a\sigma}T_{ab}(t)\hat{\phi}_{b\sigma} + \hat{H}_{V}, \quad 1$$

where italic letters $a$, $b$ denote sets of site and orbital indexes; $\hat{H}_V$ accounts for on-site interactions between the electrons. The matrix element $T_{ab}(t)$ accounts for in-



J.-Y. Bigot et al. (eds), *Ultrafast Magnetism I*, Springer Proceedings in Physics 159
DOI 10.1007/978-3-319-07743-7_57

ter-site hopping and orbital hybridization due to both the equilibrium structure and an external time-dependent electric field, switched on at time $t_0$. Here we do not consider relativistic effects nor external magnetic fields, so $T_{ab}(t)$ is spin-independent.

We write the partition function for the electronic system in the Kadanoff-Baym formalism and we derive an effective action that describes the system in equilibrium for $t \le t_0$, as well as out of equilibrium for $t > t_0$. We assume spontaneous symmetry breaking (SSB) along the direction *z*, specified by the unit vector $\vec{u}_z$, and we study the spin excitations on the top of the collinear equilibrium ground state. Thus, we apply time- and site- dependent Holstein-Primakoff spin rotations to map the old fermionic fields $\phi_{a\sigma}$ into new fields $\psi_{a\sigma}$ having their spins aligned with *t*-dependent vectors $\vec{e}_a(t)$, which are the unit vectors of the (classical) magnetic moments. The deviations of the $\vec{e}_a(t)$'s from $\vec{u}_z$ are described by auxiliary bosonic fields; in the low-energy sector we assume such deviations to be small, obtaining an action quadratic in the bosons. Integrating out the fermionic fields $\psi_{a\sigma}$ gives an effective bosonic action, and we finally map the bosons to the $\vec{e}_a(t)$'s to obtain the spin-spin interactions.

## Results

Our results are presented in Ref.[8]. Here we summarize some of the main points.

We find that there are two forms of quadratic interactions between magnetic moments. The first form, $\propto \vec{e}_a \cdot \vec{e}_b$, is exchange. The second form, $\propto (\vec{e}_a \times \vec{e}_b) \cdot \vec{u}_z$, looks similar to an effective Dzyaloshinskii-Moriya interaction (DMI), but this interpretation is not correct since relativistic effects are absent in our model. We note, instead, that an *effective three-body* interaction of the form $(\vec{e}_a \times \vec{e}_b) \cdot \vec{e}_c$ generates precisely this term when up to second-order deviations of the magnetic moments from $\vec{u}_z$ are considered. We have proposed the name *twist exchange* for this new interaction [9]. Our method provides both the equilibrium and the non-equilibrium formulas for the coefficients describing exchange and twist-exchange, which are obtained after the fermionic field integration as convolutions of electronic Green functions and self-energies. These can be computed by the means of non-equilibrium dynamical mean-field theory. The formulas for exchange parameters reproduce, in the particular case of the single-band Hubbard model in equilibrium, the results of Ref.[6].

We find that the non-equilibrium spin-spin interactions are not, in general, local in time, i.e., the system has *memory*: the parameters depend on two times. We here present the formula for non-equilibrium exchange parameters in the Hartree-Fock approximation (Ref.[8] contains the more general formulas, also for twist exchange):

$$J_{ab}(t,t') = \frac{1}{4}\delta(t-t')\mathrm{Re}\left\{\int_{-\infty}^{+\infty}\frac{\mathrm{d}\omega}{2\pi}f(\omega,t)\left[\overline{\Sigma}_{ab}^{\downarrow}(t)A_{ba}^{\uparrow}(\omega,t)+A_{ab}^{\downarrow}(\omega,t)\overline{\Sigma}_{ba}^{\uparrow}(t)\right]\right\}$$

$$+\frac{\mathrm{sign}(t_r)}{8}\mathrm{Im}\left\{\int_{-\infty}^{+\infty}\frac{\mathrm{d}\omega}{2\pi}\int_{-\infty}^{+\infty}\frac{\mathrm{d}\omega'}{2\pi}e^{-i(\omega-\omega')t_r}\left[f(\omega,t_c)-f(\omega',t_c)\right]\right. \qquad 2$$

$$\cdot\left\{\left[\overline{\Sigma}^{\downarrow}(t)\cdot A^{\downarrow}(\omega,t_c)\right]_{ab}\left[\overline{\Sigma}^{\uparrow}(t')\cdot A^{\uparrow}(\omega',t_c)\right]_{ba}-A_{ab}^{\downarrow}(\omega,t_c)\left[\overline{\Sigma}^{\uparrow}(t')\cdot A^{\uparrow}(\omega',t_c)\cdot\overline{\Sigma}^{\uparrow}(t)\right]_{ba}\right.$$

$$\left.\left.+\left[A^{\downarrow}(\omega,t_c)\cdot\overline{\Sigma}^{\downarrow}(t')\right]_{ab}\left[A^{\uparrow}(\omega',t_c)\cdot\overline{\Sigma}^{\uparrow}(t)\right]_{ba}-\left[\overline{\Sigma}^{\downarrow}(t)\cdot A^{\downarrow}(\omega,t_c)\cdot\overline{\Sigma}^{\downarrow}(t')\right]_{ab}A_{ba}^{\uparrow}(\omega',t_c)\right\}\right\}.$$

Here $t_r = t-t'$, $t_c = (t+t')/2$, and $\overline{\Sigma}^{\sigma}(t)$ is the Hartree-Fock self-energy (because of SSB, $\overline{\Sigma}^{\downarrow} \neq \overline{\Sigma}^{\uparrow}$), while $f(\omega,t)$ and $A^{\sigma}(\omega,t)$ are, respectively, the time-dependent occupation number and spectral function, related to the non-equilibrium Green functions via the relation $G_{ab}^{<}(\omega,t) \equiv if(\omega,t)A_{ab}(\omega,t)$. As an application, in Ref.[8] we also derive the time-dependent spin stiffness tensor for a ferromagnet, which may be relevant for the interpretation of experiments where spin dynamics is induced by a laser beam of macroscopic size [7].

**Acknowledgments** The authors acknowledge useful discussions with Johan Mentink, Martin Eckstein, Laszlo Szunyogh, Oksana Chubykalo-Fesenko, Ulrich Nowak, Peter Oppeneer, Theo Rasing and Roy Chantrell. This work is supported by the European Union ERC Grant Agreement No. 281043 (FEMTOSPIN) and SFB925 (Germany).

# References

[1] E. L. Nagaev, "Physics of Magnetic Semiconductors" (Mir Publishers, Moscow, 1983).
[2] M. I. Auslender, M. I. Katsnelson, "The effective spin Hamiltonian and phase separation instability of the almost half-filled Hubbard model and the narrow-band s-f model" Solid State Communications 44, 387-389 (1982).
[3] V. Y. Irkhin, M. I. Katsnelson, "Spin waves in narrow-gap ferromagnetic semiconductors" Sov. Phys. JETP 61, 306-311 (1985).
[4] S. A. Turzhevskii, A. I. Lichtenstein, M. I. Katsnelson, "Degree of localization of magnetic moments and the non-Heisenberg character of exchange interactions in metals and alloys" Sov. Phys. Solid State 32, 1138-114 (1990).
[5] A. I. Liechtenstein, M. I. Katsnelson, V. P. Antropov, V. A. Gubanov, "Local spin density functional approach to the theory of exchange interactions in ferromagnetic metals and alloys" J. Magn. Magn. Mater. 67, 65-74 (1987).
[6] M. I. Katsnelson, A. I. Lichtenstein, "First-principles calculations of magnetic interactions in correlated systems" Phys. Rev. B 61, 8906-8912 (2000).
[7] A. Kirilyuk, A. V. Kimel, Th. Rasing, "Ultrafast optical manipulation of magnetic order" Rev. Mod. Phys. 82, 2731-2784 (2010).

[8] A. Secchi, S. Brener, A. I. Lichtenstein, M. I. Katsnelson, "Non-equilibrium magnetic interactions in strongly correlated systems" Ann. Phys. 333, 221-271 (2013).
[9] L. P. Kadanoff, G. Baym, "Quantum Statistical Mechanics" (W. A. Benjamin, Inc., New York, 1962).

# Study of the X-ray-plasma interaction for high intensity laser pulses

**O. Morandi, J. Zamanian, G. Manfredi, P.-A. Hervieux**

Institut de Physique et Chimie des Matériaux de Strasbourg

23, rue du Loess, F-67034 Strasbourg, France.

omar.morandi@ipcms.unistra.fr

**Abstract** The optical response of a plasma under strong X-ray laser excitation is investigated. We focus on the effects of the spin-dependent semi-relativistic corrections on the polarization of the reflected and transmitted light.

## Introduction

The study of the magneto-optical properties of a material exposed to an intense laser irradiation is nowadays an active field of research. High intensity X-ray lasers could induce strong modifications of the magnetic order of a nanometric system [1]. Furthermore, by irradiating a thin metallic slab with high intensity laser pulse, a hot relativistic plasma of electrons whose speed approaches the speed of light could be generated [2]. The theoretical study of the light-matter interactions in such high-field regimes should include relativistic effects and quantum corrections induced by the excitation of the negative-energy component of the particle wave function. The simplest example of quantum relativistic phenomena is the spin-orbit interaction. Particle confinement and high speed are the ingredients that induce strong interaction between particle spin and the orbital motion. As an example, valence electrons of a solid state material experience the strong Coulomb field of the ions and are accelerated at relativistic speed around the nuclei. In the same way a strong external X-ray field can induce a direct coupling between the electron plasma spin and its orbital momentum.

## Modeling relativistic quantum plasmas

The plasma is described by using a quantum hydrodynamic formalism. In this framework, the plasma is defined in terms of the macroscopic charge ($n$), velocity ($\vec{u}$), and spin ($\vec{s}$) densities. The plasma evolution equations are obtained in the

 
J.-Y. Bigot et al. (eds), *Ultrafast Magnetism I*, Springer Proceedings in Physics 159
DOI 10.1007/978-3-319-07743-7_58

semi-classical and semi-relativistic limit. The particle motion is governed by the so called Foldy-Wouthuysen Hamiltonian which is derived by expanding the Dirac Hamiltonian in terms of the Plank constant ħ and the semi-relativistic parameter $\beta = v/c$, where $v$, $c$ are respectively the particle and the speed of light.

$$\partial_t n + \partial_z (u_z n) = 0 \qquad 1$$

$$\partial_t \vec{s} + (u_z - v_c)\partial_z \vec{s} = \vec{s} \times \left( \frac{q}{mc}\vec{B} - \frac{q}{2mc^2}\vec{u} \times \vec{E} + \frac{\hbar}{2nm}\partial_z (n\partial_z \vec{s}) + \vec{s} \times \vec{c} \right) \qquad 2$$

$$\partial_t \vec{u} + \left( u_z + \frac{v_c}{2} \right)\partial_z \vec{u} = \frac{q}{m}\left( \vec{E} + \frac{1}{c}\vec{u} \times \vec{B} \right) + \frac{\hbar}{2m}\left[ \frac{q}{mc}\left( \vec{s} \cdot \partial_z \vec{B} \right) - \frac{\hbar}{2nm}\partial_z \left( n|\partial_z \vec{s}|^2 \right) - \frac{q\hbar}{4m^2c^2}\partial_z^2 E_z \right]\hat{e}_z \qquad 3$$
$$- \frac{q\hbar}{4m^2c^2}\left[ \frac{q}{m}\vec{s} \times \left( \vec{E} \times \vec{B} \right) + \vec{s} \times (\partial_t + u_z \partial_z)\vec{E} \right].$$

Here, $\hat{e}_z$ is the unit vector and we defined $v_c = \frac{q\hbar}{2m^2c^2}\left(\vec{E} \times \vec{S}\right)_z$ and $\vec{c} = \frac{q\hbar}{4nm^2c^2}\nabla \times \left(n\vec{E}\right)$. Details on the derivation of the equations will be given in a forthcoming publication [3]. The plasma equations of motion are rather complex to analyze. In our model we keep second order corrections with respect to β and ħ. This takes into account the direct coupling between the electromagnetic wave and the spin (last term of Eq. 2) and the spin-dependent corrections to the plasma speed (the terms of Eq. 3 proportional to ħ). The vectors $\vec{E}$ and $\vec{B}$ denote the electric and magnetic fields respectively. They include the laser field and the self-consistent fields generated by the plasma current and density.

We focus on the spin properties of the plasma and on their implications on the laser-plasma interaction. Important light-matter interaction phenomena like the Kerr and Faraday effects are usually explained in terms of the interplay between the spin and the orbital motion of the particles in a solid, gas or plasma. Spin oscillations could induce a current along the transverse plane and generate an electromagnetic field, which is in general rotated with respect to the polarization plan of the laser and can modify the Kerr-Faraday angles. In particular, the modification of the light polarization plane is accentuated by the increase of the laser frequency. We study the evolution of a plasma slab with a thickness $l_0$ (uniform in the transverse plane) and density $n_0$ excited by an intense source of X-ray linearly polarized light propagating along the $z$ direction. The intensity of the laser is equal to $10^{14}$ W cm$^{-2}$ and corresponds to electric fields with maximum amplitude of $1.6 \cdot 10^9$ V/m. X-ray lasers with similar intensity are currently used in experiments of spin excitation in metals and organic systems [4-5]. In our simulations, we choose $l_0$= 40 nm and $n_0$= 1.5 $10^{-4}$ nm$^{-3}$. In fig. 1(a) we depict the trajectory of the transverse components of the electric field of the incident (left panel) and reflected (right panel) waves. These quantities are particularly relevant since they could be directly measured in pompe-probe experiments. For sake of comparison, we depict (red dashed line) the same quantities when all the spin effects are neglected (i.e. by setting $s \equiv 0$ everywhere in the equations). Thus our simulations show that strong laser excitation of the plasma could lead to spin semi-relativistic effects which have a clear signature in the polarization of the propagating light. The study of the variation of the Kerr angle with respect to the laser energy is presented in Fig. 1.

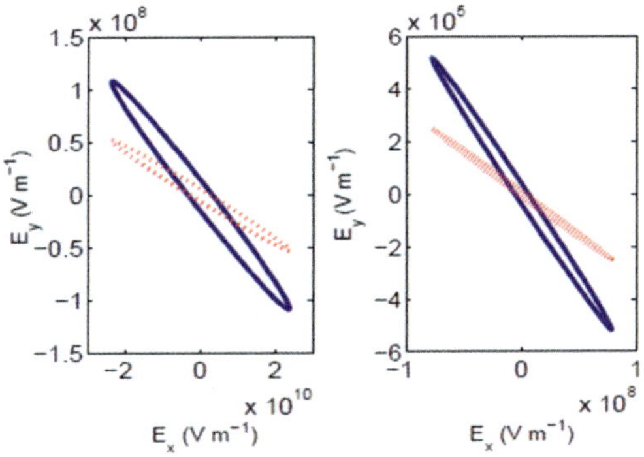


**Fig. 1.** Trajectory in the transverse plane of the electric field of the transmitted (left panel) and reflected light (right panel). The simulations are pre-formed for spin (blue continuous curve) and spinless (red dotted curve) plasma. Laser frequency: $4.5 \cdot 10^4$ eV.

In conclusion, a signature of the quantum relativistic corrections could be observed on the Kerr and Faraday angles of a plasma excited by an intense laser pulse. Our results help on the interpretation of the origin of the light polarization resulting from pompe-probe experiments in thin films.

**Acknowledgments** This work was partially funded by the European Research Council through the Advanced Grant "Atomag" (ERC-2009-AdG-20090325 247452).

## References

[1] Tianhan Wang et al., “Femtosecond Single-Shot Imaging of Nanoscale Ferromagnetic Order in Co/Pd Multilayers Using Resonant X-Ray Holography” Phys. Rev. Lett. **108**, 267403 (2012).

[2] D. Kiefer, M. Yeung, T. Dzelzainis, P.S. Foster, S.G. Rykovanov, C.LS. Lewis, R.S. Marjoribanks, H. Ruhl, D. Habs, J. Schreiber, M. Zepf & B. Dromey, "Relativistic electron mirrors from nanoscale foils for coherent frequency upshift to the extreme ultraviolet", Nature Communications **4**, 1763 (2013).

[3] O. Morandi, J. Zamanian, G. Manfredi, P.-A. Hervieux, *in preparation*.

[4] B. Dromey et al., "Coherent synchrotron emission from electron nanobunches formed in relativistic laser–plasma interactions", Nature Physics **8,** 804 (2012).

[5] M. Marvin Seibert, "Single mimivirus particles intercepted and imaged with an X-ray laser" Nature **470**, 178 (2011).

# Part V
# Ultrafast Coherent Magnetism

# Femtosecond Opto-Magnetism

**Alexey V. Kimel**

Radboud University Nijmegen, Institute for Molecules and Materials, P.O. Box 9010 6500 GL Nijmegen, The Netherlands Author affiliation and full address

a.kimel@science.ru.nl

**Abstract** The action of electric field of light on electronic dipoles, being the largest perturbation in physics of light-matter interaction, conserves the spin of electron. Nevertheless, an effective optical control of magnetism becomes possible due to spin-orbit interaction. Here we review the fundamentals and recent progress in the area of femtosecond opto-magnetism.

## Introduction

The ever increasing demand for faster information processing has triggered an intense search for ways to manipulate magnetically stored bits at the short time-scale. However, there are many fundamental questions concerning the mechanisms of ultrafast control of magnetism which are still poorly understood. This is mainly because an ultrashort stimulus brings a medium into a strongly non-equilibrium state where a conventional description of magnetic phenomena in terms of thermodynamics is no longer valid. Obviously, if the pulse duration is much shorter than the time of thermal equilibration in the spin system (~100 ps), such a pulse brings the medium into a strongly non-equilibrium state, where a conventional description of magnetic phenomena in terms of thermodynamics is no longer valid. Generation of magnetic field pulses much shorter than 100 ps and strong enough to reverse magnetization (>1T) is an extremely challenging technical problem. As the result the dynamics of the magnetization reversal at the sub-100 ps time-scale remains to be a very intriguing and rather unexplored area of modern magnetism. To solve the problem of generation of an ultrashort and strong pulse of effective magnetic field we suggest to employ a femtosecond laser pulse.

In order to reveal the mechanism allowing an effective control of magnetism by light, we have to discuss the interaction between photons and spins using energy considerations. It can be shown, for instance, that the thermodynamical potential $\Phi$ of a non-dissipating, magnetically ordered medium with static magnetization $\mathbf{M}(0)$ and in a monochromatic light field $\mathbf{E}(\omega)$ includes a term:

 
J.-Y. Bigot et al. (eds), *Ultrafast Magnetism I*, Springer Proceedings in Physics 159
DOI 10.1007/978-3-319-07743-7_59

$$\Phi^{(m)} = \varepsilon_0 \chi_{ijk} E_i(\omega) E_j^*(\omega) M_k(0) \qquad 1$$

where $\varepsilon_0$ is the fundamental constant, $\chi_{ijk}$ is the magneto-optical susceptibility. In the electric dipole approximation the linear optical response of a medium to a field $\mathbf{E}(\omega)$ is defined by the optical polarization $\mathbf{P}(\omega) = \delta\Phi/\delta\mathbf{E}(\omega)^*$. From Eq. (1) one can easily see that the optical polarization $\mathbf{P}(\omega)$ should have a contribution $\mathbf{P}^{(m)}$ proportional to the magnetization $P^{(m)}_i(\omega) = \varepsilon_0 \chi_{ijk} E_i(\omega) M_k(0)$. This contribution to the optical polarization results in a family of magneto-optical effects linear with respect to the magnetization. The magneto-optical Faraday and Kerr effects are examples of such phenomena. However, from the same Eq. (1) one can also see that an electric field of light at the frequency $\omega$ should act on the magnetization as an effective magnetic field $\mathbf{H}^{\text{eff}}$ directed along the wave-vector of the light

$$H_k^{(eff)} = \frac{\varepsilon_0}{\mu_0} \chi_{ijk} E_i(\omega) E_j^*(\omega) \qquad 1$$

For an isotropic medium $\chi_{ijk}$ in this equation can be substituted by a scalar $\chi$. From the equation one can see that that right- and left-handed circularly polarized waves should act as magnetic fields of opposite sign. It is well known that due to the spin-orbit interaction the magneto-optical susceptibility $\chi$ in magnetic materials can be very large.

In my lecture I will discuss how to generate such ultrashort pulses of effective magnetic field and employ these pulses for investigation of ultrafast magnetization dynamics. It will be shown that circularly polarized femtosecond laser pulses can excite and coherently control the spins in magnetic materials [1]. The effect of this optical pulse on a magnetic system was found to be equivalent to the effect of an equally short magnetic field pulse with strengths up to several Tesla. I will demonstrate that the mechanisms of magnetic switching triggered by such pulses can be very counterintuitive. One example is illustrated in Figure 1. The figure shows single-shot magneto-optical images obtained in (Sm,Pr)$FeO_3$ at different delays after excitation with a 100 fs pulse of effective magnetic field. Intuitively one expects that a magnetic domain is written during the action of magnetic field pulse. Unlike the expectations, the system hardly shows any sign of spin reorientation the first 2 ps and actual formation of a magnetic domain occurs long after the action of the field pulse [2].

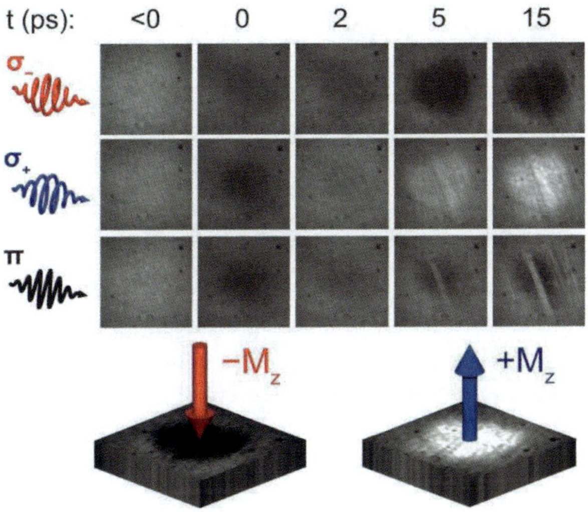


**Fig. 1.** Writing magnetic domains with 60 fs polarized laser pulses. The out-of-plane component of the magnetization can point "up" (black) or "down" (white). Circularly polarized pulses with opposite helicities (σ+ and σ-) create magnetic domains with opposite orientation of the magnetization. Linearly polarized (π) pulses produce a multidomain state. Each picture shows an area of approximately 70 ×70 $\mu m^2$ [2].

**Acknowledgments** The author thanks The Netherlands Organization for Scientific Research (NWO), the Foundation for Fundamental Research on Matter (FOM), the European Union's Seventh Framework Program (FP7/2007-2013) Grants No. 214469 (UltraMagnetron), No. 246102 (IFOX), and No.280555. (GoFast) as well as the European Research Council Grant Agreement No. 257280 (Femtomagnetism).

## References

[1] A. V. Kimel, A. Kirilyuk, P. A. Usachev, R. V. Pisarev, A. M. Balbashov, R. V. Pisarev, and Th. Rasing, "Ultrafast non-thermal control of magnetization by instantaneous photomagnetic pulses" Nature **435** 655 (2005).

[2] J. de Jong, I. Razdolski, A. M. Kalashnikova, A. Balbashov, R. V. Pisraev, A. Kirilyuk, Th. Rasing and A. V. Kimel "Coherent control of the route of an ultrafast magnetic phase transition via low-amplitude spin precession" Phys. Rev. Lett. **108**, 157601 (2012).

# Optical magnetization control in EuO films

**Manfred Fiebig**

Department of Materials, ETH Zurich, Wolfgang-Pauli-Strasse 10, 8093 Zurich, Switzerland

manfred.fiebig@mat.ethz.ch

**Abstract** Optical second and third harmonic generation (SHG and THG) on epitaxial ferromagnetic EuO films couple linearly to the ferromagnetic order parameter. This is exemplified by investigating the effect of Gd-doping of EuO by SHG and the magnetization of very thin EuO films by the third-order magneto-optical rotation associated to THG. In contrast to the Faraday rotation the latter does not depend on the thickness of the film.

The ferromagnetic semiconductor EuO possesses a variety of remarkably pronounced magneto-optical properties [1]. Since nonlinear optics has a great potential to disclose the microscopic mechanisms of light-matter interaction and to get unique information about the crystallographic, electronic, and magnetic structure, including states that are inaccessible by linear optics, the investigation of the nonlinear optical properties of EuO is presently of special interest. Here we report on the second- and third-order optical response of epitaxial EuO(001) films. SHG and THG probe different directions of the three-dimensional magnetization. We apply SHG to investigate the effect of Gd doping on the in-plane magnetic order of EuO. The magnetization of out-of-plane-magnetized EuO films is probed by the third-order magneto-optical rotation because, in contrast to the linear Faradayeffect, it does not depend on the film thickness. Thus, a magnetic rotation of up to 90° is obtained on films of 10 nm.

EuO has a centrosymmetric cubic rocksalt structure. It undergoes a ferromagnetic transition $T_C$=69 K which retains the inversion symmetry. In centrosymmetric systems, SHG is only allowed if higher-order multipole contributions in the expansion of the electromagnetic light field are involved [2, 3]. In EuO this leads to $P_i(2\omega) = \varepsilon_0 (\chi^i_{ijk}+\chi^c_{ijk}) E_j(\omega) H_k(\omega)$, where the time-invariant tensor $\chi^i_{ijk}$ and the time-noninvariant tensor $\chi^c_{ijk}$ are the second-order nonlinear susceptibilities related to, in the leading order, crystallographic SHG and magnetization-induced SHG, respectively. $E(\omega)$ and $H(\omega)$ denote the j-polarized electric-dipole field and the k-polarized magnetic-dipole field of the fundamental light, respectively, that induce the i-polarized SHG polarization $P(2\omega)$. In contrast, electric-dipole-type THG is allowed in centrosymmetric materials. This is expressed by $P_i(3\omega) = \varepsilon_0(\chi^i_{ijkl}+\chi^c_{ijkl}) E_j(\omega) E_k(\omega) E_l(\omega)$ where $\chi^i_{ijkl}$ and $\chi^c_{ijkl}$ are related to crystallographic and magnetic THG, respectively.



J.-Y. Bigot et al. (eds), *Ultrafast Magnetism I*, Springer Proceedings in Physics 159
DOI 10.1007/978-3-319-07743-7_60

The epitaxial EuO(001) films of 10-100 nm are probed with laser pulses of 120 fs incident along the z axis with a setup described elsewhere [4, 5]. The EuO films were grown on a two-side polished $YAlO_3(110)$ substrate by molecular-beam epitaxy and protected by an amorphous silicon (a-Si) cap layer of 10 nm [6]. The direction of the film magnetization M was set by magnetic fields of 0.1 T (M||x, M||y) or 3 T (M||z).

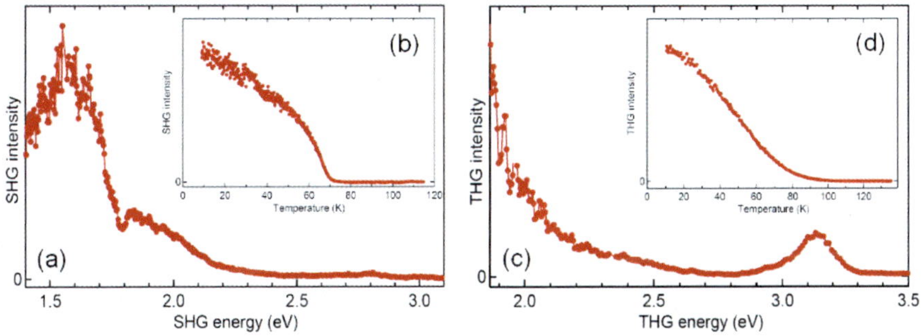


**Fig. 1.** Magnetically induced SHG and THG on EuO(100). (a) SHG spectrum at 10 K and $\mu_0 H_y$ = 0.1 T with light from $\chi^c_{xxy}$. (b) Temperature dependence at $2\hbar\omega$ = 2.6 eV. (c) THG spectrum at 10 K and $\mu_0 H_z$ = 3 T with light from $\chi^c_{yxxx}$. (d) Temperature dependence at $3\hbar\omega$ = 2.0 eV.

In Fig. 1 we see the spectral and temperature dependence of the SHG and THG signal obtained on the EuO(001) films. The SHG signal is gained from $\chi^c_{xxy}$which probes the y component of the magnetization. The emergence of the SHG signal at 69 K evidences its magnetic origin. The resonant enhancement of the SHG signal below $2\hbar\omega$=2.5 eV coincides with the $4f^7 \rightarrow 4f^6 5d^1(t_{2g})$ two-photon-resonant transitions of the $Eu^{2+}$ ion [4]. The THG signal is gained from $\chi^c_{yxxx}$which probes the z component of the magnetization. Again, the emergence of the nonlinear signal at about 69 K evidences its magnetic origin. The transition is less sharp than in the SHG measurement because of the larger magnetic field applied in order to orient the magnetization out-of-plane. The THG spectrum reveals two-and three-photon-resonant $4f^7 \rightarrow 4f^6 5d^1(t_{2g})$ transitions of the $Eu^{2+}$ ion [5].

After Fig. 1 has established SHG and THG as probe of the ferromagnetic order of EuO, Figs. 2 and 3 show applications of the two techniques. In Fig. 2 the temperature dependence of the SHG amplitude (i.e. the square root of the SHG intensity) from $\chi^c_{xxy}$ at 2.60 eV is shown for a series of $Eu_{1-x}Gd_xO$ samples with $0 < x <$ 19.5%. We see that at x = 0.64% the Curie temperature begins to increase up to 120 K at x = 2.65%. The temperature dependence becomes distinctly non-Brillouin-like with increasing doping. Figure 2 is in striking coincidence with magnetization measurements on the $Eu_{1-x}Gd_xO$ system reported in [7]. The double-dome-like temperature dependence is related to complementary Eu- and Gd-driven ferromagnetic ordering processes, possibly supplemented by magnetic polaron formation, as explained in detail elsewhere [8, 9].

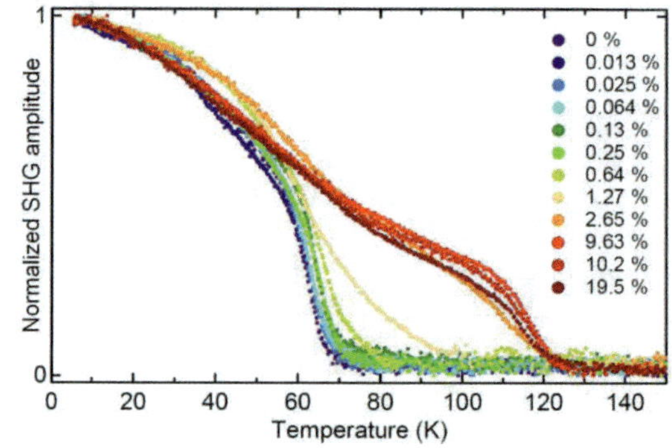


**Fig. 2.** Temperature dependence of the SHG amplitude from $\chi^{c}_{xxy}$ at 2.60 eV for $Eu_{1-x}Gd_xO(001)$.

Figure 3 shows the polarization analysis of the THG signal emitted from a EuO(100) film which was subjected to magnetic fields of 0 and ±3 T applied along the z axis. At 0 T the maximum of the THG signal is observed at $\varphi_A = 0°$. The magnetization along the z axis is 0 and, accordingly, no magneto-optical rotation occurs: The polarization of the incident light at ω and the emitted light at 3ωboth point along the x axis. At ±3 T the maximum of the THG signal is shifted to $\varphi_A = \pm 80°$. The magnetic field rotates the magnetization into the z direction and this rotates the polarization of the THG wave by +80° or −80°, depending on the sign of the magnetization. The impressively large value of this third-order magneto-optical rotation is independent of the thickness of the EuO film. Even in films of 10 nm the value of the rotation is fully retained. In higher magnetic fields the rotation can be increased up to 90° [5].

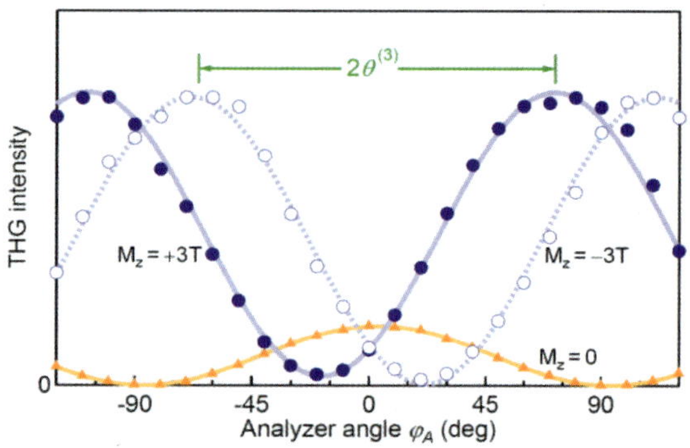


**Fig. 1.** Polarization dependence of the THG signal at 10 K and $3\hbar\omega = 2.0$ eV with light from $\chi^{c}_{yxxx}$. The angle $\theta^{(3)}$ refers to the third-order rotation, i.e., the rotation of the polarization of light at 3ω with respect to the polarization of the light at ω.

We thus see that nonlinear optics is a versatile probe for the magnetic order of EuO films. While this order can also be probed by magnetization measurement, the nonlinear optical approach opens new degrees of freedom. Spatial resolution can be used to image sample inhomogeneities and domain structures. Most of all, the investigation with pulsed lasers gives access to dynamical processe down to <1ps in pump-probe experiments. In comparison to linear optics, SHG and THG allows one to excite additional electronic transitions and distinguish them. In addition magneto-optical effects are often larger in nonlinear than in linear optics. In particular, magneto-optical rotation by up to 90° is obtained on films of 10 nm whereas Faraday rotation would be <<1°.

**Acknowledgments** The author thanks M. Matsubara, C. Becher, A. Schmehl, D. G. Schlom, and J. Mannhartwho are the authors of the papers on which this talk, presented at the Ultrafast Magnetism Conference 2013, is based.

## References

[1] J. Feinleib, W. J. Scouler, J. O. Dimmock, J. Hanus, T. B. Reed, C. R. Pidgeon, "Spin-Polarized Splittings in the Temperature-Dependent Reflectance of EuO", Phys. Rev. Lett. **22**, 1385 (1969).

[2] P. S. Pershan, "Nonlinear Optical Properties of Solids: Energy Considerations" Phys. Rev. **130**, 919 (1963).

[3] M. Fiebig, V. V. Pavlov, R. V. Pisarev, "Review: Second Harmonic Generation as a Tool for Studying Electronic and Magnetic Structures of Crystals" J. Opt. Soc. Amer. B **22**, 96 (2005).

[4] M. Matsubara, A. Schmehl, J. Mannhart, D. G. Schlom, M. Fiebig, "Large Nonlinear Magneto-Optical Effect in the Centrosymmetric Ferromagnetic Semiconductor EuO" Phys. Rev. B **81**, 214447 (2010).

[5] M. Matsubara, A. Schmehl, J. Mannhart, D. G. Schlom, M. Fiebig, "Giant Third-Order Magneto-Optical Rotation in Ferromagnetic EuO" Phys. Rev. B **86**, 195127 (2012).

[6] A. Schmehl, V. Vaithyanathan, A. Herrnberger, S. Thiel, C. Richter, M. Liberati, T. Heeg, M. Röckerath, L. F. Kourkoutis, S. Mühlbauer, P. Böni, D. A. Muller, Y. Barash, J. Schubert, Y. Idzerda, J. Mannhart, D. G. Schlom, "Epitaxial Integration of the Highly Spin-Polarized Ferromagnetic Semiconductor EuOwith Silicon and GaN"Nature Mater. **6**, 882 (2007).

[7] T. Mairoser, A. Schmehl, A. Melville, T. Heeg, L. Canella, P. Böni, W. Zander, J. Schubert, D. E. Shai, E. J. Monkman, K. M. Shen, D. G. Schlom, J. Mannhart, "Is There an Intrinsic Limit to the Charge-Carrier-Induced Increase of the Curie Temperature of EuO?" Phys. Rev. Lett. **105**, 257206 (2010).

[8] M. Arnold, J. Kroha, "Simultaneous Ferromagnetic Metal-Semiconductor Transition in Electron-Doped EuO" Phys. Rev. Lett. **100**, 046404 (2008).

[9]N. Tsuda, K. Nasu, A. Yanase, K. Siratori, "Electronic Conduction in Oxides", Springer Series in Solid-State Sciences Vol. 94, (Springer, 1991).

# Non-thermal light-induced spin dynamics in YIG:Co films via the photomagnetic effect

**A. Stupakiewicz[1*], M. Pashkevich[1,2], A. Maziewski[1]**

[1]Laboratory of Magnetism, Faculty of Physics, University of Bialystok, 15-424, Bialystok, Poland

[2]Scientific-Practical Materials Research Centre of the NASB, 220072, Minsk, Belarus

[*] and@uwb.edu.pl

**Abstract** We observed different frequencies of the magnetization precession within GHz-THz range in Co-substituted yttrium iron garnet films by exciting linearly polarized femtosecond pulses. The photomagnetic effect can be useful for non-thermally controlling of spin systems and ultrafast switching in magnetic devices.

## Introduction

Previous studies have revealed the fascinating role of electron-spin-based phenomena in spintronic and magnonic devices operating in a wide frequency range. The functionality of garnet systems has been shown to be very broad: for example, the excitation of surface plasmons, the propagation of nonlinear spin-waves, the photomagnetic effect [1], the observation of the inverse Faraday effect induced by an ultrafast laser pulse [2], and many others. Optical non-contact methods offer special ways of studying local spin-density effects with high spatial and temporal resolution. A recent study on a Co-substituted yttrium-iron garnet (YIG:Co) thin-film reported a 20° angle of the magnetization precession through non-thermal excitation by linearly polarized femtosecond laser pulses via photoinduced magnetic anisotropy [3]. Interesting light-induced magnetic properties have been observed in an ultrathin Co layer deposited on a garnet film. In this bilayer, damped oscillations in the Faraday rotation transients were demonstrated, representing precessional motion of the magnetization vector in both an ultrathin Co layer and garnet of the bilayer, with a modulation resulting from two distinct frequencies [4].

We observed in YIG:Co films the magnetization precession with different frequencies: an unexpected field-independent high-frequency mode and a field-dependent low-frequency ferromagnetic mode within 1-10 GHz values. The polarization dependencies in a wide spectral range were measured.

 
J.-Y. Bigot et al. (eds), *Ultrafast Magnetism I*, Springer Proceedings in Physics 159
DOI 10.1007/978-3-319-07743-7_61

## Results and discussion

Experimental studies were performed on garnet films with a composition of $Y_2CaFe_{3.9}Co_{0.1}GeO_{12}$ on a gadolinium-gallium-garnet (001)-oriented substrate. The surface of the garnet thin film was treated with a low energy oxygen ion beam [5]. The saturation magnetization of garnets at room temperature was about 7 Gs. We used ferromagnetic resonance with an X-band microwave source and magneto-optical magnetometer to study the magnetic anisotropy of our samples at room temperature. The easy magnetization axis was slightly inclined from a <111>-type crystallographic direction relative to the sample plane. To investigate the spin precession using femtosecond laser excitation we performed Faraday rotation measurements on garnets using time-resolved tools with a magneto-optical pump-probe method. The polarization plane of the pump beam was along the [100] and [010] crystallographic axes, when we have maximal amplitude of magnetization precession in the garnet film. On the contrary, the polarization plane of the probe beam was along [110] axis. We measured the Faraday rotation angle (proportional to the out-of-plane component of the magnetization) of the probe pulses as a function of the delay time and external in-plane magnetic field H.

Figure 1 shows the results for the magnetization precession as a function of the delay time at H=0.75 kOe with different polarizations of pump light. We observed that the precession of the magnetization with an opposite phase is triggered as non-thermal characterization of ultrafast dynamics in garnets.

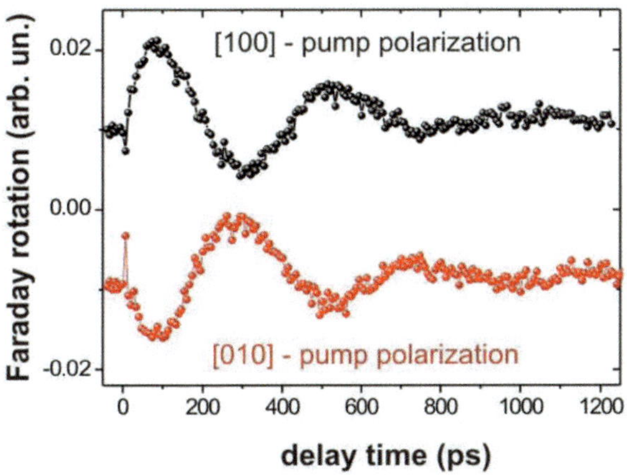


**Fig. 1.** Time-resolved Faraday rotation for linear pump polarizations along [100] and [010] axes of pump laser pulses with in-plane magnetic field H=0.75 kOe.

At the initial timescale, up to few picoseconds, the high-frequency mode was observed. This mode was independent from the external magnetic field at a wide spectral range (800-1600 nm) of pump light. The low frequency mode increased linearly with the amplitude of the applied magnetic field. In this case, the observations agree with the behavior for ferromagnetic spin resonance modes in garnet films [4]. During pulses of about 20 ps, we simultaneously observed the relaxation of the high-frequency precession and the excitation of the low-frequency precession of the magnetization. This excitation is consistent with an optically induced magnetic anisotropy change in a garnet film [3]. Though qualitatively similar in behavior, the ultrafast dynamic modification of the anisotropy differs from the static one in the orientation of the light polarization, which raises a question about the origins of the two modifications. In this case, one can expect that the ultrafast photomagnetic anisotropy at YIG:Co may be connected with femtosecond excitation of magnetic ions in tetrahedral sites, using light excitation of the exchange resonance mode.

The experimental studies and analysis as demonstrated show the possibility of non-thermally exciting and controlling the magnetization of ferrimagnets using femtosecond laser pulses by the ultrafast photomagnetism. Our result allows for the possibility of tunable manipulations of the magnetization switching with frequencies from GHz up to THz values.

**Acknowledgments** Supported by SYMPHONY project operated within the Foundation for Polish Science Team Programme co-financed by the EU European Regional Development Fund, OPIE 2007–2013.

## References

[1] A. Stupakiewicz, A. Maziewski, I. Davidenko, V. Zablotskii, "Light-induced magnetic anisotropy in Co-doped garnet films" Phys. Rev. B **64**, 644405 (2001).

[2] A. Kirilyuk, A.V. Kimel, and T. Rasing, "Ultrafast optical manipulation of magnetic order" Rev. Mod. Phys. **82**, 2731 (2010).

[3] F. Atoneche, A. M. Kalashnikova, A. V. Kimel, A. Stupakiewicz, A. Maziewski, A. Kirilyuk, and Th. Rasing, "Large ultrafast photoinduced magnetic anisotropy in a cobalt-substituted yttrium iron garnet" Phys. Rev. B **81**, 214440 (2010).

[4] A. Stupakiewicz, M. Pashkevich, A. Maziewski, A. Stognij, N. Novitskii, "Spin precession modulation in a magnetic bilayer" Appl. Phys. Lett. **101**, 262406 (2012).

[5] M. Pashkevich, A. Stupakiewicz, A. Kirilyuk, A. Maziewski, A. Stognij, N. Novitskii, A. Kimel, Th. Rasing, "Tunable magnetic properties in ultrathin Co/garnet heterostructures" J. Appl. Phys. **111**, 023913 (2012).

[6] A. H. M. Reid, A. Kimel, A. Kirilyuk, Th. Rasing, "Optical Excitation of a Forbidden Magnetic Resonance Mode in a Doped Lutetium-Iron-Garnet Film via the Inverse Faraday Effect" Phys. Rev. Lett. **105**, 107402 (2010).

# Ultrafast charge contribution to magneto-optics in strong correlated magnetic oxides

**C. Piovera[1*], F. Boschini[1], H. Hedayat[1], C. Dallera[1], M. Münzenberg[2], A. Gupta[3], E. Carpene[4]**

[1] Dipartimento di Fisica, Politecnico di Milano, 20133 Milan, Italy

[2] I. Physikalisches Institut, Georg-August-Universität Göttingen, 37077 Göttingen, Germany

[3] Department of Chemistry, University of Alabama, Tuscaloosa, Alabama 35487, USA

[4] IFN-CNR, Dipartimento di Fisica, Politecnico di Milano, 20133 Milan, Italy

[*]Christian.piovera@polimi.it

**Abstract** Wavelength- and time-resolved magneto-optical Kerr measurements on $CrO_2$ and LSMO show a strong dependence on transient optical parameters leading to a not obvious interpretation of MOKE data. Comparing the complex Kerr angle and the transient reflectivity allows us to reveal the true spin dynamics in strong correlated ferromagnetic oxides.

## Introduction

Time-resolved magneto-optical Kerr (TR-MOKE) effect has been exploited to measure ultrafast laser induced magnetic dynamics for almost 20 years. Since the the pioneering work of Beaurepaire et al. (Ref. 1), the reliability of the technique has been questioned especially in the sub-picosecond timescale. Many experiments demonstrated that in simple ferromagnetic (FM) metals, e.g. Cobalt and Nickel, the Kerr signal follows the real magnetic dynamics in a perturbative regime and for pump-probe delays longer than the electronic thermalization time. Although that, the problem has not been as much debated in other kind of FM materials like magnetic oxides where spin, orbital and charge orders are strongly correlated.

In this paper we exploited broadband pump-probe spectroscopy in order to study laser induced MOKE dynamics in $CrO_2$ and $La_{0.67}Sr_{0.33}MnO_3$ compounds. Detecting the complex Kerr angle in a wide spectral region allows us to avoid bleaching and state blocking effects. By a careful characterization of the MOKE signal we are able to distinguish the pump induced evolution of the diagonal and off-diagonal elements of the dielectric tensor which are strictly related to the electronic structure and the magnetic order.

 
J.-Y. Bigot et al. (eds), *Ultrafast Magnetism I*, Springer Proceedings in Physics 159
DOI 10.1007/978-3-319-07743-7_62

## Experimental results

The studied samples are half metallic $CrO_2$ and $La_{0.67}Sr_{0.33}MnO_3$ (LSMO) epitaxial films. They are characterized by a high spin polarization at the Fermi level due to the presence of an energy gap in the minority electronic density of states and by a double exchange mediated ferromagnetic order.

Time-resolved MOKE are performed in a temporal window of about 50 ps exploiting ultrashort (50 fs) laser pulses generated by an amplified Ti:sapphire laser centered at 800 nm (1.55 eV). Part of the beam energy is focused in a Sapphire plate to generate a white light supercontinuum employed as probe beam. The probe spectrum ranges between 400 and 750 nm corresponding to 2.75 and 1.65 eV. This energy window is particularly important in FM oxides since it includes several electronic features as minority band gap and insulator to metal transition. Kerr rotation $\theta$ and ellipticity $\eta$, together with transient differential reflectivity $\Delta R/R$, have been measured in longitudinal MOKE geometry with a *p*-polarized probe light (Ref. 2). The normalized modulus of the complex Kerr angles $|\Theta| = (\theta^2 + \eta^2)^{1/2}$ are reported in Fig. 1 as a function of the pump-probe delay and probe wavelength for $CrO_2$ (panel a) and LSMO (panel c). In a spectral window of about 200 nm the MOKE signals show different dynamics. In $CrO_2$ a fast (1 ps) decrease is clearly visible close to the UV region while a slower change is observable at lower energies. In LSMO (panel c) the Kerr angle enhances up to 20% at about 540 nm while reduces to about 15% of its static value at 650 nm. The experimental results sort out that TR-MOKE cannot genuinely reveal the spin dynamics in such systems. The complex Kerr angle is a function of the diagonal $\varepsilon_{xx}$ and off-diagonal $\varepsilon_{xy}$ components of the dielectric tensor. While $\varepsilon_{xy}$ is proportional to the sample magnetization, the transient $\varepsilon_{xx}$ depends only on charge evolution. As demonstrated by Carpene et al. (Ref. 3), TR-MOKE can be largely affected by pump induced dynamics of $\varepsilon_{xx}$ and only a careful analysis of the transient rotation, ellipticity and reflectivity can univocally bring to $\varepsilon_{xy}$.

Following the same method we have disentangled the charge contribution from $|\Theta|$. The results are reported in Fig. 1 (panels b and d). In panel b, the normalized modulus of $\varepsilon_{xy}$ shows the same dynamics for all wavelengths in $CrO_2$ disclosing a slow demagnetization taking place in more than 50 ps (Ref. 3). On the other hand, in LSMO similar spin dynamics are present for wavelength higher than 600 nm. In the spectral region spreading from 450 to 540 nm, $|\varepsilon_{xy}|$ drastically increases as if an optically activated enhancement of the magnetization occurs. This behavior of the dielectric tensor cannot be completely attributed to a pure spin dynamics since $\varepsilon_{xy}$ strongly depends on the electronic structure and the state filling close to the Fermi level. It is interesting to notice that 540 nm (2.3 eV) matches the estimated values of the half-metallic gap for minority spin electrons. The experiments are still in progress and different stoichiometries of LSMO will be studied to find out the correlation between the observed effects and the strongly correlated electronic, magnetic and crystalline structures.

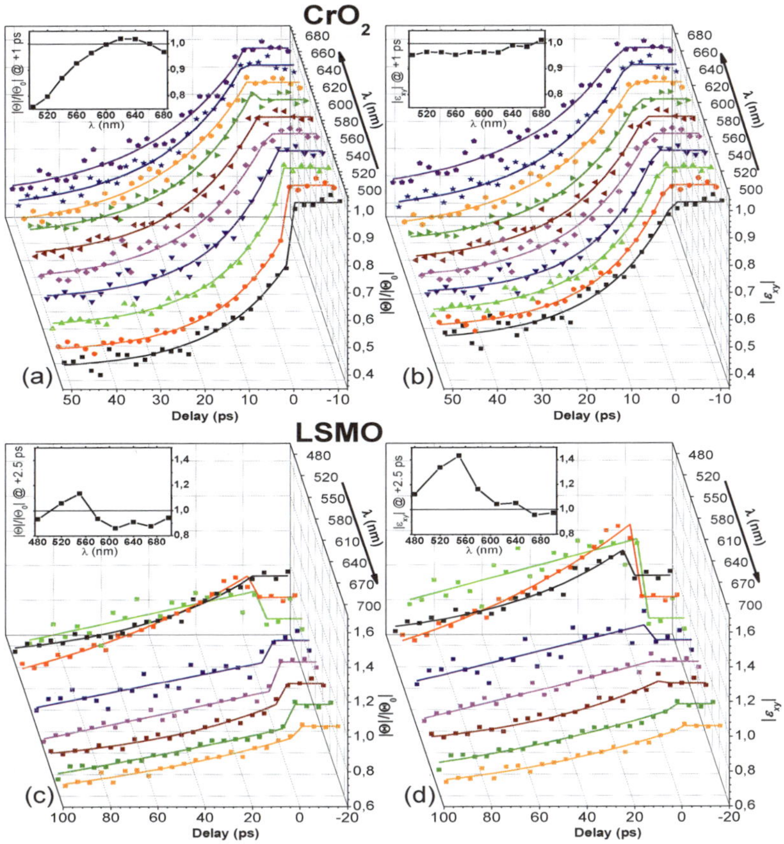


**Fig. 1.** Normalized modulus of the complex Kerr angle $\Theta$ and $\varepsilon_{xy}$: (a-b) $CrO_2$ and (c-d) LSMO. Insets: value of $|\Theta|/|\Theta_0|$ and $|\varepsilon_{xy}|$ at +1 ps ($CrO_2$) and +2.5 ps (LSMO) delay.

**Acknowledgments** Fondazione Cariplo is gratefully acknowledged for financial support.

# References


[1] E. Beaurepaire, J.-C. Merle, A. Daunois, and J.-Y. Bigot, "Ultrafast Spin Dynamics in Ferromagnetic Nickel" Phys. Rev. Letters **76**, 22 (1996).

[2] E. Carpene, F. Boschini, H. Hedayat, C. Piovera, C. Dallera, E. Puppin, M. Mansurova and M. Münzenberg, X. Zhang and A. Gupta, "Measurement of the magneto-optical response of Fe and $CrO_2$ epitaxial films by pump-probe spectroscopy: Evidence for spin-charge separation" Phys. Rev. B. **87**, 174437 (2013).

[3] G. M. Müller et al., "Spin polarization in half-metals probed by femtosecond spin excitation" Nature Materials **8**, 56 - 61 (2009).

# Heat Assisted Magnetic Recording

**Tim Rausch, Ed Gage and John Dykes**

Seagate Technology, 1280 Disc Drive, Shakopee, MN 55379, USA

tim.s.rausch@seagate.com

**Abstract** Heat Assisted Magnetic Recording (HAMR) is being staged as the next technology to replace perpendicular magnetic recording (PMR) in the hard disk drive industry. In this paper we will introduce HAMR and discuss some of the challenges and opportunities for this new technology.

## Introduction

As shown in Figure 1, in 2012 the hard disk drive industry shipped 569 million hard disk drives with an average capacity of 702 GB per drive. In other words, the industry shipped a combined 400 Exabytes of storage capacity. This is a 10% increase in exabytes shipped over 2011 and a 46x increase over the last ten years. As the chart shows, much of the total capacity shipped has been due to the increase in the capacity of the drives. Accounting for the increase in the number of heads and media per drive during this time frame, the areal density has increased by 23x during the last 10 years.

Due to limitations in the size of the field a magnetic recording head can generate and scaling rules in the media, it will not be possible to continue areal density growth using current magnetic recording technology. In a HAMR system, the media is temporarily heated during the recording process. This magnetically softens the recording media so that it can be recorded on with conventional magnetic writers. By heating the media we can use high coercivity materials, such as FePt, to extend the areal density capability.

 
J.-Y. Bigot et al. (eds), *Ultrafast Magnetism I*, Springer Proceedings in Physics 159
DOI 10.1007/978-3-319-07743-7_63

## HAMR Recording Process

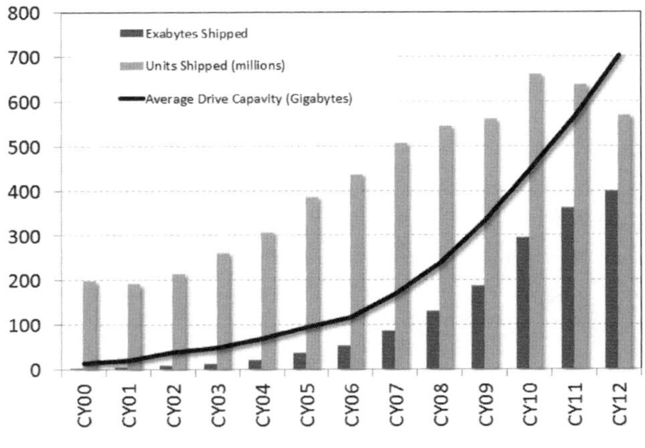


**Fig. 1.** Exabyte, unit and average drive capacity industry growth trends

In currently shipping hard disk drives the track pitch is already well below the diffraction limit of light. To confine light beyond the diffraction limit a near field transducer is required. Figure 2 shows an SEM image of one type of NFT[1] and some modeling results showing the scalability of this design. Light incident on the NFT is converted intoa surface plasmonand propagates down the edges of the NFT disk. When the surface plasmonsreach the peg at the bottom of the disk, theyinteract with the mediacausing it to heat up. This causes a drop in coercivity so that a nearby field source can change the orientation of the magnetization of the media enabling the storage of data.

## HAMR Challenges and Opportunities

Recent demonstrations of HAMR have exceeded 1TBPSI[2] areal density on the spin stand and early prototype HAMR drives have been built[3]. The main impediment to full commercialization of HAMR is the reliability of the NFT and head-disk mechanical interface. During the recording process the media is heated above the Curie temperature of the medium. For FePt, the Curie temperature is between 350C and 450C necessitating peak temperatures in the media between 450C and 550C. Developing head and media overcoats that can withstand these kinds of temperatures over the 5+ year operable life of a typical hard drive are challenging. In addition, the NFT is a highly resonant structure and any changes in its shape or physical properties over the life of the drive can lead to a reduction in coupling efficiency into the media or outright failure of the NFT. To solve this problem new NFT designs and/or materials capable of withstanding higher temperatures but still possess the necessary physical properties to support surface plasmons

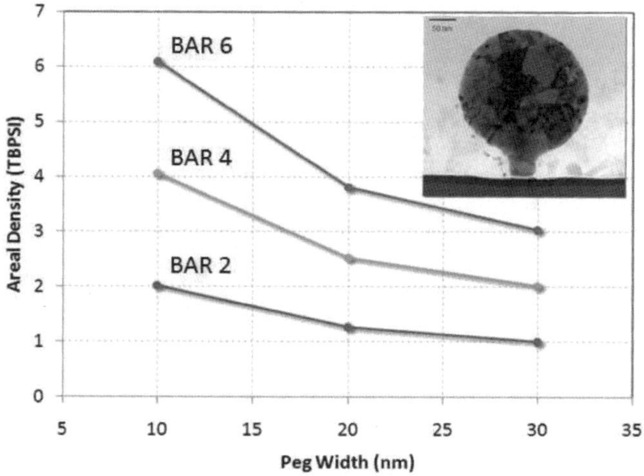


**Fig. 2.** Scalability of NFT with various bit-aspect ratios (BAR) and a TEM image of an actual NFT

## Conclusion

HAMR has demonstrated the necessary areal density capability to meet the every growing needs of the world-wide demand for data storage. Reliability of the NFT and the head-disk mechanical interface still needs further improvement before full scale commercialization of the technology can begin.

**Acknowledgments** Over the years far too many people have contributed to the success of HAMR at Seagate to list here. The authors are grateful to all our friends and colleagues at Seagate who have taken HAMR from a simple idea to a technology which will satisfy the data storage needs for billions of people in the future.

## References

[1] W. A. Challener et al, "Heat-assisted magnetic recording by a near-field transducer with efficient optical energy transfer" Nature Photonics **3**, 220 - 224 (2009)
[2] A. Q. Wu et al, "HAMR areal density demonstration of 1+ Tbpsi on spinstand" in IEEE Transactions on Magnetics Volume **49** Issue 2, 779-782 (2013).
[3] T. Rausch et al, "HAMR drive performance and integration challenges" in IEEE Transactions on Magnetics Volume **49** Issue 2, 730-733 (2013).

# Photo-induced ferromagnetic resonance in systems incorporating magnetic junctions

**H. Munekata**

Imaging science and engineering laboratory, Tokyo Institute of Technology 4259-J3-15 Nagatsuta, Midori-ku, Yokohama 226-8503, Japan

hiro@isl.titech.ac.jp

**Abstract** Careful analysis of experimental data obtained from photo-induced ferromagnetic resonance (ph-FMR) for systems incorporating magnetic junctions shed light on spin dynamics associated with interface magnetism. Examples are grate field-dependent magnetic damping in a Fe/(Ga,Mn)As and enhanced precession amplitude in a Co/Pd multi-layered structure. In the former, we discuss a scenario of spin pumping induced by a domain wall across the interface. For the latter, we discuss a scenario of efficient energy transfer from orbital (electronic) to spin sub-systems at the Co/Pd interface. Those examples suggest importance of magnetic interfaces on ultrafast magnetic phenomena.

## Introduction

Photo-induced precession of magnetization in ordered-spin systems, call hereafter the photo-induced ferromagnetic resonance (ph-FMR), is an interesting phenomenon in that the energy of ultra-short laser pulses is stored in part for relatively long duration in a spin sub-system in the form of precession with natural damping. This phenomenon was studied in various magnetic materials, including metals, semiconductors, and insulators, with the aim to directly extract parameters associated with magnetization dynamics. In this paper, we are concerned with two new topics in ph-FMR: a spin current during the magnetization precession and efficiency of energy transfer at the initial state of ph-FMR. As to the spin current, I show grate field-dependent magnetic damping in a Fe/(Ga,Mn)As heterojunction, and discuss spin pumping induced by a domain wall across the Fe/(Ga,Mn)As heterojunction. As for the efficient energy transfer, I show ph-FMR in an ultra-thin Co/Pd multi-layered structure in the regime of low energy excitation as low as 0.25 μJ/cm$^2$ per a laser pulse of 150 fs, and suggests a mechanism of efficient energy transfer from orbital (electronic) to spin subsystems in the vicinity of the interface region at which contribution of orbital angular momentum is supposed to be large.



J.-Y. Bigot et al. (eds), *Ultrafast Magnetism I*, Springer Proceedings in Physics 159
DOI 10.1007/978-3-319-07743-7_64

## A Fe/(Ga,Mn)As system

We have studied the ph-FMR for three samples with different structures; a (Ga,Mn)As single layer, a Pt/(Ga,Mn)As heterostructure, and a Fe/(Ga,Mn)As heterostructure [1,2]. The Mn content $x$ is $x = 0.045$ for all cases. In the single (Ga,Mn)As, precession frequency $\omega$ increases and precession lifetime $\tau$ decreases with increasing an external field applied along the easy axis. The resultant $\omega\tau$ products remain almost constant in this case. Similar trend is observed in Pt/(Ga,Mn)As, except that the $\omega\tau$ product is smaller than that of (Ga,Mn)As. The value of inverse $\omega\tau$ product, so called the effective Gilbert damping $\alpha$, is $\alpha = 0.07$ and 0.15, respectively, for (Ga,Mn)As and Pt/(Ga,Mn)As. The enhanced magnetic damping in Pt/(Ga,Mn)As can be understood qualitatively in terms of the spin pumping. To our surprise, in Fe/(Ga,Mn)As, the value of inverse $\omega\tau$ product, which is larger than that of the Pt/(Ga,Mn)As ($\alpha \sim 0.23$) at H = 0, varies with an external field, reflecting the relative magnetic configuration between Fe and (Ga,Mn)As layers: the $\alpha$ value maximizes ($\alpha \sim 0.29$ at H = 70 Oe) when the crossed magnetic configuration is realized between the two layers, whereas the $\alpha$ value minimizes at the parallel configuration ($\alpha \sim 0.07$ at H $\geq$ 400 Oe). The $\alpha$ - H curves follow well with the corresponding $M^2$ - H curves. We infer that the great field-dependent change in damping could be attributed to the spin-wave excitation in the domain well appearing at the Fe/(Ga,Mn)As interface.

## A Co/Pd multi-layered structure

As for the ultra-thin Co/Pd multi-layered structure, which has been known for the interface-induced perpendicular magnetic anisotropy [3], a change in the amplitude A and frequency $f$ of ph-FMR occurs with alternating the structural parameters. When the Co layer thickness is varied ($t_{Co} = 0.3 \sim 1.4$ nm) with the constant Pd layers thickness ($t_{Pd} = 1.62$ nm), a change in the amplitude can be explained within the framework of a standard gyromagnetic model; $A \propto 1/f$. On the other hand, when the Pd layer thickness is varied ($t_{Pd} = 0.1 \sim 2.6$ nm) with the constant Co layer thickness ($t_{Co} = 0.78$ nm), the ph-FMR amplitude increases significantly by a factor of fifty with reducing the $t_{Pd}$ value despite of a very moderate change in the static magnetic characteristics [4]. Consequently, ph-FMR with the excitation as weak as 0.25 μJ/cm$^2$ per fs-laser pulse has been observed in the $[Co/Pd]_5$ sample with $t_{Pd} = 0.2$ and $t_{Co} = 0.78$ nm. The data have been analyzed using a model which connects the amplitude and parameters associated with spin dynamics, through which we have concluded that the observed enhancement in the FMR amplitude is due to enhanced efficiency of energy transfer between orbital and spin sub-systems. A microscopic mechanism on the basis of super-lattice model will be discussed at the time of presentation.

## Conclusion

Experimental results of ph-FMR for two different layered-structures have been reviewed, through which important roles of magnetic interfaces have been pointed out. Tailoring interface would lead us to the development of new magnetic materials that would not be achievable with single layers, especially in view of spin dynamics in ultra-short time scale.

**Acknowledgments** This work is supported in part by the Advanced Photon Science Alliance Project from MEXT, and Grand-in-Aid for Scientific Research No. 22226002 from JSPS.

## References

[1] S. Kobayashi, K. Suda, J. Aoyama, D. Nakahara, and H. Munekata, "Photo-Induced Precession of Magnetization in Metal/(Ga, Mn)As Systems" IEEE Trans. Mag. **46**, 2470-2473 (2010).

[2] S. Kobayashi, K. Suda, and H. Munekata, "A grate field-dependent change of magnetic damping in Fe/(Ga,Mn)As" APS March Meeting 2011 (Dallas, Texas, U.S.A), presentation given on March 22nd, Session J15:

[3] N. Nakajima, et al., "Perpendicular magnetic anisotropy caused by interfacial hybridization via enhanced orbital moment in Co/Pt multilayers: magnetic circular x-ray dichroism study" Phys. Rev. Lett. **81**, 5229-5232 (1998).

[4] K. Yamamoto, T. Matsuda, K. Nishibayashi, Y. Kitamoto and H. Munekata, "Low-power photo-induced precession of magnetization in ultra-thin Co/Pd multi-layer films" IEEE Trans. Mag. **49** (2013), in press.

# Nonlinear Spin Waves in Two-Dimensional Arrays of Magnetic Nanodots

**Konstantin Guslienko[1,2], Yuri Kobljanskyj[3], Gennady Melkov[3], Valentyn Novosad[4], Samuel D. Bader[4], Michael Kostylev[5], and Andrei Slavin[6]**

[1] Depto. Física de Materiales, Universidad del País Vasco, 20018 San Sebastián, Spain

[2] IKERBASQUE, The Basque Foundation for Science, 48011 Bilbao, Spain

[3] Faculty of Radiophysics, T. Shevchenko National Univ. of Kyiv, Kyiv 01601, Ukraine

[4] Materials Science Division, Argonne National Laboratory, Argonne, IL 60439, USA

[5] School of Physics, University of Western Australia, Crawley, WA 6009, Australia

[6] Department of Physics, Oakland University, Rochester, MI 48309, USA

kostyantyn.gusliyenko@ehu.es

**Abstract.** It is shown experimentally that in a two-dimensional array of Permalloy nanodots the lifetime of the electromagnetic microwave radiation at the sub-harmonic frequency originated from the parametrically excited ferromagnetic resonance mode is increased by *two orders* of magnitude compared to the case of a continuous magnetic film.

## Introduction

When magnetic elements are reduced to a sub-micrometer size, their spin wave (SW) spectra are modified substantially [1]. In particular, in a flat cylindrical magnetic dot of radius $R$ the long-wavelength spin waves having wave number $k < 1/R$ are excluded from the SW spectrum and the frequency degeneracy between the quasi-uniform ferromagnetic resonance (FMR) mode and higher SW modes is removed. This spectral modification qualitatively changes all nonlinear dynamic properties of the dots, as they become insusceptible to the four-wave nonlinear processes that limit the FMR precession amplitude. Due to reduction of the dot sizes the lifetime of the electromagnetic radiation at a sub-harmonic frequency originated from the parametrically excited FMR mode is essentially increased.



J.-Y. Bigot et al. (eds), *Ultrafast Magnetism I*, Springer Proceedings in Physics 159
DOI 10.1007/978-3-319-07743-7_65

## Experiment and discussion

We investigated two-dimensional square arrays of cylindrical Permalloy dots having the radius R = 1000 nm, distance d=1000 nm between the dot edges and two different thicknesses: L1=100 nm (array #1) and L2=12 nm (array #2) formed on GaAs substrate. As a "control" sample we used continuous Py film of the thickness of 100 nm formed on the same substrate.

We studied the process of parametric generation of a subharmonic of external microwave signal (having frequency twice of the FMR frequency) in a dot array (Fig. 1). The sample was subjected to simultaneous action of the long (9 μs), powerful (10-100 W) pulse of microwave parallel pumping ($\omega_p/2\pi$=9.4 GHz) and short (30 ns) and weak (10 μW) pulse of synchronizing signal at half of the pumping frequency, and output microwave signal $P_{out}$ was received by an antenna. The output signal appeared only when the pumping amplitude exceeded the threshold of parametric excitation of the FMR mode. The maximum value of $P_{out}$ was obtained by tuning the in-plane bias magnetic field $H_0$ to achieve the resonance condition $\omega_0 = \omega_p/2$ of the FMR mode with the pumping subharmonic [2].

The measured time dependences of the power of subharmonic radiation $P_{out}$ are presented in Fig. 2. If the parametric pumping is switched on, the output power $P_{out}$ (intensity of the radiated subharmonic signal) starts to increase exponentially. A maximum level determined by a four-magnon mechanism of power limitation is reached after $t$=200-300 ns, and the output power in continuous film grows faster than in both patterned samples due to higher threshold of parametric excitation. The temporal evolution of $P_{out}$ in continuous film and dot arrays is different. In continuous film $P_{out}(t)$ decreases exponentially $P_{out} \sim exp(-t/\tau)$, where $\tau \approx$200 ns. This decrease is caused by competitive four-magnon interaction of the FMR mode with the parametrically excited SWs having frequencies close to FMR frequency. These SWs have the lowest threshold of parametric excitation in continuous magnetic film. For the dot arrays patterning leads to exclusion of the majority of four-wave processes of magnon relaxation limiting the amplitude of the quasi-uniform FMR mode in continuous film and bulk magnetic samples. This exclusion resulted in substantial enhancement of nonlinear properties of the magnetic dot arrays. In particular, the patterning resulted in increase of the characteristic time of parametrically induced subharmonic radiation from the arrays in two orders of magnitude in comparison to continuous film. The slow decrease of the radiated subharmonic power observed in Fig. 2 is caused by the microwave heating of magnetic dots and can be substantially reduced by reducing the dot thickness [2].

## Conclusions

Nano-structuring of magnetic material by fabricating 2D dot arrays leads to development of a novel metamaterial with nonlinear microwave properties that are substantially superior to that of magnetic films and traditional bulk magnetic materials. These properties can be controlled by the geometric parameters of materials' nanostructure. This novel metamaterial can be useful for applications in microwave signal processing devices operating at high levels of microwave power.

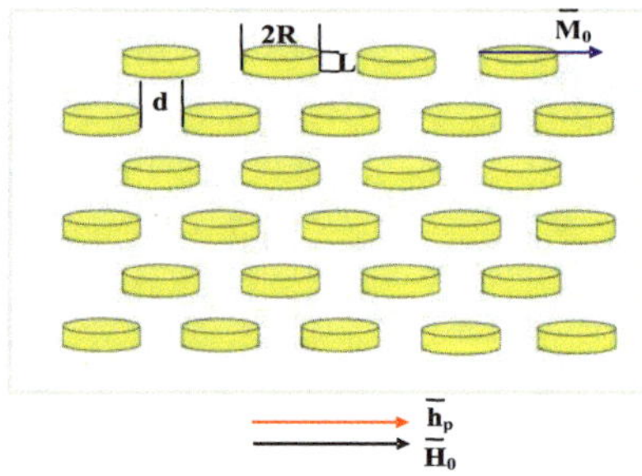


**Fig. 1.** Two-dimensional array of Permalloy dots: *R* is the dot radius, *L* is the dot thickness, *d* – is the distance between the dot edges, $\mathbf{M_0}$ is the saturation magnetization, $\mathbf{H_0}$ is the in-plane bias magnetic field, and $\mathbf{h_p}$ is the magnetic field of microwave parallel pumping.

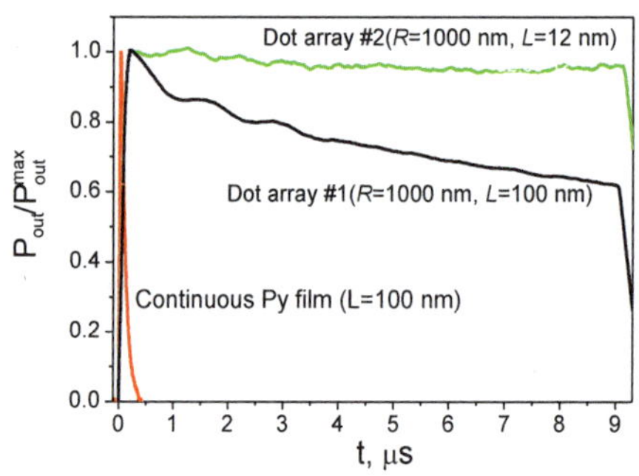


**Fig. 2.** Measured dependence of the normalized output power $P_{out}$ on the time of action of the microwave pumping: red line – continuous film, black line – dot array #1, green line – array #2.

**Acknowledgments** K.G. acknowledges support by IKERBASQUE (the Basque Foundation for Science). This work was partially supported by the Spanish MEC grants PIB2010US-00153, FIS2010-20979-C02-01, by the National Sci. Foundation of USA (grant DMR-1015175), by the State Fund of Fundamental Res. of Ukraine (project #UU34/008), by the MES of Ukraine (grant No. M/90-2010), by the Australian Res. Council, and by the U.S. DOE Office of Science (contract DE-AC02-06CH11357).

## References

[1] S.O. Demokritov (Ed.), "Spin wave confinement" (Pan Stanford Publ., Singapore, 2009).
[2] Y. Kobljanskyj, G.A. Melkov, K.Y. Guslienko, V. Novosad, S.D. Bader, M. Kostylev, and A.N. Slavin, "Nano-structured magnetic metamaterial with enhanced nonlinear properties", *Sci. Reports* **2**, 478 (2012).

# Ultrafast photoinduced linear and circular anisotropy in multiferroic manganite $YMnO_3$

**M. Pohl,[1] V.V. Pavlov,[2] I.A. Akimov,[1,2] V.N. Gridnev,[2] R.V. Pisarev,[2] D.R. Yakovlev,[1,2] and M. Bayer[1]**

[1] Experimentelle Physik 2, Technische Universität Dortmund, 44221 Dortmund, Germany

[2] Ioffe Physical-Technical Institute, Russian Academy of Sciences, 194021 St. Petersburg, Russia

Author e-mail address: pisarev@mail.ioffe.ru

**Abstract** Ultrafast photo-induced rotation and ellipticity are observed in multiferroic hexagonal manganite $YMnO_3$, which we attribute to the optical spin alignment and orientation within the charge transfer transition $A_1$-$E_1$ at ~1.6 eV. The recombination decay time of ~500 fs and the Raman coherence time of ~5 fs have been determined. We attribute the ultrashort coherence time to the strong electron Coulomb interaction.

## Introduction

Ultrashort pulsed lasers have opened new opportunities for implementing coherent control of electronic and magnetic states in matter, which is important for applications in multidisciplinary fields such as information processing, opto-electronics, and spintronics [1]. Ultrafast phenomena are widely studied in various materials. However the important problem of the light-matter interaction on the shortest time scales of tens of femtoseconds remains essentially unresolved. In this paper, we report on the photo-induced ultrashort phenomena observed in strongly-correlated multiferroic manganite $YMnO_3$ which possesses an intense band at ~1.6 eV.

## Results

Undoped 3*d*-transition-metal oxides are typically good insulators with a large optical gap in the range of 1-4 eV. The material of choice for our study is hexagonal manganite $YMnO_3$ (P.G. *6mm)*, which is widely studied by various techniques. Besides interesting multiferroic properties, this material demonstrates an unusual optical spectrum with an isolated charge-transfer transition at ~1.6 eV [2]. Moreover, it exhibits large second and third order optical nonlinearities [3,4].

 
J.-Y. Bigot et al. (eds), *Ultrafast Magnetism I*, Springer Proceedings in Physics 159
DOI 10.1007/978-3-319-07743-7_66

The pump-probe setup for time-resolved transient measurements including the lock-in detection method has been described earlier [5]. The modulation of the pump beam is accomplished by a photo-elastic modulator. The λ/2 and λ/4 mode modulations create linearly and circularly polarized pump light and are used for the measurements of photo-induced linear and circular anisotropy, respectively.

During the excitation, an electron moves from the oxygen $O^{2-}(2p^6)$ states to the $Mn(3d(z^2))$ state. This process is shown in Fig. 1(a) for the electronic transition $A_1 \rightarrow E_1$. The lifetime of this transition was measured via pump induced reflectivity using a conventional chopper modulation working at 2 kHz (see Fig. 1(b)). The excited carriers relax with a recombination decay time of ~500 fs, which is distinguished from coherent electron dynamics, as we show below.

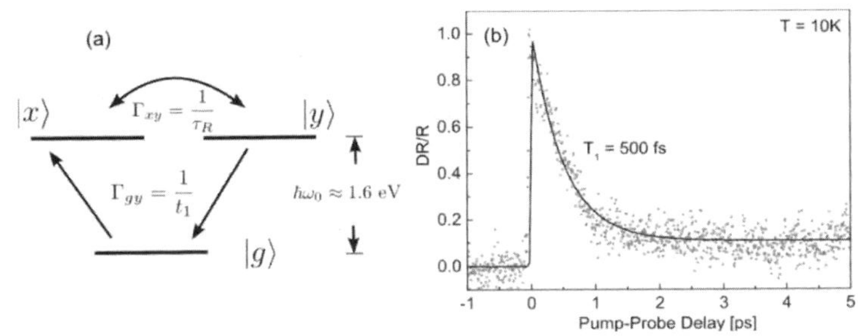


**Fig. 1.** (a) Schematic energy-level diagram for the dipole-allowed optical transition $A_1 \rightarrow E_1$ between the hybridized $O^{2-}(2p^6)$ states (|g> is the ground state) and the $Mn^{3+}(3d^4)$ states (|x>, |y> are the excited degenerate states). (b) Dynamics of the photo-induced reflectance in $YMnO_3$, points are experimental results, the line is a fit by an exponential function.

In the pump-probe experiment a linearly polarized pump beam generates a non-equilibrium electron polarization. The less intense linearly polarized probe beam with an angle of 45° between the polarization planes of pump and probe beams experiences different phase velocities and absorptions for its $x$- and $y$-components, resulting in changes of optical rotation and ellipticity. This effect will be referred to as the photo-induced linear anisotropy. Upon circular excitation by the pump beam, the angular momentum of the absorbed light is transferred to the medium creating a non-equilibrium electron polarization with a nonzero angular momentum. In the following, this effect will be referred to as photo-induced circular anisotropy. The left and right circularly polarized components of the linearly polarized probe beam consequently experience different phase velocities and absorptions, giving rise to the circular birefringence and dichroism.

Results on the ultrafast photoinduced linear and circular anisotropy in $YMnO_3$ for the different pulse durations of 17 and 34 fs are shown in Fig. 2. The dynamical response here is definitely faster than 100 fs, i.e. the electron relaxation dynamics accessed in this experiment are at least one order of magnitude faster than the carrier relaxation dynamics shown in Fig. 1(b). The signals of linear-induced ellipticity and circular induced rotation have a strongly asymmetric shape with a sign change for both pulse durations. The photoinduced rotation θ and ellipticity ϵ have been fitted with the phenomenological expression

$$\theta + i\epsilon = A\exp\left(-\frac{t^2}{\sigma^2}\right) + B\exp\left(\frac{\sigma^2}{4\tau_R^2} - \frac{t}{\tau_R}\right)\left[1 - \mathrm{erf}\left(\frac{\sigma}{2\tau_R} - \frac{t}{\sigma}\right)\right] \quad 1$$

where $t$ is the pump-probe time delay, σ is the width of the Gaussian pulse, erf is the Gauss error function.

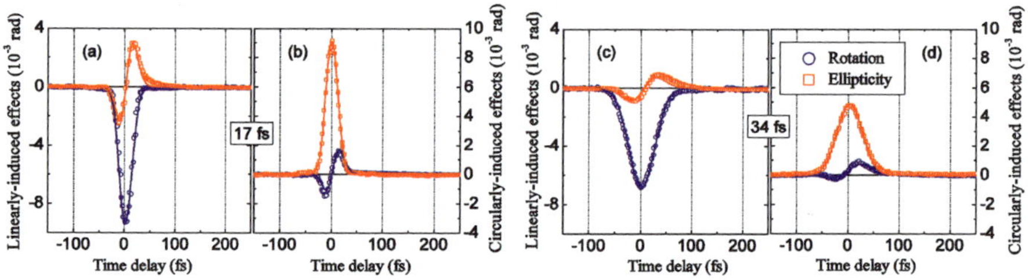


**Fig. 2.** Experimental data on linear- and circular-induced phenomena for pulse duration times of ~17 fs and ~34 fs (dots). The calculated dependencies of temporal behavior of the photo-induced phenomena on the basis of Eq. (1) are shown by solid lines.

Complex coefficients A and B are functions of the diagonal and non-diagonal elements of the dielectric tensor. The Raman coherence time $\tau_R$ was obtained by a fitting procedure taking into account the complex character of the photoinduced phenomena with linear and circular excitations.

The Raman coherence time $\tau_R = 5$ fs has been determined for two experiments with pulse durations of 17 and 34 fs. The calculated dependencies of temporal behavior of the photoinduced phenomena on the basis of Eq. (1) are shown in Fig. 4 by lines, which are in good quantitative agreement with the experimental data on linear- and circular-induced phenomena.

**Acknowledgments** This work was supported by the Deutsche Forschungsgemeinschaft, the Russian Foundation for Basic Research (13-02-00754), the Government of the Russian Federation (14.B25.31.0025), V.V.P. thanks the Alexander-von Humboldt Foundation.

## References

[1] A. Kirilyuk, A. V. Kimel, and Th. Rasing, "Ultrafast optical manipulation of magnetic order" Rev. Mod. Phys. **82**, 2731 (2010).

[2] A. M. Kalashnikova and R.V. Pisarev, "Electronic Structure of Hexagonal Rare-Earth Manganites $RMnO_3$" JETP Lett. **78**, 143 (2003).

[3] D. Fröhlich, St. Leute, V.V. Pavlov, and R.V. Pisarev, "Nonlinear Optical Spectroscopy of the Two-Order-Parameter Compound $YMnO_3$" Phys. Rev. Lett. **81**, 3239 (1998).

[4] A.V. Kimel, R.V. Pisarev, F. Bentivegna, and Th. Rasing, "Time-resolved nonlinear optical spectroscopy of $Mn3^{+}$ ions in rare-earth hexagonal manganites $RMnO_3$ (R=Sc, Y, Er)" Phys. Rev. B **64**, 201103 (2001).
[5] V. V. Pavlov, R. V. Pisarev, V. N. Gridnev, E. A. Zhukov, D. R. Yakovlev, and M. Bayer, "Ultrafast Optical Pumping of Spin and Orbital Polarizations in the Antiferromagnetic Mott Insulators $R_2CuO_4$" Phys. Rev. Lett. **98**, 047403 (2007).

# Magneto-optical wave mixing in Garnets

**M. Barthelemy[1], M. Sanches Piaia[1], H. Vonesh[1], M. Vomir[1], P. Molho[2], B. Barbara[2] and J.-Y. Bigot[1]**

1. Institut de Physique et Chimie des Matériaux de Strasbourg, Université de Strasbourg - CNRS, 23 rue du Loess, BP 43, 67034, Strasbourg, France.

2. Institut Louis Néel, CNRS - Université Joseph Fourrier, 25 avenue des Martyrs, BP 166, 38042 Grenoble, France.

barthelemy@ipcms.unistra.fr

**Abstract** We report on magnetic field dependent four wave mixing signals emitted from an iron bismuth Garnet film. The coherent and population contributions to the magneto optical response are separately measured in a three beams configuration. Advantages of this technique over the commonly used pump probe magneto-optical configurations are discussed.

## Introduction

Since ultrafast demagnetization was observed [1], several approaches are used to explain the loss of angular momentum induced by ultrashort pulses in magnetic materials. One of them considers the coupling between photon and spin momenta giving rise to coherent magnetism. Previous experiments based on Kerr and Faraday pump probe experiments show that the coherent contribution plays a non-negligible role in the demagnetization process [2]. On the other hand, the transfer of angular momentum between the orbital and spin momenta has been recently unraveled using femtosecond X-ray pulses [3]. In this work, we report on the magneto optical coherent emission measurement based on four wave mixing configuration.

## Coherent magnetism

The physical origin of ultrafast coherent magnetism can be understood by taking into account the spin photon interaction in the framework of short pulse excitation of a magnetic system. The following interaction Hamiltonian considers a single electron of mass m, charge e, momentum p and spin s:

 
J.-Y. Bigot et al. (eds), *Ultrafast Magnetism I*, Springer Proceedings in Physics 159
DOI 10.1007/978-3-319-07743-7_67

$$H_{\text{int}} = \frac{e}{m}\vec{\pi}_{\pm}\vec{A}_L \qquad \text{with } \vec{\pi}_{\pm} = \left(\vec{p} + \frac{e}{c}\vec{A}_M + \vec{s}\times\vec{\nabla}V_{ion}\right)_{\pm} \qquad 1$$

Where $\mathbf{A_L}$ and $\mathbf{A_M}$ are the potential vectors of the laser electromagnetic field and of the static magnetic field; $V_{ion}$ is the scalar potential of the ion; $\pi_{\pm}$ is the generalized momentum for right or left circularly polarized light. We will show that the Hamiltonian of equation 1, applied to a simplified 8 levels spins system, allows understanding the dynamics of the coherences and populations of spins [4]. In particular, the spin orbit coupling term $\vec{s}\times\vec{\nabla}V_{ion}$ gives rise to a coherent magneto optical component in the time scale of the laser pulse.

## Three beams Magneto Optical Four Waves Mixing

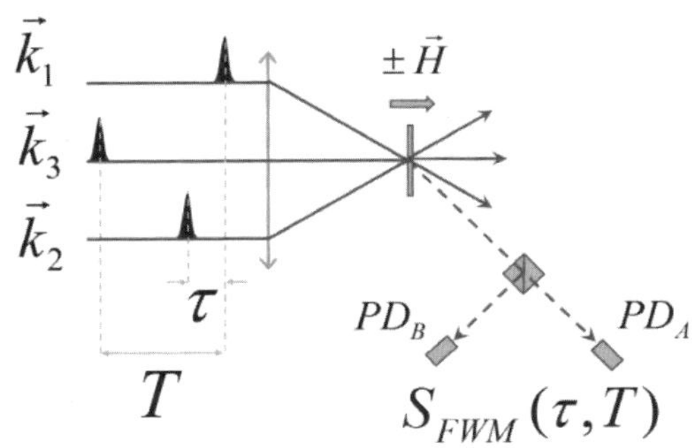


Fig. 1. Three beams magneto-optical four waves mixing configuration

The three beams four waves mixing technique has proved to be well adapted to separate the coherent and population dynamics of charges. In our experiment, it is transposed to a magneto optical configuration (MOFWM) to separate the coherent from population dynamics of the spins. We used the magneto-optical configuration shown in Fig.1. Three fields $E_1$, $E_2$ and $E_3$ propagate along $k_1$, $k_2$ and $k_3$. In this case, the MOFWM process phase-matches along both directions $k_3 \pm (k_1-k_2)$.

The experiment was performed using amplified femtosecond laser pulses (48 fs; 795 nm; 5 kHz) and a polarization bridge (photodiodes $PD_A$ and $PD_B$ in Fig. 1). A static magnetic field H=±3kOe is applied at an angle $\phi$ with respect to the normal to the sample. The MOFWM signals are detected in the $k_3+(k_1-k_2)$ direction by changing the delays between pulses 1 and 2 (delay $\tau$) or pulses 1 and 3 (delay T).

The sample is a 7 μm thick bismuth doped garnet $(GdThPtBi)_3(FeGa)_5O_{12}$ film deposited on a 1mm GGG substrate by liquid phase epitaxy. It is known to have a strong magneto optical response and its transparency at 800nm makes it a good candidate for degenerating four wave mixing measurements at this wavelength.

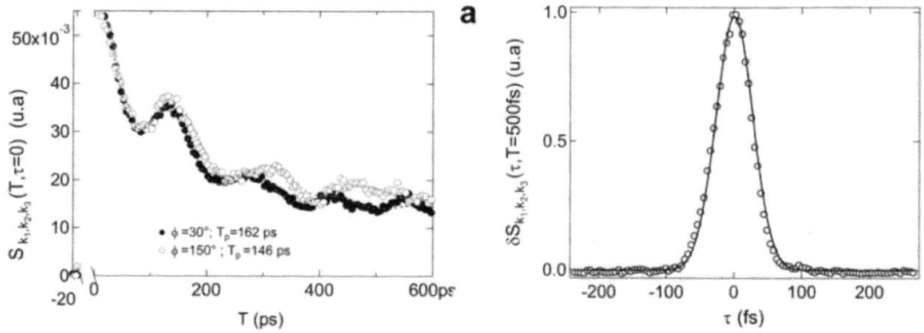


**Fig. 2.**. a) Population magneto optical contribution as a function of delay T for $\tau$=0. Full symbols: $\phi$=30° with a corresponding precession period $T_p$=162ps. Empty symbols: $\phi$=150° , $T_p$=146ps. b) Coherent Magneto Optical contribution measured by varying the delay $\tau$ and setting T=500fs.

As shown in Figure 2, this experimental configuration allows a selective meas*urement of coherent and population contributions involved in the magnetization* dynamics. Figure 2.a) illustrates the MOFWM signal $S_{k1k2k3}(\tau=0,T)$ measured as a function of the delay T. The oscillations correspond to the precession of the magnetization around the effective magnetic field. The period of the oscillations depends on the magnetic field orientation. The double exponential decay is related to the population of spins involving the spin-lattice interaction. In contrast, the dynamical signals $S_{k1,k2,k3}$ measured as a function of $\tau$ with a fixed T=500fs for two opposite directions of the magnetic field $\phi$=90° and $\phi$=-90° gives information on the coherent spins dynamics. The difference $\delta S_{k1,k2,k3}(\tau,T=500fs)= S_{k1,k2,k3}(\tau,T=500fs)|_{H+} - S_{k1,k2,k3}(\tau,T=500fs)|_{H-}$ corresponds to the magneto optical coherent contribution as shown in Figure 2b). Details concerning the respective role of the coherent and population dynamics of the spins in the magneto-optical response will be reviewed and discussed.

**Acknowledgments:** This work was supported by the European Research Council, project ERC Advanced Grant ATOMAG (ERC-2009-AdG-20090325 247452).

## References

[1] E. Beaurepaire, J.-C. Merle, A. Daunois, & J.-Y. Bigot, “Ultrafast Spin Dynamics in Ferromagnetic Nickel” Phys. Rev. Lett. **76**, 4250-4253 (1996).
[2] J.-Y. Bigot, M. Vomir, & E. Beaurepaire, “Coherent ultrafast magnetism induced by femtosecond laser pulses” Nature Phys. **5**, 515-520 (2009).

[3] C. Boeglin, E. Beaurepaire, V. Halté, V. López-Flores, C. Stamm, N. Pontius, H. A. Dürr and J.-Y. Bigot, "Distinguishing the ultrafast dynamics of spin and orbital moments in solids" Nature **465**, 458-461 (2010).
[4] H. Vonesh and J.-Y. Bigot, "Ultrafast spin-photon interaction investigated with coherent magneto-optics" Phys. Rev. B **85**, 180407(R) (2012).

# Quantum Femtosecond Magnetism in a Strongly Correlated Manganese Oxide

**Tianqi Li[1,2], Aaron Patz[1,2], Leonidas Mouchliadis[3,4], Jiaqiang Yan [2], Thomas A. Lograsso[2], Ilias E. Perakis [3,4], and JigangWang[1,2]**

[1.] Department of Physics and Astronomy, Iowa State University, Ames, Iowa 50011 U.S.A.

[2.] Ames Laboratory – USDOE, Ames, Iowa 50011, USA

[3.] Department of Physics, University of Crete, Heraklion, Crete 71003, Greece

[4.] Institute of Electronic Structure and Laser, Foundation for Research and Technology-Hellas, Heraklion, Crete 71110, Greece.

jwang@iastate.edu; jwang@ameslab.gov

**Abstract** We show a photoinduced magnetic phase transition from antiferromagnetic to ferromagnetic ordering in a strongly correlated manganite during ~100 fs laser pulses when optical polarization still interacts with spins. This reveals an initial quantum coherent regime of magnetism, which is driven by fast quantum spin-flip fluctuations correlated with a coherent superposition of many-body electronic states.

There is growing evidence that femtosecond transient polarization of condensed matter systems during a laser pulse can be used to manipulate spin or change magnetic order [1]. Non-adiabatic coherent photo-excitation during very early times can be used to control subsequent slower dynamics driven by the free energy or excited potential surface. This research provides fascinating opportunities to address outstanding and cross-cutting challenges of condensed matter/chemical/biological physics: how to achieve *quantum control* of magnetism and reveal highly non-equilibrium, *"thermodynamically hidden"* orders during femtosecond timescales? Recently we explored a new paradigm, referred as to *quantum femtosecond magnetism*—fs magnetic phase transitions driven by quantum spin flucations and laser-excited inter-atomic coherences [2]. Such far-from-equilibrium many-body problems involve an extremely poorly-understood interplay between *quantum coherence*, *strong correlation* and *nonlinearity*. While thermodynamic, transport and linear optical properties do not depend as critically on these properties, we show they dominate ultrafast nonlinear spin and charge dynamics.

The *quantum fs magnetism* provides a general quantum many–body framework for fs manipulation of ground states with magnetic orders and strong correlation. We use as example a prototype insulating ground state with CE–type

 
J.-Y. Bigot et al. (eds), *Ultrafast Magnetism I*, Springer Proceedings in Physics 159
DOI 10.1007/978-3-319-07743-7_68

antiferromagnet-ic (AFM) and CO/OO orders, shown schematically in Fig.1. This phase consists of one-dimensional (1D) zig–zag chains with alternating Mn4+/Mn3+ sites (CO) and ferromagnetically–aligned (FM) spins. Neigboring chains are AFM–ordered. The Mn $t_{2g}$–orbitals are fully occupied by 3 electrons forming local S=3/2 spins $S_i$ on every site *i*. We denote the $Mn^{3+}$ atomic states, with the additional electron of spin $s_i$ populating a Mn $e_g$–orbital. Our understanding of dynamics could build on classical or quantum spin scenarios. Classical scenarios neglect spin–flips. Electron hopping from site to site diminishes with increasing angle between the local spin moments. Applied to our system, inter–chain $e_g$–electron hopping is suppressed by AFM spin alignment, which results in electronic confinement within the 1D chains (white arrows). This confinement suppresses fs spin dynamics, as $e_g$ and $t_{2g}$ spins remain parallel within the same chain, consistent with the slow ps magnetization changes observed thus far. In the case of quantum spin picture, photoinduced electron hopping to sites with anti–parallel local spin is possible when correlated with accompanying quantum spin–flips (red arrows). Photoexcitation thus populates degenerate non–equilibrium quantum states of photoexcited and local electron spins, shown schematically in Fig. 1:

$$|i\alpha M\rangle = \sqrt{\frac{S+M+\frac{1}{2}}{2S+1}}c^{+}_{i\alpha\uparrow}\left|iS, M-\frac{1}{2}\right\rangle + \sqrt{\frac{S-M+\frac{1}{2}}{2S+1}}c^{+}_{i\alpha\downarrow}\left|iS, M+\frac{1}{2}\right\rangle \quad (1)$$

where α is shorthand for orbital, lattice, and spin quantum numbers. The eigenvalues (J, M) characterize the total spin $J_i = s_i + S_i$. The quantum coherent superposition Eq.(1) allows for fs spin dynamics, driven by the off–diagonal Hund's rule magnetic interaction, during timescales shorter than the phonon oscillation period. It is suppressed in the classical spin limit $S \rightarrow \infty$.

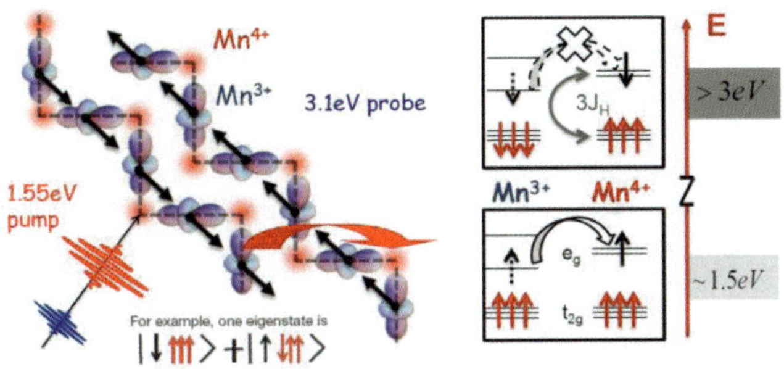


**Fig. 1.** A quantum many-body scheme for fs manipulation of a CE-type AFM order. Quantum spin–flip fluctuations allow $e_g$ electrons to hop on sites with opposite local $t_{2g}$ spin, by forming non–equilibrium total spin state as illustrated (Eqn. 1 in the text).

Below we show femtosecond photo-induced switching from antiferromagnetic to ferromagnetic ordering in $Pr_{0.7}Ca_{0.3}MnO_3$, by observing the establishment (within about 120 femtoseconds) of a huge temperature-dependent magnetization with photo-excitation threshold behavior absent in the optical reflectivity. Typical data showing the general temporal profile of the photoinduced ellipticity change $\Delta\eta$, are plotted in Fig. 2, for the first 25 ps. The sign of $\Delta\eta$ is positive, indicating tran-sient photoinduced magnetization enhancement. Distinct temporal regimes can be identified with different pump fluence dependence: at low pump fluence, there is a slow, ps gradual enhancement (up to ~30 ps). In strong contrast to this, at high pump fluence above a threshold behavior is seen ~ 2.5 mJ/cm$^2$, an initial fast, fs increase (~120 fs) in magnetization is followed by a slow ps enhancement with periodical oscillations. This distinctive threshold, "step-function-like" en-hance-ment in the magnetization is indicative critical photoinduced antiferro- to ferro-magnetic phase transitions.

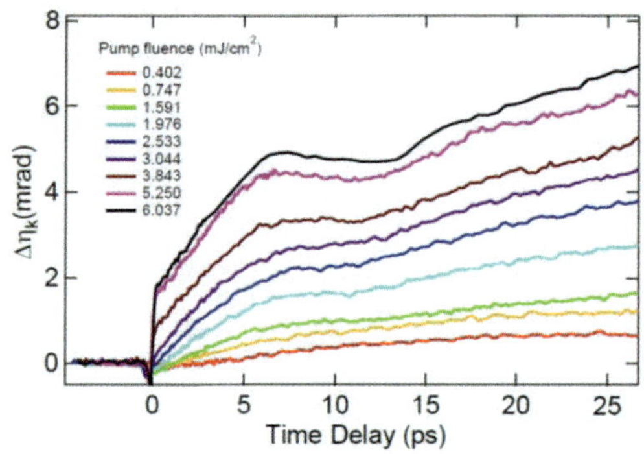


**Fig. 2.** Photoinduced transient magnetization changes obtained by two-color, NIR pump/UV probe MCD spectroscopy (1.5 eV /3 eV). A threshold behavior is seen ~ 2.5 mJ/cm$^2$ for significant magnetization enhancement within the first 120 fs, suggesting a photoinduced magnetic phase transition.

The development of ferromagnetic correlations during the femtosecond laser pulse reveals an initial quantum coherent regime of magnetism, distinguished from the picosecond lattice-heating regime characterized by phase separation without threshold behaviour. Our simulations [2] reproduce the nonlinear femtosecond spin generation and further corroborate fast quantum spin-flip fluctuations corre-lated with coherent superpositions of electronic states to initiate local ferromagnetic correlations.

**Acknowledgments** This work was supported by the National Science Foundation (contract no. DMR-1055352) and by the U.S. Department of Energy No. DE-AC02-7CH11358.

## References

[1] J.-Y Bigot, M. Vomir, E. Beaurepaire, "Coherent ultrafast magnetism induced by femtosecond laser pulses". Nature Phys. **5**, 515–520 (2009).

[2] T. Li et al., "Femtosecond switching of magnetism via strongly correlated spin–charge quantum excitations" Nature, **496**, 69 (2013)

# Ultrafast opto-magnetism in KNiF3

**D. Bossini[1], A.M. Kalashnikova[2], R.V. Pisarev[2], Th. Rasing[1] and A.V. Kimel[1]**

[1]Radboud University Nijmegen, Institute for Molecules and Materials, Heyendaalseweg 135, Nijmegen, The Netherlands

[2]Ioffe Physical-Technical Institute, Russian Academy of Sciences, 194021 St. Petersburg, Russia

d.bossini@science.ru.nl

**Abstract** Optical control of spins in the highly symmetric Heisenberg antiferromagnet $KNiF_3$ is demonstrated. It is shown that Impulsive Stimulated Raman Scattering of light by magnons, combined with time-resolved magneto-optical detection, allows the generation and detection of antiferromagnetic resonances with unprecedented high sensitivity.

## Introduction

Since the pioneering observation of ultrafast demagnetization in Ni by an intense femtosecond laser pulse [1], intense research efforts have been focused on the control of magnetism on ever shorter time-scales. Though a lot of the early works involved ferromagnets [2-3], the magnetization dynamics in antiferromagnets is intrinsically faster [4].

Recently, selective excitation of spins has been achieved using THz radiation in the antiferromagnet NiO [6]. The same magnetic mode in the material was also accessed [5] through the opto-magnetic effect, which is described in terms of Impulsive Stimulated Raman Scattering (ISRS) [4]. Although the energy of the optical stimulus lay in the gap, it corresponded to a weak localized d-d transition. Thus real electronic transitions were excited and energy dissipation did occur. We also note that NiO has a complex strongly anisotropic magnetic structure, which originates from the significantly unquenched orbital momentum [7] resulting in strong spin-orbit coupling.

These results raise some fundamental questions: which of the optical or the THz stimulus provides the most efficient excitation of ultrafast spin dynamics? And, are magnetic anisotropy and absorption necessary requirements for opto-magnetism?

We address these questions, by an experimental investigation of the ultrafast optically induced spin dynamics in $KNiF_3$. This cubic crystal has the perovskite

 
J.-Y. Bigot et al. (eds), *Ultrafast Magnetism I*, Springer Proceedings in Physics 159
DOI 10.1007/978-3-319-07743-7_69

type crystal structure (point group m3m) [8], while the magnetic structure consists of two sublattices, ordered up to $T_N$ = 246 K [9]. This material is known as the *perfect Heisenberg antiferromagnet* because of its cubic magnetic structure, due to a very weak magnetic anisotropy. This aspect, combined with the high transparency in the optical range (band gap $E_g \approx$ 6.2 eV [8]), makes $KNiF_3$ a very suitable system to investigate the role of dissipation and anisotropy in the opto-magnetic effect.

In a pump-probe experiment we excited ultrafast spin dynamics via ISRS and we studied the dependence of the dynamics on an external magnetic field. The results are summarized in Figure 1. The photon energies of the pump and probe were 2.6 eV and 1.5 eV, respectively. The fluence of the pump beam was set to ≈ 14.5 mJ/cm$^2$.

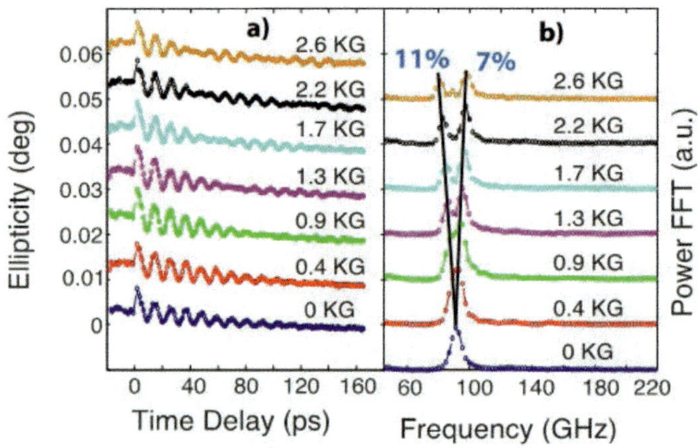


**Fig. 1.** Field dependence of the opto-magnetic effect in the time domain (a) and in the frequency domain (b). The magnetic field was applied in plane. The frequency shifts relative to the zero-field data induced by the 2.6 KG field, which amount to 7% and 11%, are reported.

In the experiment we measured changes in the ellipticity of the probe polarization, acquired by the light upon propagation through the sample. The ellipticity represents the magnetic linear birefringence in the material. Oscillations at a frequency of ≈ 90 GHz are observed and must be assigned to the antiferromagnetic resonance (AFM). In an external field beatings appear in the dynamics, indicating that at least two modes are excited. From the Fourier transforms in panel (b) two peaks are clearly visible. The magnitude of the magnetic field determines the splitting between these two spectral features. Considering the field dependence of the AFM frequency [10] in $KNiF_3$, a relative shift of 7.5 % of the zero-field frequency is expected for an applied field of 2.6 KG. Our data show an excellent agreement with this prediction, revealing a 7% and 11% frequency shift. We could excite and detect spin waves with such a high sensitivity that we observed the split of the two AFM modes in a field as small as 1.3 mT. To the best of our knowledge, this is the lowest field at which the splitting of the antiferromagnetic resonance has ever been observed.

Our experiment shows that absorption and magnetic anisotropy are not necessary requirements for an optical generation of coherent magnons. We also note that the ISRS mechanism provides a broader excitation bandwidth, given the indirect coupling between light and spins, than in the case of resonant THz pumping.

**Acknowledgments** The authors thank A. van Roij and T. Toonen for the technical support. This research was partially supported by de Nederlandse Organisatie voor Wetenschappelijk Onderzoek (NWO), de Stichting voor Fundamenteel Onderzoek der Materie (FOM), the European Union's Seventh Framework Program (FP7/2007-2013) Grants No. NMP3-LA-2010-246102 (IFOX), No. 280555 (Go-Fast), No. 214810 (FANTOMAS), the European Research Council under the European Union's Seventh Framework Program (FP7/2007-2013)/ERC Grant Agreement No. 257280 (Femtomagnetism) as well as the program "Invited Scientist" funded by the Russian Ministry of Education and Science (Grant Agreement 14.B37.21.0899). R.V.P. and A.M.K. acknowledge the support from the Russian Ministry for Education and Science (grant No. 14.B25.31.0025). A.M.K. thanks RFBR for support under the grant No. 12-02-31508.

## References

[1] E. Beaurepaire, J.-C. Merle, A. Daunois, and J.-Y. Bigot, "Ultrafast Spin Dynamics in Ferromagnetic Nickel", Phys. Rev. Lett. **76**, 4250 (1996).

[2] J.-Y. Bigot, "Femtosecond magneto-optical processes in metals", C. R. Acad. Sci. Ser. IV 2, 1483-1504 (2001).

[3] J.-Y. Bigot, L. Guidoni, E. Beaurepaire, and P. N. Saeta, "Femtosecond Spectrotemporal Magneto-optics" Phys. Rev. Lett. **93,** 077401 (2004).

[4] Andrei Kirilyuk, Alexey V. Kimel, and Theo Rasing, "Ultrafast optical manipulation of magnetic order" Rev. Mod. Phys. **82**, 2731 (2010).

[5] T. Satoh, S.-J. Cho, R. Iida, T. Shimura, K. Kuroda, H. Ueda, Y. Ueda, B. A. Ivanov, F. Nori, and M. Fiebig, "Spin Oscillations in Antiferromagnetic NiO Triggered by Circularly Polarized Light" Phys. Rev. Lett. **105**, 077402 (2010).

[6] T. Kampfrath, A. Sell, G. Klatt, A. Pashkin, S. Mährlein, T. Dekorsy, M. Wolf, M. Fiebig, A. Leitenstorfer & R. Huber, "Coherent terahertz control of antiferromagnetic spin waves" Nature Photonics **5**, 31 (2010).

[7] V. Fernandez, C. Vettier, F. de Bergevin, C. Giles, and W. Neubeck, "Observation of orbital moment in NiO" Phys. Rev. B. **57**, 7870 (1998).

[8] Y. Ivanov, E. A. Zhurova, V. V. Zhurov, K. Tanaka and V. Tsirelson, "Electron density and electrostatic potential of KNiF3: multipole, orbital and topological analyses of vacuum-camera-imaging plate and four-circle diffractometer data." Acta. Cryst. B **55**, 923 (1999).

[9] J. Nonet, A. Zarembowitch, R. V. Pisarev, J. Ferre, and M. Lecomte, "Determination of TN for KNiF3 through elastic, magneto-optical, and heat capacity measurements" Appl. Phys. Lett. **21**, 161 (1972).

[10] C. Kittel, "Theory of Antiferromagnetic Resonance" Phys. Rev. **82**, 565 (1951).

# Classical Modeling of Coherent Ultrafast Demagnetization Experiments

**Y. Hinschberger and P.-A. Hervieux**

Institut de Physique et Chimie des Matériaux de Strasbourg, CNRS and University of Strasbourg, 23 rue du Loess, BP 43, 67034 Strasbourg Cedex 2, France

Yannick.Hinschberger@ipcms.u-strasbg.fr and Hervieux@unistra.fr

**Abstract** A classical model based on the anharmonic Drude-Voigt model and the Maxwell's equations is developed for modeling ultrafast nonlinear coherent magneto-optical experiments performed on ferromagnetic thin films. Our theoretical predictions of the Faraday angles are in good agreement with available experimental data.

## Introduction

A spin-orbit like mechanism resulting from the direct interaction between the external field of the laser pulse and the electron spins of the material sample has been recently evoked for explaining the nonlinear Faraday measurements performed on Ni thin films [1]. After recalling the principle of the single-femtosecond-pulse Faraday experiment realized in [1], we present a nonlinear extension of the usual circular refractive indices based on a classical model which includes the spin-orbit interaction and needed for the modeling of the Faraday rotation angles. Comparisons between our theoretical predictions and the experimental data of [1] are presented.

## Nonlinear Faraday experiment and theoretical modeling

Faraday rotation $\theta$ and ellipticity $\eta$ can be related to the complex phase $\Phi$ acquired during the propagation of a polarized light of pulsation $\omega$ through a magnetized medium of thickness $d$. It represents the phase shift between the right $(+)$ and the left $(-)$ circular polarization states of the incident light and can be expressed as

$$\Phi = \frac{\omega d}{2c}(n_+ - n_-) = \theta + i\eta \,, \qquad 1$$

where $n_\pm^2 = \varepsilon_{xx} \pm i\varepsilon_{xy}$ are the refractive indices of the circular states $(\pm)$, $\varepsilon_{xx}$ and $\varepsilon_{xy}$ are respectively, diagonal and non-diagonal elements of the dielectric tensor $[\varepsilon_r]$ defined by $\boldsymbol{D} = \varepsilon_0 \boldsymbol{E} + \boldsymbol{P} = \varepsilon_0[\varepsilon_r]\boldsymbol{E}$. In [1], a 48 femtosecond laser pulse is focused on a ferromagnet-

 
J.-Y. Bigot et al. (eds), *Ultrafast Magnetism I*, Springer Proceedings in Physics 159
DOI 10.1007/978-3-319-07743-7_70

ic nickel thin film ($d$=7.5 nm) and, the rotation and ellipticity of the transmitted beam are measured. It was observed that these latter quantities exhibit a nonlinear dependence in the absorbed energy (or in the amplitude of the incident electric field). In the linear regime, the well-known expressions of the optical indices $n_\pm$ are incomplete because they do not exhibit any dependence in the laser electric field amplitude. In order to extend the modeling to the nonlinear regime, we have derived nonlinear circular refractive indices from the Maxwell's equation (similarly to the linear case) by adding a third-order polarization current $\boldsymbol{j} = \partial_t \boldsymbol{P}^{(3)}$ and a magneto-optical current coming directly from the spin-orbit interaction. Thus, we have been able to obtain new expressions of the refractive indices depending on the square and the fourth power of the amplitude of the incident electric field and on the third-order dielectric response functions

$$n_\pm^2 = f\left(\varepsilon_{xxxx}, \varepsilon_{xxxy}, E^2, E^4\right).$$

This expression is used to calculate the rotation $\theta$ of Eq. (**1**). The tensor elements are determined by employing a classical anharmonic Drude-Voigt model and by calculating the first- and third-order dielectric polarizations of the bound and free charge carriers of nickel. Finally, magnetic properties of the sample are incorporated in the model by using the Weiss molecular field approach.

## Results and Conclusion

The normalized rotation $\theta/\theta^{min}$ ($\theta^{min}$corresponds to the lowest absorbed energy) as a function of the absorbed energy has been calculated and compared to the experimental data reported in [1]. Since nickel has a band structure, the calculations have been performed by substituting the electron mass $m_e$ by an effective mass $m_{\text{eff}}$. Our theoretical predictions of $\theta/\theta^{min}$ depicted in **Fig. 1** exhibit a nonlinear shape very similar to the one which is experimentally observed.

The main result of the present work is that a classical nonlinear theory seems to be well-adapted to describe nonlinear magneto-optical effects and anisotropic phenomena. The present model can also be improved by calculating the dielectric response functions within a quantum mechanical framework or/and by adding a magnetization dynamics which is presently missing. The difference between the experimental measurements and our theoretical predictions could be attributed to these improvements.

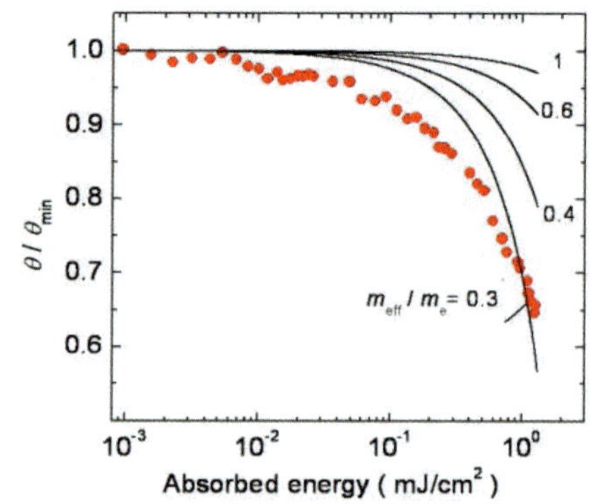


**Fig. 1.** Normalized rotation angle as a function of the absorbed energy for different values of the effective mass: $m_{eff}/m_e$=1, 0.6, 0.4 and 0.3. Red circles are the experimental data reported in [1].

**Acknowledgments** The authors would like to thank J. -Y. Bigot, H. Vonesch and M. Vomir.

# References

[1] J.-Y. Bigot, M. Vomir and E. Beaurepaire, "Coherent ultrafast magnetism induced by femto-second laser pulses" Nature Physics **5**, 515 (2009).

# Part VI
# Ultrafast Magnetism Control

# Sub-nanosecond Heat Assisted Magnetic Recording of FePt Media

**D. Weller, O. Mosendz, H.J. Richter, G. Parker, S. Pisana, T.S. Santos, J. Reiner, O. Hellwig, B. Stipe and B. Terris**

HGST a Western Digital Company, San Jose Research Center, San Jose, CA USA

dieter.weller@hgst.com

**Abstract** Chemically ordered and textured L10 FePtX-Y (001) granular media with high perpendicular anisotropy are being developed for future high areal density heat-assisted magnetic recording (HAMR) applications. Proper heat sink layers optimize the recording time window down to 0.1-0.2 ns and achievable areal densities beyond1Tb/in2.

## Introduction

Current perpendicular magnetic recording (PMR) based on CoCrPt media is supposed to achieve areal densities (AD) of 1-1.5 $Tb/in^2$. Heat-assisted magnetic recording (HAMR) based on FePt media is regarded to be the most promising option to extend AD beyond 1.5 $Tb/in^2$ [1-3]. To write bits in HAMR, laser light is focused with a near field transducer (NFT) down to spot sizes below 50 nm in diameter. The media grains experience very short heat pulses at temperatures of ~ 850K, which are ~100K beyond the Curie temperature ($T_C$). During this heat pulse a magnetic field orients the magnetization. Typical data rates of 1-2 Gbits/s require short time recording below one nanosecond at elevated temperatures necessitating proper understanding of ultrafast magnetization reversal [4].The research on HAMR media is focused on chemically ordered $L1_0$ FePtX-Y (001) alloys with sufficient magnetic anisotropy to allow small thermally stable grains down to $D_G$=4-5nm diameter relative to $D_G$=7-8nm in CoCrPt based alloys [5-10]. Seagate reported a high HAMR areal density of 1$Tb/in^2$ with linear density=1975 kfci and track density=510 ktpi in 2012 [2, 3]. They used, highly textured FePtX-Y (001) media with an average grain size $\langle D_g \rangle$=9 nm and grain pitch $\langle D_p \rangle$=10.5 nm. Going forward, HAMR heads, media, head-media-spacing (HMS) and read-back channel technologies need to improve to extend AD to 1.5-5 $Tb/in^2$ [11]. Alternative technologies to HAMR are Microwave Assisted Magnetic Recording (MAMR) [12, 13], Bit Patterned Magnetic Recording (BPMR) [14, 15] and two-dimensional magnetic recording (TDMR) [16, 17]. Extensions beyond 3-5 $Tb/in^2$ will likely combine HAMR with BPMR [18].



J.-Y. Bigot et al. (eds), *Ultrafast Magnetism I*, Springer Proceedings in Physics 159
DOI 10.1007/978-3-319-07743-7_71

## Media Structure

The basic media structure for HAMR recording is shown in Fig. 1. Analogous to conventional recording media, the magnetic layer needs to be grown on a suitable seedlayer, which ensures proper texturing of the magnetic grains, where the magnetically easy axes should be oriented perpendicularly to the film plane. The magnetic layers need to be protected by a carbon overcoat and a lubrication layer. A new feature for HAMR media is the heatsink, which has the function to prevent "blooming" of the heatspot.

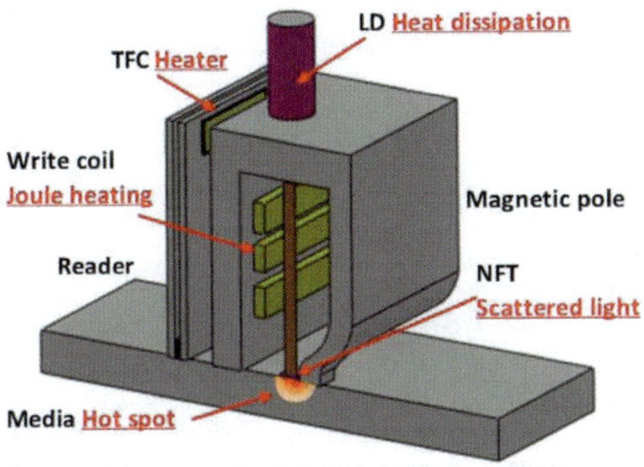


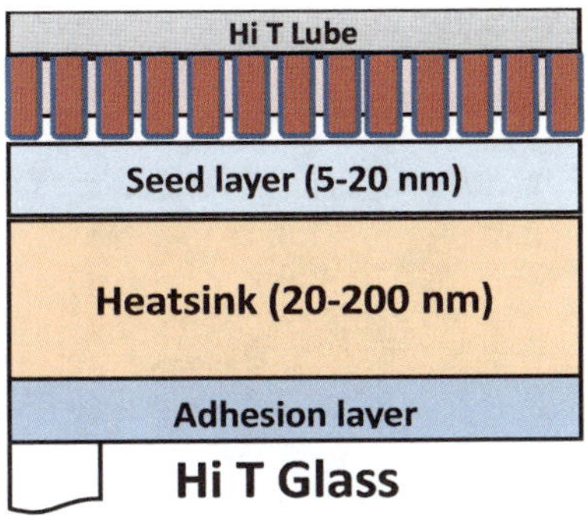


**Fig. 1.** HAMR head (left) [19] and media (right) [10]. Laser diode (LD) and near field transducer (NFT) induce a small hot spot in the media. Thermal fly-height control (TFC) is critical. The media are based on granular FePt (001) separated by segregants (e.g. Carbon) on texture (e.g. MgO), heatsink (e.g. Cr) and adhesion (e.g. NiTa) layers. High T lube and overcoat is needed.

To be technologically useful, HAMR recording must be able to sustain sufficiently high data rates. The presence of the high temperature during recording results in a new set of challenges. It is well known that high-density recording requires narrow magnetization transitions, which can only be created with a high effective field gradient during writing. The high write field gradient is created by the temperature profile of the heat spot. Too high temperature gradients may be detrimental for HAMR. As an example, consider a recorded track of 50 nm width, where for simplicity, we assume a circular heat spot. As a consequence of the temperature dependence of the anisotropy field [20], the recording temperature is only 1-2% lower than $T_C$. From spin-stand measurements [21], the thermal gradient is determined to be of the order of 5-10K/nm and consequently the distance between the contours of the writing and the Curie temperatures is estimated to be between 0.7 to 3nm. With a typical linear velocity of 15m/s, this gives the magnetization about 0.05 to 0.2 ns to be recorded meaning that the higher the thermal gradient, the less time the magnetization has to orient along the recording field. In other words, it is possible to freeze in the (almost) random state that naturally occurs when the temperature exceeds $T_C$. On the other hand, poor thermal gradients result in wide recording zones, which also deteriorate recording quality.

A good design for HAMR media requires careful thermal engineering, where a heat-sink is located underneath the magnetic and the seed layer (Fig.1). The heat-sink concentrates the heat on a small area and enhances the thermal gradient. In addition, proper thermal conductivity of the magnetic layer is important. Fig. 2 shows plan view and cross sectional TEM images of a $L1_0$ FePt medium. A small mean grain size of 6.3 nm but a rather wide grain size distribution of $\sigma_D/\langle D\rangle=26\%$ is found [10]. So far, a smaller grain size distribution has been achieved only for larger grains e.g. $\sigma_D/\langle D\rangle=16\%$ and $D_G=7.2$ nm [7]. A coercivity as high as $H_C=5.2$T and anisotropy beyond $K_u>4.5\times10^7$ erg/cm$^3$ is possible. Hence, the small grains exhibit sufficient thermal stability for recording media applications [7-10].

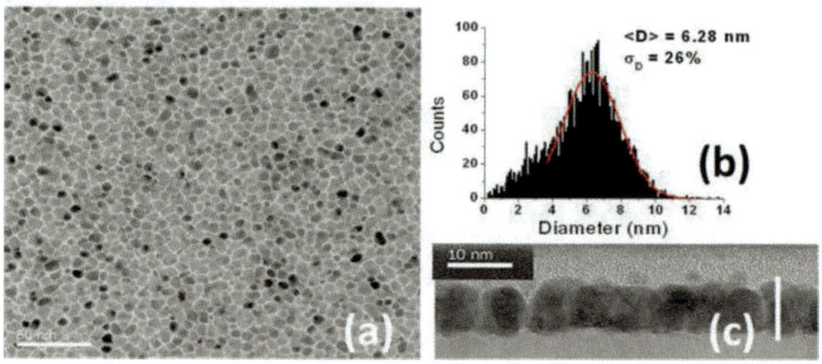


**Fig. 2.** (a) Plan view TEM image, (b) histogram of the grain size distribution, and (c) cross-sectional TEM image of 10 nm thick FePt-Y granular media. The grain aspect ratio is $\delta/D=1.5$ (from ref. [10]).

Going forward the roadmap among other parameters requires many improvements to enhance AD beyond 1.5 Tb/in2. It is particularly important to reduce distributions, in particular of TC and grain size of the media [11].

## References

[1] M. H. Kryder, E. C. Gage, T. W. McDaniel, W. A. Challener, R. E. Rottmayer, G. Ju, Y.-T. Hsia, and M. F. Erden, "Heat Assisted Magnetic Recording" Proc. IEEE vol **96**, 1810 (2008).

[2] A. Q. Wu, Y. Kubota, T. Klemmer, T. Rausch, C. Peng , Y. Peng, D. Karns, X. Zhu, Y. Ding, E. KC Chang, Y. Zhao, H. Zhou, K. Gao, J.-U. Thiele, M. Seigler, G. Ju, and E. Gage," HAMR Areal Density Demonstration of 1+ Tbpsi on Spinstand" IEEE Trans Mag. **49**, 779 (2013).

[3] X. Wang, K. Gao, H. Zhou, A. Itagi, M. Seigler, E. Gage, "HAMR Recording Limitations and Extendibility" IEEE Trans Mag. **49**, 686 (2013).

[4] W. A. Challener , C. Peng, A. V. Itagi, D. Karns, W. Peng, Y. Peng, XM. Yang, X. Zhu, N. J. Gokemeijer, Y.-T. Hsia, G. Ju, R. E. Rottmayer, M. A. Seigler and E. C. Gage, "Heat-assisted magnetic recording by a near-field transducer with efficient optical energy transfer" Nat. Photonics **3**, 220 (2009).

[5] R. E. Rottmayer, S. Batra, D. Buechel, W. A. Challener, J. Hohlfeld, Y. Kubota, Lei Li, Bin Lu, C. Mihalcea, K. Mountfield, K. Pelhos, C. Peng, T. Rausch, M. A. Seigler, D. Weller, and XM Yang "Heat-Assisted Magnetic Recording" IEEE Trans. Mag. **42**, 2417 (2006).

[6] D. Weller and T. McDaniel ""Media for extremely high density recording" Advanced Magnetic. Nanostructures, eds D. Sellmyer and R. Skomski, Springer 2007, Chapter 11

[7] O. Mosendz, S. Pisana, J. W. Reiner, B. Stipe, and D. Weller, "Ultra-high coercivity small-grain FePt media for thermally assisted recording" J. Appl. Phys. **111**, 07B729 (2012).

[8] B.S.D.Ch.S.Varaprasad, M. Chen, Y.K. Takahashi and K. Hono, "L10 ordered FePt based perpendicular recording media for heat assisted magnetic recording", IEEE Trans Mag, **49**, 718 (2013).

[9] S. Pisana, O. Mosendz, G.J. Parker, J.W. Reiner, T.S. Santos, A.T. McCallum, H.J. Richter, D.Weller, "Effects of grain microstructure on magnetic properties in FePtAg-C media for temperature assisted recording", J. Appl. Phys. **113**, 043910 (2013).

[10] D. Weller, O. Mosendz, G. Parker, S. Pisana and T. S. Santos, "L10 FePtX-Y media for heat assisted magnetic recording" Phys. Status Solidi A, **210**, 1245 (2013).

[11] D. Weller, G. Parker, O. Mosendz, E. Champion, B. Stipe, X. Wang, T. Klemmer, G. Ju, A. Ajan, "The HAMR Media Technology Roadmap to an Areal Density of 4 Tb/in2" (submitted for publication).

[12] J.-G. Zhu and Y. Wang, "Microwave Assisted Magnetic Recording Utilizing Perpendicular Spin Torque Oscillator With Switchable Perpendicular Electrodes" IEEE Trans. Mag. **46**, 751 (2010).

[13] J.-G. Zhu, M. Mallary, S. Hinata, S. Saito, M. Takahashi, "Ferromagnetic resonance analysis of internal effective field of classified grains by switching field for granular perpendicular recording media", J. Appl. Phys. **111**, 07B722 (2012).

[14] H.J. Richter, A.Y. Dobin, R.T. Lynch, D .Weller "Recording potential of bit-patterned media" Appl Phys Lett **88,** 222512, 119, (2006).

[15] T. R. Albrecht, D. Bedau, E. Dobisz, He Gao, M. Grobis, O. Hellwig, D. Kercher, J. Lille, E. Marinero, K. Patel, R. Ruiz, M. E. Schabes, L. Wan, D. Weller and Tsai Wei-Wu, "Bit Patterned Media at 1 Tdot/in2 and Beyond" IEEE Trans Mag, **49**, 773 (2013).

[16] R. Wood, M. Williams, A. Kavcic and J. Miles "The Feasibility of Magnetic Recording at 10 Terabits Per Square Inch on Conventional Media" IEEE Trans. Mag. **45**, 917 (2009).

[17] Roger Wood, "Future Hard Disk Drive Systems", J. Magn. Magn. Mater. 321, 555 (2009).

[18] B. C. Stipe, T.C. Strand, Chie C. Poon, H. Balamane, T. D. Boone, J. A. Katine, J.-L. Li, V. Rawat, H. Nemoto, A. Hirotsune, O. Hellwig, R. Ruiz, E. Dobisz, D. S. Kercher, N. Robertson, T. R. Albrecht & B. D. Terris "Magnetic recording at 1.5 Pb m-2 using an integrated plasmonic antenna" Nat. Photonics **4**, 484 (2010).

[19] Lidu Huang, B. Stipe, M. Staffaroni, J.-Y. Juang, T. Hirano, E. Schreck and F.-Y. Huang, "HAMR Thermal Modeling Including Media Hot Spot", IEEE Trans. Mag. **49**, No. 6, June 2013.

[20] J.-U. Thiele, K. R. Coffey, M. F. Toney, J. A. Hedstrom, and A. J. Kellock, "Temperature dependent magnetic properties of highly chemically ordered Fe Ni Pt L10 films," J. Appl. Phys., **91**, 6595 (2002).

[21] H.J. Richter, C.C. Poon, G. Parker, M. Staffaroni, O. Mosendz, R. Zakai, and B.C. Stipe, "Direct Measurement of the Thermal Gradient in Heat Assisted Magnetic Recording ", IEEE Trans. Mag. 2013 (accepted for publication).

# Controlling ultrafast transport in magnetic heterostructures

**A.J. Schellekens and B. Koopmans**

Department of Applied Physics, Center for NanoMaterials (cNM), Eindhoven University of Technology, P.O. Box 513, 5600 MB Eindhoven, The Netherlands

a.j.schellekens@tue.nl

**Abstract** We report on recent experiments addressing the interplay between magnetic ordering and spin/heat transport after femtosecond pulsed laser excitation. First, we demonstrate ultrafast magnetic control of heat flow through a metallic spin valve. Second, we show an experimental indication that superdiffusive spin currents can exert a spin-transfer torque on unprecedented timescales.

## Introduction

The field of ultrafast magnetism has come a long way since the first observation of sub-picosecond demagnetization of Ni after femtosecond (fs) laser heating [1]. Where first it was only possible to create disorder in ferromagnetic materials, new phenomena like reversal of ferrimagnetic alloys [2,3] and the generation of large non-equilibrium spin currents [4] have shown to be possible by pulsed laser excitation. Further exploring magnetism and spintronics on the ultimate timescale is of utmost importance to both fundamental understanding of these processes as to improve the speed and efficiency of future spintronic devices.

Here we present measurements on two phenomena in the field of spintronics, being giant thermal magnetoresistance (GMTR) and spin-transfer torque (STT), which yet are unexplored at (sub-)ps timescales after pulsed laser excitation. First, we show that the demagnetization of a ferromagnetic thin film can be altered by a neighboring spin-valve, providing magnetic control of heat transport on unprecedented timescales. Second, we discuss a pilot study of the control of the orientation of a non-collinear magnetic bilayer by ultrafast STT.

## Ultrafast Giant thermal magnetoresistance

In metals thermal conductivity is governed by the free electrons, implying that thermal and electrical conductivity are related. In a magnetic bilayer separated by

 
J.-Y. Bigot et al. (eds), *Ultrafast Magnetism I*, Springer Proceedings in Physics 159
DOI 10.1007/978-3-319-07743-7_72

a conductive spacer layer the electrical conductivity is related to the mutual orientation of the magnetization in the magnetic layers, which is called giant magnetoresistance (GMR). The related effect, i.e. that the thermal conductivity of such a bilayer also depends on the mutual orientation, has already been demonstrated [5]. Here, we investigate this so called GMTR on ultrafast timescales after fs laser excitation, paving the way for ultrafast magnetic control of heat flow.

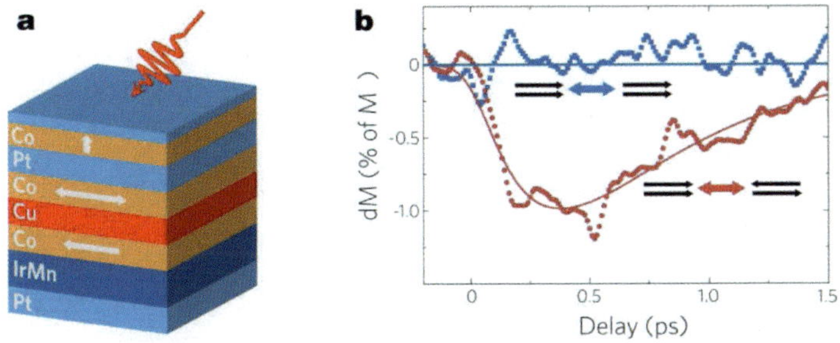


**Fig. 1.** (a) Sample grown for investigating ultrafast GMTR. (b) Change in demagnetization of the perpendicular Pt/Co/Pt layer by changing the orientation of the spin valve by an in-plane magnetic field, including a reference measurements where the applied field was marginally smaller than the switching field.

The idea behind the experiments is depicted in Fig. 1(a). An intense fs laser pulse heats a perpendicularly magnetized Pt/Co/Pt layer. The amount of heat transport is controlled by the Co/Cu/Co GMR stack below the top layer. In Fig. 1(b) the change in demagnetization between the parallel and anti-parallel configuration dM is plotted as a function of delay time as measured by time-resolved MOKE. When the in-plane field is too small to switch the GMR stack, no signal is observed. However, when the applied field exceeds a critical value, a clear change of about 1% in the demagnetization efficiency is measured. A negative value of dM means that the demagnetization is smaller in the parallel alignment. This is expected, since a parallel orientation means a large electrical and thermal conductivity, leading to faster transport of heat away from the top layer, finally resulting in a smaller demagnetization.

## Ultrafast spin-transfer torque

Although the presence of ultrafast spin currents after fs laser pulse excitation is well established [6], these currents have not yet been used to actually control the orientation of magnetic layers on ultrafast timescales. Here, we show preliminary experimental evidence that ultrafast spin currents can induce a torque on a magnetic layer on unprecedented timescales.

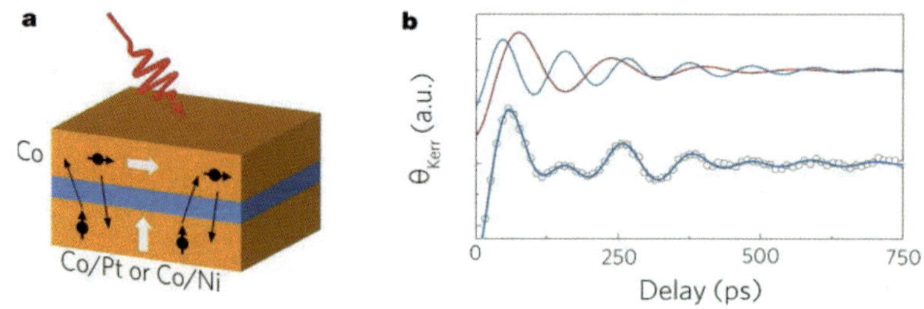


**Fig. 2.** (a) Sample grown to investigate ultrafast STT. (b) Time resolved Kerr measurements, revealing a precession of both the top and bottom magnetic layer. The lines through the data are fits with a double exponentially decaying sine, and the separate lines are the fitted individual precessions plotted with a slight offset for clarity.

In Fig. 2(a) the sample to investigate STT after pulsed laser excitation is depicted. Two non-collinear magnetic layers are separated by a conductive spacer layer, allowing for exchange of angular momentum. Due to the non-collinear alignment the transferred spins exert a torque on the magnetization, bringing the system out of equilibrium. After the ultrafast spin currents, the magnetization will precess back to equilibrium.

In Fig. 2(a) an example of the precessions induced by ultrafast STT is depicted. Two oscillations, corresponding to the top and bottom layer, can be clearly distinguished. The properties of these oscillations as a function of applied field, field angle, and spacer layer thickness suggest that superdiffusive spin currents from the bottom to top layer are at the origin of the precession of the top layer. The STT caused by superdiffusive spin currents can provide a method to switch the magnetization of magnetic bits on unprecedented timescales.

## References

[1] E. Beaurepaire, J.C. Merle, A. Daunois, J.-Y. Bigot, "Ultra-fast spin dynamics in ferromagnetic Nickel" Phys. Rev. Lett. **76**, 4250(1996).

[2] I. Radu *et al.*, "Transient ferromagnetic-like state mediating ultrafast reversal of antiferromagnetically coupled spins" Nature **472**, 205-208 (2011).

[3] A.J. Schellekens and B. Koopmans, "Microscopic model for ultrafast magnetization dynamics of multisublattice magnets" Phys. Rev. B **87**, 020407(R) (2013).

[4] M. Battiato, K. Carva, P.M. Oppeneer, "Superdiffusive Spin Transport as a Mechanism of Ultrafast Demagnetization" Phys. Rev. Lett. **105**, 027203 (2010).

[5] D. Daniela, J.F. Gregg, S.M. Thompson, J.M.D. Coey, A. Fagan, K. Ounadjela, C. Fermon, G. Saux, "Thermal conductivity of a giant magnetoresistive mechanical alloy" J. Magn. Magn. Mater. **140,** 493 (1995).

[6] G. Malinowski et al., "Control of speed and efficiency of ultrafast demagnetization by direct transfer of spin angular momentum" Nature Physics **4**, 855-858 (2008).

# Ultrafast Magnetoacoustics in Nickel

**Ji-Wan Kim, Mircea Vomir, and Jean-Yves Bigot**

Institut de Physique et Chimie des Matériaux de Strasbourg, UMR 7504, CNRS, Université de Strasbourg, BP 43, 23 rue du Loess, 67034 Strasbourg Cedex 02, France

bigot@ipcms.u-strasbg.fr

**Abstract** We report that picosecond acoustic pulses generated by femtosecond laser pulses can induce a change of the magnetization in ferromagnetic nickel films. We show that this effect can be used to manipulate the amplitude of the magnetization precession via a change in the magnetic anisotropy. The conditions for realizing magneto-acoustic devices will be discussed.

## Introduction

Intense researches are being carried out for manipulating the spins in magnetic materials in the fastest and most efficient manners. The way of using femtosecond laser pulses, named "femtomagnetism", is at the heart of ultrafast spin control [1-5]. Recently, several groups have shown an alternative way of manipulating the magnetization using ultrafast acoustic pulses in various materials [6-9]. In the present talk we will describe these methods, focusing on the magnetoacoustics induced in Ni thin films by femtosecond laser pulses.

## Results and discussion

The experiment was performed using the time-resolved magneto-optical Kerr effect. The pump pulses (150 fs, 397 nm) excite the front side of the sample and the variation in the reflectivity $\Delta R/R(t)$, the magneto-optical Kerr rotation $\Delta\theta/\theta(t)$, and ellipticity $\Delta\varepsilon/\varepsilon(t)$ are measured with two probe pulses (120 fs, 794 nm) on both the front and the rear sides of the film. The sample is a polycrystalline film of Ni (200 nm) deposited on a glass substrate.

Figure 1(a) shows the calculated strain profile $\eta$ at $t = 30$ ps induced by a pump pulse with an energy density $I_p = 1.7$ mJcm$^{-2}$. The shape of the strain pulse is determined by the temperature profiles calculated by three coupled baths (electrons, spins, and lattice) and a one-dimensional wave equation [6] that models the acoustic echoes propagating back and forth in the Ni film. The solid line in the in-

 
J.-Y. Bigot et al. (eds), *Ultrafast Magnetism I*, Springer Proceedings in Physics 159
DOI 10.1007/978-3-319-07743-7_73

set of Fig. 1(a) shows a fit of the reflectivity on the front side of the film using this model. The magneto-optical responses $\Delta\theta/\theta_s$ (closed circle) and $\Delta\varepsilon/\varepsilon_s$ (open circle) measured at the rear side of Ni are displayed in Fig. 1(b) for the magnetic field angle $\phi = 25°$. They show that the magnetization is first brought out of equilibrium by the strain pulse at time $T_R/2$ ($T_R$: round trip time of acoustic pulse), therefore initiating a motion of precession. The amplitude and the frequency of the precession motion have similar values for $\Delta\theta/\theta_s$ and $\Delta\varepsilon/\varepsilon_s$. The amplitude of precession is increased after $t = 3T_R/2$. In contrast, $\Delta\theta/\theta_s$ and $\Delta\varepsilon/\varepsilon_s$ slightly differ at $t = T_R/2$. We attribute this difference to the magneto-optic spectral shape. Indeed, at the photon energy 1.5 eV, the static rotation $\theta(\omega)$ is almost flat while $\varepsilon(\omega)$ has an abrupt variation. The polarities of the signals depend on the external field and can even have an opposite sign as shown for $\phi = 0°$ in Fig. 1(c).

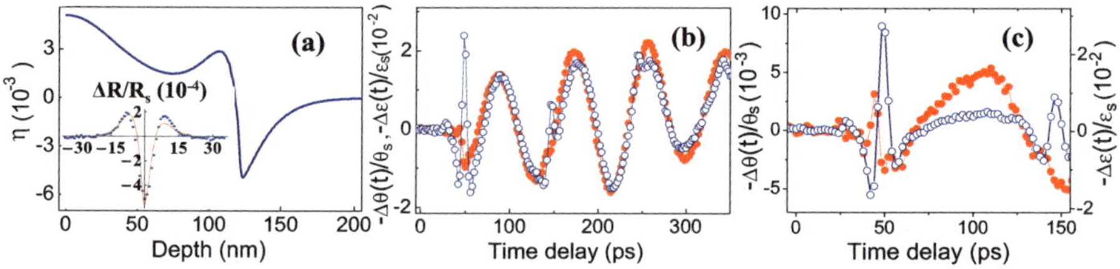


**Fig. 1.** (a) Model of the strain profile at $\boldsymbol{t}$ =30 ps. Inset: Model fit (solid curve) of the acoustic echo on the front side of the film. (b) Differential Kerr rotation $\boldsymbol{\Delta\theta/\theta_s}$ (closed circle) and $\boldsymbol{\Delta\varepsilon/\varepsilon_s}$ (open circle) probed on the rear side of the film with $\boldsymbol{\phi = 25°}$. (c) Same as (b) for $\boldsymbol{\phi = 0°}$.

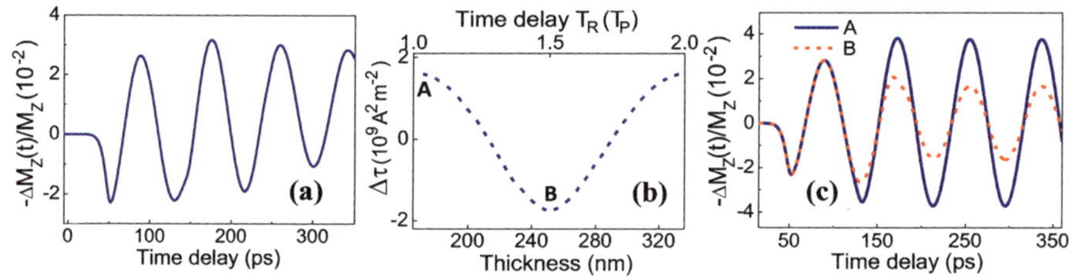


**Fig. 2.** Simulation of the magnetoacoustic response. (a) $\boldsymbol{\Delta M_z/M_z}$ for $\boldsymbol{\phi}$ =25°. (b) $\boldsymbol{\Delta\tau}$ calculated as a function of $\boldsymbol{l}$ (bottom abscissa) or $\boldsymbol{T_R}$ (top abscissa). (c) $\boldsymbol{\Delta M_z/M_z}$ for two $\boldsymbol{T_R}$ values.

To understand the magnetoacoustic dynamics quantitatively, we calculate the perturbation induced by the strain pulse $\eta$ considering the magnetoelastic energy in the Landau-Liftshitz-Gilbert equation (see [6] and references therein). Figure 2(a) shows the calculated motion of precession triggered by the strain pulse for $\phi = 25°$. Since the spectral variation of the Kerr signals, due to the strain, is not taken into account, the simulation curve is similar to the measured signal $\Delta\theta/\theta_s$ (fig. 1(b)) which has no sharp spectral feature. Importantly, the precession amplitude is also increased after the arrival of the 1st order echo as pointed out in Fig. 1(b). To account for this effect, we define $\Delta\tau$ as the difference between the torque value $\tau = |M \times H_{eff}|$ after the 0th and 1st order echoes. As shown in Fig. 2(b), the variation of $\Delta\tau$ as a function of the film thickness $l$ is then calculated assuming a Gilbert damping parameter $\alpha = 0$ for simplicity. For $l = 200$ nm, the fact that $\Delta\tau > 0$ indicates that the precession amplitude increases after the 1st order echo. In Fig. 2(c), the precession motion is shown for two points (A and B) marked in Fig. 2(b). We clearly see that the maximum (minimum) value of $\Delta\tau$ produces a maximum (minimum) change of the precession amplitude. By introducing the precession period $T_P$, we can find a simple relation for the variation of the precession amplitude and the strain echoes: $T_R = mT_P$, corresponding to constructive (destructive) effect for even (odd) number of m. This simple relation is useful for the design of magnetoacoustic devices as will be discussed in the presentation.

In conclusion we have shown that acoustic pulses generated by femtosecond laser pulses in magnetic metallic films induce a motion of precession of the magnetization which can be controlled by matching the round trip time of the strain pulse echoes with the period of precession. These results open a new way of controlling magneto-acoustic devices at room temperature.

**Acknowledgments** The authors acknowledge the financial support of the European Research Council with the ERC Advanced Grant ATOMAG (ERC-2009-AdG-20090325 247452).

## References

[1] E. Beaurepaire, J.-C. Merle, A. Daunois, and J.-Y. Bigot, "Ultrafast Spin Dynamics in Ferromagnetic Nickel" Phys. Rev. Lett. **76**, 4250 (1996).

[2] G. Zhang, W. Hübner, E. Beaurepaire, and J.-Y. Bigot, "Laser-induced ultrafast demagnetization: femtomagnetism, a new frontier?" in "Spin Dynamics in Confined Magnetic Structures I", (Springer, 2002), Chap. 8, pp. 245-290.

[3] J.-Y. Bigot, M. Vomir, and E. Beaurepaire, "Coherent ultrafast magnetism induced by femtosecond laser pulses" Nature Phys. **5**, 515 (2009).

[4] M. Vomir, L. H. F. Andrade, L. Guidoni, E. Beaurepaire, and J.-Y. Bigot, "Real Space Trajectory of the Ultrafast Magnetization Dynamics in Ferromagnetic Metals" Phys. Rev. Lett. **94**, 237601 (2005).

[5] F. Hansteen, A. Kimel, A. Kirilyuk, and Th. Rasing, "Femtosecond Photomagnetic Switching of Spins in Ferrimagnetic Garnet Films" Phys. Rev. Lett. **95**, 047402 (2005).

[6] J.-W. Kim, M. Vomir, and J.-Y. Bigot, "Ultrafast Magnetoacoustics in Nickel Films" Phys. Rev. Lett. **109**, 166601 (2012).

[7] A. V. Scherbakov, A. S. Salasyuk, A. V. Akimov, X. Liu, M. Bombeck, C. Brüggemann, D. R. Yakovlev, V. F. Sapega, J. K. Furdyna, and M. Bayer, "Coherent Magnetization Precession in Ferromagnetic (Ga,Mn)As Induced by Picosecond Acoustic Pulses" Phys. Rev. Lett. **105**, 117204 (2010).

[8] L. Thevenard, E. Peronne, C. Gourdon, C. Testelin, M. Cubukcu, E. Charron, S. Vincent, A. Lemaître, and B. Perrin, "Effect of picosecond strain pulses on thin layers of the ferromagnetic semiconductor (Ga,Mn)(As,P)" Phys. Rev. B **82**, 104422 (2010).

[9] O. B. Wright and V. E. Gusev, "Ultrafast generation of acoustic waves in copper" IEEE Trans. Ultrason. Ferroelectr. Freq. Control **42**, 331 (1995).

# Thermally assisted all-optical helicity dependent switching of ferrimagnetic amorphous $Fe_{100-x}Tb_x$ thin films

**A. Hassdenteufel[1], B. Hebler[1], C. Schubert[1], A. Liebig[1], M. Teich[2], J. Schmidt[2], M. Helm[2], M. Aeschlimann[4], M. Albrecht[1] and R. Bratschitsch[1, 4]**

[1]Institute of Physics, Chemnitz University of Technology, D-09107 Chemnitz, Germany

[2]Helmholtz Zentrum Dresden-Rossendorf, P.O. Box 510119, D-01314 Dresden, Germany and TU Dresden, D-01062 Dresden, Germany

[3]Department of Physics and Research Center OPTIMAS, University of Kaiserslautern, D-67663 Kaiserslautern, Germany

[4]Institute of Physics, University of Münster, Wilhelm-Klemm-Str. 10, D-48149 Münster, Germany

alexander.hassdenteufel@physik.tu-chemnitz.de

**Abstract** We observe all-optical switching (AOS) in ferrimagnetic $Fe_{100-x}Tb_x$ alloy films below, above, and in samples without a magnetic compensation point. AOS is linked to a low remanent magnetization $M_R$ and associated with laser heating up to the Curie temperature. Above $M_R$ = 220 emu/cc AOS is replaced by pure thermal demagnetization.

## Introduction

Magnetization switching is at the heart of both modern information storage technology and fundamental science. Ultrafast laser pulses are promising to explore and finally reach the ultimate speed limit of this process without applying an external magnetic field. The process of ultrafast all-optical switching (AOS) has first been observed in thin GdFeCo alloy films [1]. A multitude of experimental and theoretical investigations have been performed with this material system [2]. However, for technological applications and for eventually gaining insight into the underlying physical mechanism, finding alternative AOS materials is of utmost importance.

We present a new material system exhibiting all-optical switching [3]. Twelve samples of Fe100-xTbx alloys with Tb content between 19 at.% and 38.5 at.% are fabricated under ultrahigh vacuum conditions by magnetron co-sputtering and carefully characterized by SQUID-VSM measurements, Rutherford backscattering, and X-ray diffraction. All of them reveal a strong uniaxial perpendicular



J.-Y. Bigot et al. (eds), *Ultrafast Magnetism I*, Springer Proceedings in Physics 159
DOI 10.1007/978-3-319-07743-7_74

magnetic anisotropy and an amorphous structure. Two of the samples (x = 24 at.% and 27 at.%) exhibit a magnetic compensation point at $T_{comp} = (218.5 \pm 1.0)$ K and $T_{comp} = (345.5 \pm 1.0)$ K, respectively.

For the switching investigations an ultrafast regenerative laser amplifier system with a pulse repetition rate of 250 kHz and a pulse duration of 100 fs at the sample position is used. The orientation of sample magnetization is observed in situ by a Faraday microscope, where opposite magnetic states appear as black ($M^-$) and gray ($M^+$) areas (Fig. 1 a, b). The polarization of the switching pulses is altered by a zero order quarter waveplate. The AOS experiments are performed as follows: first, the sample is homogeneously magnetized in one direction by applying an external magnetic field (Fig. 1a). Second, the switching pulses are focused onto the sample surface, and the sample is subsequently moved below the laser spot in one direction, resulting in the formation of a narrow stripe domain with opposite magnetization direction compared to the surrounding material (Fig. 1b). Finally, by applying the contrary light polarization this written domain structure may also be "erased" (Fig 1c). Interestingly, AOS is found below, above, and in samples without magnetic compensation point. If the erasing process fails, a helicity independent pure thermally demagnetized (PTD) state is witnessed, as observed below 22 at.% and above 34 at.% Tb. The reason for this observation is the increase of remanent sample magnetization $M_R$ and hence the presence of strong magnetic stray fields. Figure 1 d summarizes this finding: the AOS region ($M_R$ below 220 emu/cc) is clearly separated and shows a different slope than the PTD region. Using the two temperature model, we calculated the equilibrium temperature of the electronic system and the lattice after laser heating and found them to be identical to the Curie temperature in this material. Therefore, we conclude that heating up to the Curie temperature is necessary for all-optical switching. But heating alone is not sufficient for AOS in $Fe_{100-x}Tb_x$ thin films [3]. The striking difference of excitation with linearly (pure heating) and circularly polarized light is depicted in Fig. 2. Thermally induced laser depinning of the domain wall is witnessed for linearly polarized light. Circularly polarized light induces AOS.

In summary, the impetus for helicity dependent AOS in FeTb is determined by the properties of the laser pulses, but the ability for AOS is an intrinsic material property, given by the remanent sample magnetization $M_R$.

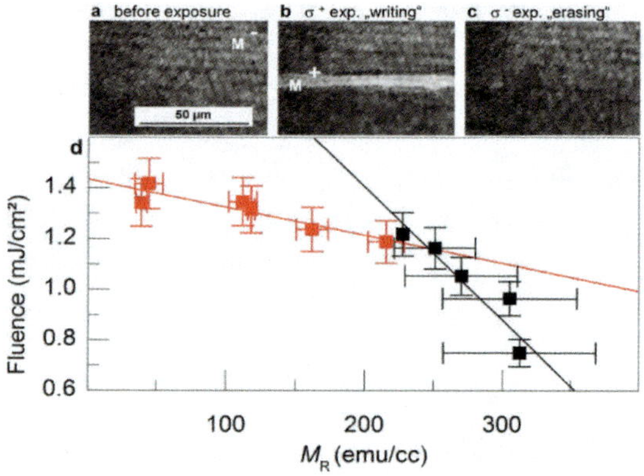


**Fig. 1.** All-optical switching in $Fe_{70}Tb_{30}$. **a** Homogeneously magnetized sample with magnetization pointing into the sample plane ($M^-$). **b** Irradiation with right handed polarized light ($\sigma^+$) at threshold fluence for AOS: $F_{AOS} = (1.20 \pm 0.08)$ mJ/cm². **c** "Erasing" by using the same fluence as in b but opposite helicity ($\sigma^-$). **d** Threshold fluence for AOS (red) and PTD (black). The slope for the AOS region is four times smaller than that in the PTD region.

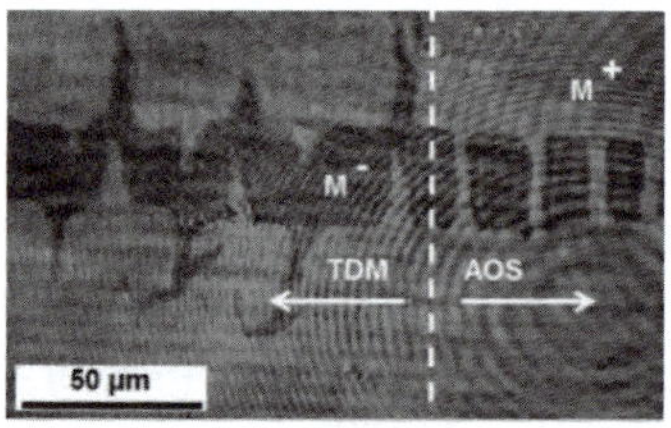


**Fig. 2.** Manipulation of the magnetization by linearly (left) and circularly (right) polarized light in $Fe_{70}Tb_{30}$. A stripe with magnetization $M^-$ is irradiated on a meandered path first, with linearly polarized light, which causes laser induced thermal depinning of the domain wall and second, with circularly polarized light, resulting in clean switching ($F_{AOS} = (1.20 \pm 0.08)$ mJ/cm²).

## References

[1] C. D. Stanciu, F. Hansteen, A. V. Kimel, A. Kirilyuk, A. Tsukamoto, A. Itoh, and Th. Rasing, "All-Optical Magnetic Recording with Circularly Polarized Light" Phys. Rev. Lett. **99**, 047601 (2007).
[2] A. Kirilyuk, A. V. Kimel, and Th. Rasing, "Laser-induced magnetization dynamics and reversal in ferromagnetic alloys" Rep. Prog. Phys. **76**, 026501 (2013).
[3] A. Hassdenteufel, B. Hebler, C. Schubert, A. Liebig, M. Teich, M. Helm, M. Aeschlimann, M. Albrecht, and R. Bratschitsch, "Thermally Assisted All-Optical Helicity Dependent Magnetic Switching in Amorphous $Fe_{100-x}Tb_x$ Alloy Films" Adv. Mater. **25**, 3122 (2013).

# Ultrafast laser-excited spin transport in Au/Fe/MgO(001): Relevance of the Fe layer thickness

**A. Alekhin[1], D. Bürstel[2], A. Melnikov[1], D. Diesing[2], and U. Bovensiepen[3]**

[1]Fritz-Haber-Institut der MPG, Phys. Chemie, Faradayweg 4-6, 14195 Berlin, Germany

[2]Universität Duisburg-Essen, Fakultät für Chemie, Universitätsstr. 5, 45117 Essen, Germany

[3]Universität Duisburg-Essen, Fakultät für Physik, Lotharstraße 1, 47057 Duisburg, Germany

uwe.bovensiepen@uni-due.de

**Abstract** Propagation dynamics of spin-dependent optical excitations is investigated by back-pump front-probe experiments in Au/Fe/MgO(001). We observe a decrease for all pump-probe signals detected at the Au surface, if the Fe thickness in increased. Relaxation processes within Fe limit the emission region of ballistic spins at the Fe/Au interface to ~1 nm.

Recently, we have established magneto-optical femtosecond back-pump front-probe experiments [1] in order to provide insight into spin-dependent (i) transport contributions in ultrafast magnetization dynamics and (ii) non-equilibrium transport in metallic films in general.

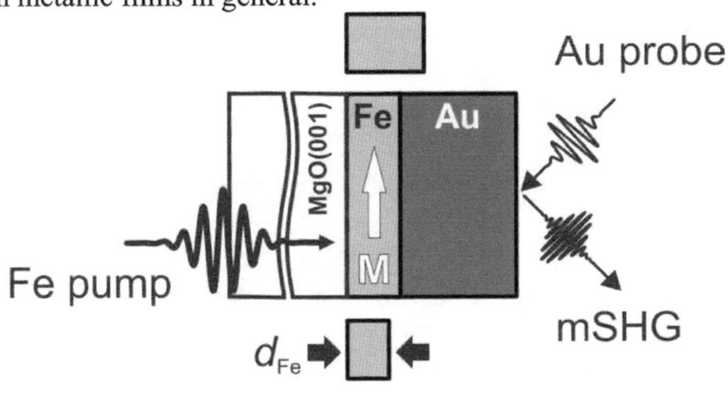


Fig. 1. Scheme of back-pump front-probe experiment. The fs pump pulse at 800 nm is absorbed in the metallic bilayer after transmission through MgO(001) depending on the investigated Fe layer thickness dFe, here 2, 11, and 17 nm. The dynamics of excitations propagating through the bilayer stack to the Au surface is probed by magneto-induced second harmonic generation (mSHG) as a function of time delay between pump and probe pulses.



J.-Y. Bigot et al. (eds), *Ultrafast Magnetism I*, Springer Proceedings in Physics 159
DOI 10.1007/978-3-319-07743-7_75

As depicted in Fig. 1, the investigated Fe/Au bilayer stack, which is grown epitaxially on MgO(001) [1], is photo-excited by femtosecond laser pump pulses of 14 fs pulse duration and at 800 nm wave length, which are transmitted through the transparent substrate. The probe is sent to the Au surface in order to analyze the propagation dynamics of the excitations through the bilayer stack - similar to a time-of-flight experiment. Here, we use second harmonic generation to detect the excitations arriving at the Au surface for different Fe thickness.

To become highly sensitive to spin-dependent excitations and their propagation dynamics we detect the second harmonic (SH) intensity for opposite Fe magnetization directions. The SH intensity $I_{2\omega}(\mathbf{M},t) \propto |\mathbf{E_{even}}+\mathbf{E_{odd}}|^2$, $E_{even}(t)$ is independent of $\mathbf{M}$ and $E_{odd}(t) \propto M(t)$. Below, we discuss the SH magnetic contrast $\rho(t)$ and the pump-induced change $\Delta_{even}$ of $E_{even}$; $E^0_{even}$ is the value before optical excitation.

$$\rho(t) = \frac{I^{\uparrow}_{2\omega}(t) - I^{\downarrow}_{2\omega}(t)}{I^{\uparrow}_{2\omega}(t) + I^{\downarrow}_{2\omega}(t)} \propto M(t) \ ; \quad \Delta_{even}(t) = \frac{E_{even}(t) - E^0_{even}}{E^0_{even}} .$$

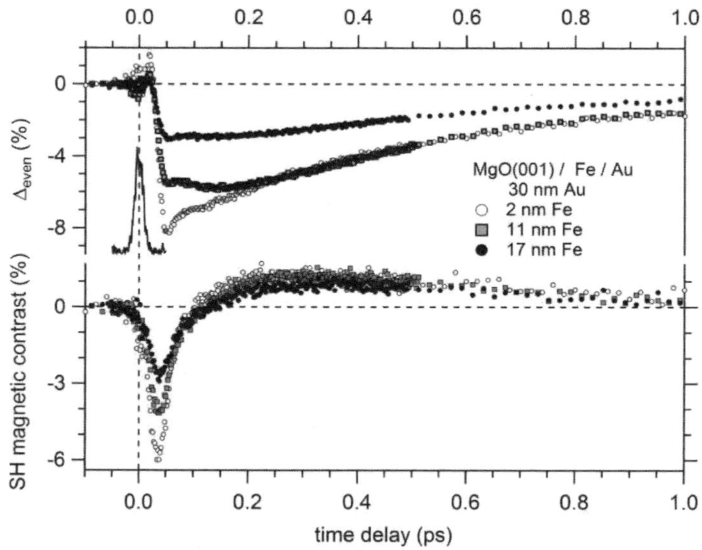


**Fig. 2.** Results of back-pump front-probe experiments employing SH generation for detecting the propagation of excitations. The top panel shows the relative pump-induced change in the even SH field, the lower panel depicts the relative change of the magnetic contrast $\rho(t)$ for three Fe layer thicknesses at a fixed Au layer thickness of 30 nm. The solid line on top panel represents the second harmonic cross correlation which indicates overlap of pump and probe pulses.

Fig. 2 shows the experimental results for three investigated layer stacks of 2, 11, and 17 nm Fe thickness combined with a constant Au thickness of 30 nm. The top panel depicts $\Delta_{even}(t)$ as a function of pump-probe delay (symbols) and the SH cross correlation of pump and probe pulses (solid line), which indicates time zero. The data exhibit a pronounced reduction at 40 fs delay and a recovery of the initial value within few ps. The variation with the Fe layer thickness $d_{Fe}$ is characterized by a weaker change for larger $d_{Fe}$.

The delay of the onset of the change originates from propagation of the fastest excitation through the Au layer [1]. In comparison to a variation of the Au layer thickness reported in [1] the variation in $d_{Fe}$ leaves the delay of initial drop fixed. We therefore conclude that the propagation through Au determines the 40 fs delay of initial drop.

In the magnetic contrast we identify a negative and a positive contribution to $\rho(t)$, which were reported to be dominated by ballistic and diffusive carrier propagation, respectively [1]. The isosbestic point near 90 fs close to zero magnetic contrast indicates that the two probed characteristic excitations can be well separated. The minimum value at 40 fs is more pronounced for smaller $d_{Fe}$. Although weaker, a comparable dependence is identified for the positive change in $\rho(t)$. Both signal contributions imply therefore a larger probability to excite the spin excitations probed at the Au surface for thinner Fe films. The short spin-dependent mean free path (<3 nm) and lifetime (<10 fs) in Fe [1] provides a strong argument that for $d_{Fe}>3$ nm considerable relaxation occurs in the Fe layer, before the excitation can be injected into Au. In other words, the injection of ballistic spin currents into Au is limited to a thin Fe region at the Fe/Au interface. Due to the dominant contribution of injected majority holes [1] with a ballistic mean free path <1 nm which we estimated from electronic lifetimes [2] the active injection region is constricted to a Fe slice of 1 nm thickness at the interface. Thus, the observed effects weaken for larger $d_{Fe}$ since less energy is deposited in the interface region.

The time-dependence of $\Delta_{even}$ is regarded as the direct consequence of the carrier redistribution among Fe and Au. The weaker $\Delta_{even}$ for larger $d_{Fe}$ is therefore explained by processes in the Fe layer, rather than by light absorption in Au, although the latter occurs for small enough $d_{Fe}$.

**Acknowledgment** Funding by the DFG through ME 3570/1 and by the EU 7-th framework program (CRONOS) is gratefully acknowledged.

## References

[1] A. Melnikov, I. Razdolski, T. Wehling, E. Papaioannou, V. Roddatis, P. Fumagalli, O.A. Aktsipetrov, A. Lichtenstein, U. Bovensiepen, "Ultrafast transport of laser-excited spin polarized carriers in Au/Fe/MgO(001)" Phys. Rev. Lett. **107**, 076601 (2011).

[2] V. P. Zhukov, E. V. Chulkov, P. M. Echenique, "Lifetimes and inelastic mean free path of low-energy excited electrons in Fe, Ni, Pt, and Au: Ab initio GW+T calculations" Phys. Rev. B **73**, 125105 (2006); V. P. Zhukov, E. V. Chulkov, P. M. Echenique, "Lifetimes of excited electrons in Fe and Ni: first-principles GW and the T-matrix theory" Phys. Rev. Lett. **93**, 096401 (2004).

# All-optical switching in CoTb alloys: composition and thickness dependent studies

**Ute Bierbrauer*[1], Sabine Alebrand[1], Michel Hehn[2], Matthias Gottwald[2,3], Daniel Steil[1], Daniel Lacour[2], Eric E. Fullerton[3], Stéphane Mangin[2], Mirko Cinchetti[1] and Martin Aeschlimann[1]**

[1] Department of Physics and Research Center OPTIMAS, University of Kaiserslautern, Germany

[2] IJL, Université de Lorraine, Nancy, France

[3] University of California, San Diego, USA

* bierbrauer@physik.uni-kl.de

**Abstract** We investigate all-optical switching (AOS) in TbCo alloys, demonstrating that AOS occurs in a limited Tb concentration range and above a threshold thickness of the magnetic layer. Overall, we find that the trend of the minimum laser fluence needed for light-induced domain formation depends on the sample composition.

The possibility of writing and erasing magnetic domains exclusively with circularly polarized femtosecond laser pulses [1] causes a lot of interest in applied and fundamental science. As for this effect no external magnetic field is needed for magnetization reversal, it is called "all-optical switching" (AOS). The only necessary requirement for the light-induced domain formation is reaching a minimum threshold fluence. Within the first three years the investigations of AOS focused on the material GdFeCo. However, in the last year research activities on the magnetization reversal in other rare earth transition metal alloys were intensified [2-5]. Here, we focus on the material system TbCo, which is of high technological interest due to its large magnetic anisotropy. We intensively studied the feasibility of AOS in this alloy for different Tb concentrations [2], demonstrating that AOS is possible in a Tb concentration range between roughly 20% and 30 %. Furthermore, we investigated AOS in different TbCo alloys, varying the thickness of the film. Dependent on the TbCo composition either an increasing or decreasing trend of the minimum threshold fluence needed for light-induced domain formation is observed for increasing film thickness. Additionally, we find that AOS only occurs above a certain threshold thickness of the TbCo layer.

To detect the AOS in TbCo we use a static Faraday imaging setup, which consists of a white light source, a crossed polarizer-analyzer-pair and a CCD camera.

 

J.-Y. Bigot et al. (eds), *Ultrafast Magnetism I*, Springer Proceedings in Physics 159
DOI 10.1007/978-3-319-07743-7_76

The optical pulses utilized to manipulate the magnetization in the thickness-dependent study (duration: 400 fs, repetition rate: 6 kHz, central wavelength: 780 nm) are delivered by an amplifier system. If switching only takes place for two out of the four combinations of helicity and sample magnetization we attribute the switching scenario to AOS, otherwise to pure thermal demagnetization. In both cases we determine the minimum threshold laser fluence that is needed to optically induce magnetic domains.

Our samples consist of Glass|Ta(5nm)|$Tb_xCo_{1-x}$(d nm)|Cu(2nm)|Pt(4-5nm). In the concentration-dependent study the thickness d of the TbCo film was fixed to 20 nm and the concentration was varied from 12 – 34 %. For the thickness-dependent measurements we choose three different Tb concentrations of 5, 26 and 29 % and vary the layer thickness d from ~7 - ~40 nm.

Table 1 summarizes the results of the concentration dependent study published in [2]. For the Tb concentration (x) range between 23 % and 32 % AOS occurred, while for lower or higher Tb concentrations only a pure thermal demagnetization was observed. However, note that for 23 % < x < 32 % AOS was only found for picosecond pulses, while for 23 % < x < 26 % AOS is additionally feasible for femtosecond pulses.

| Tb conc. x (%): | < 23 | 23 - 26 | 26 - 32 | >32 |
|---|---|---|---|---|
| AOS? | no | yes for fs and ps laser pulses | yes, only for ps laser pulses | no |

**Table 1.** Summary of the results of the Tb concentration-dependent study, published in [2].

We extended the previous study by varying the thickness of the TbCo layer for samples of different compositions ($Tb_{26}Co_{74}$, $Tb_{29}Co_{71}$, and $Tb_5Co_{95}$). For all compositions and layer thicknesses we measured the minimum threshold laser fluence $F_{min}$ needed for domain formation (see Figure 1). Dependent on the composition we find an increasing ($Tb_{26}Co_{74}$, $Tb_{29}Co_{71}$) or decreasing ($Tb_5Co_{95}$) trend of $F_{min}$ with thickness. We attribute this to differences in the strength of the demagnetizing field which generally favors the formation of magnetic domains. As the demagnetizing field increases with increasing layer thickness [6], its influence leads to a decreasing trend of the minimum threshold fluence with the layer thickness. Furthermore, the demagnetizing field is higher the higher the saturation magnetization $M_S$ of the sample is [6]. Therefore the influence of the demagnetizing field is much stronger for $Tb_5Co_{95}$ (where $M_S$ is eight times higher than for $Tb_{26}Co_{74}$ and $Tb_{29}Co_{71}$ [7]) compared to $Tb_{26}Co_{74}$ and $Tb_{29}Co_{71}$ causing a decreasing trend of $F_{min}$ with increasing layer thickness.

Finally we devote to the thickness-dependent AOS behavior. As can be seen in Figure 2 AOS (marked by open diamonds) was only observed for compositions which show an increasing trend of $F_{min}$ with thickness. In other words, AOS only appears if the saturation magnetization is small, which confirms the observations made by Hassdenteufel et al. [3]. Moreover, AOS only occurs above a minimum threshold layer thickness of roughly 20nm and increases with increasing layer thickness. The latter can be attributed to the decrease of the mean absorbed energy with layer thickness.

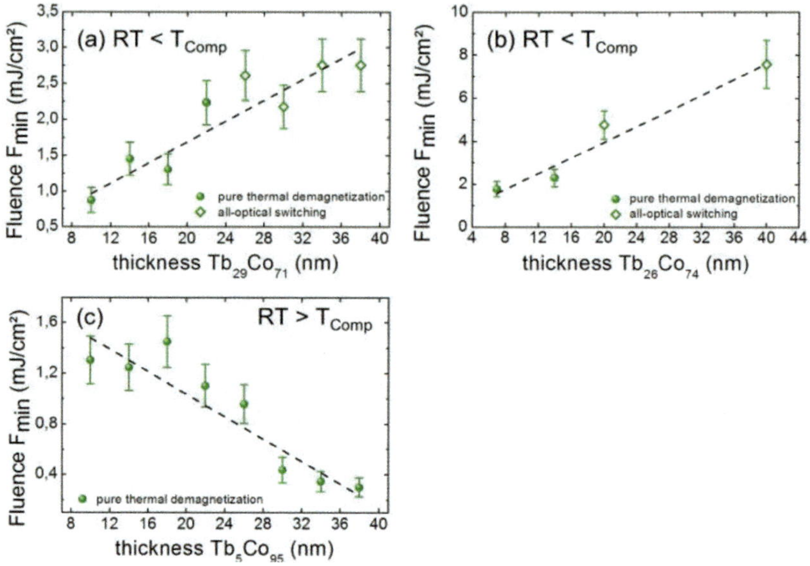


**Fig. 1.** Dependence of $F_{min}$ on the TbCo layer thickness for $Tb_{29}Co_{71}$ (a), $Tb_{26}Co_{74}$ (b) and $Tb_5Co_{95}$ (c). Green balls represent pure thermal demagnetization, while open diamonds mark AOS. The black dashed lines are guide to the eyes. $T_{Comp}$ and RT denote the magnetic compensation temperature and room temperature, respectively.

In conclusion, we demonstrated that AOS occurs in TbCo alloys for a limited Tb concentration range (~23 - 32 %) and above a threshold layer thickness (~20-30 nm). While the minimum threshold fluence needed for light-induced domain formation shows an increasing trend for $Tb_{26}Co_{74}$ and $Tb_{29}Co_{71}$, it decreases for $Tb_5Co_{95}$. We relate this to differences in the strength of the demagnetizing field, caused by the composition-dependent saturation magnetization of TbCo alloys.

## References

[1] C. D. Stanciu, F. Hansteen, A.V. Kimel, A. Kirilyuk, A. Tsukamoto, A. Itoh, and Th. Rasing, "All-Optical Magnetic Recording with Circularly Polarized Light", Phys. Rev. Lett. **99**, 1–4 (2007).

[2] S. Alebrand, M. Gottwald, M. Hehn, D. Steil, M. Cinchetti, D. Lacour, E. E. Fullerton, M. Aeschlimann, S. Mangin, "Light-induced magnetization reversal of high-anisotropy TbCo alloy films", Appl. Phys. Lett. **101**, 162408 (2012).

[3] A. Hassdenteufel, B. Hebler, C. Schubert, A. Liebig, M. Teich, M. Helm, M. Aeschlimann, M. Albrecht, R. Bratschitsch „Thermally Assisted All-Optical Helicity Dependent Magnetic Switching in Amorphous Fe100-xTbx Alloy Films", Advanced materials (Deerfield Beach, Fla.) **25**, 3122–8 (2013).

[4] A. R. Khorsand, M. Savoini, A. Kirilyuk, A. V. Kimel, A. Tsukamoto, A. Itoh, Th. Rasing, "Element-Specific Probing of Ultrafast Spin Dynamics in Multisublattice Magnets with Visible Light" Phys. Rev. Lett. **110**, 107205 (2013).
[5] V. López-Flores, N. Bergeard, V. Halté, C. Stamm, N. Pontius, M. Hehn, E. Otero, E. Beaurepaire, C. Boeglin, "Role of critical spin fluctuations in ultrafast demagnetization of transition-metal rare-earth alloys" Phys. Rev. B **87**, 214412 (2013).
[6] A. Hubert, R. Schäfer, "Magnetic Domains", (Springer Berlin Heidelberg 1998).
[7] P. Hansen, C. Clausen, G. Much, M. Rosenkranz , K.J. Witter, "" Appl. Phys. **66**, 756 (1989).

# Picosecond strain pulses for ultrafast magnetoacoustics

**O. Kovalenko[1], V. Shalagatskyi[1], T. Pezeril[1], V. Gusev[1], D. Makarov[2] and V.V. Temnov[1]**

[1]Institut des Molécules et Matériaux du Mans, UMR CNRS 6283, 72085 Le Mans, France

[2]Institute for Integrative Nanosciences, IFW Dresden, 01069 Dresden, Germany

vasily.temnov@univ-lemans.fr

**Abstract** The methods of generation and characterization of ultrashort acoustic strain pulses for ultrafast magnetoacoustics in ferromagnetic multilayer structures are discussed. Mechanisms and limitations of acoustic generation at the nano-scale are revealed.

## Introduction

Ultrafast magnetoelastic interactions at the nano-scale represent an emerging research direction, where the first experiments were performed in ferromagnetic semiconductor GaMnAs at cryogenic temperatures [1] and nickel at room temperature [2]. One of the biggest questions concerns the possibility of magnetoacoustic switching using ultrashort acoustic pulses with duration shorter than the period of ferromagnetic precession [3]. Here we review the generation and characterization of ultrashort acoustic pulses in hybrid gold-cobalt bilayer structures [4,5], which may serve as good candidates for the magnetoacoustic switching experiments.

## Generation of picosecond acoustic pulses

Ultrashort acoustic pulses can be generated by thermal expansion of metal transducer irradiated by femtosecond laser pulses [6]. Energy absorbed by the electrons may escape from the optical the skin depth $\delta_{skin} = \lambda/4\pi\kappa$ during the electron phonon-relaxation time $\tau_{e-ph} = c_e/g$ into possibly much larger electronic heat diffusion depth $\xi_{heat} = \sqrt{\chi/g}$. Here $\chi$, $g$ and $c_e = \gamma T_e$ denote the heat conductivity, electron-phonon coupling constant and the electronic heat capacity, respectively. The heat diffusion depth $\xi_{heat}$ determines the spatial extent of heated lattice region, which generates the acoustic pulses of duration $\xi_{heat}/c_s$ , where $c_s$ stands for

 
J.-Y. Bigot et al. (eds), *Ultrafast Magnetism I*, Springer Proceedings in Physics 159
DOI 10.1007/978-3-319-07743-7_77

the speed of sound in the transducer. The electron-phonon relaxation time and the electronic heat conductivity are both relatively large for noble metals leading to $\xi_{heat} \gg \delta_{skin}$. For ferromagnetic metals $\xi_{heat} \sim \delta_{skin} \sim 10$ nm and the shortest acoustic pulses can be obtained. On the other hand, despite being inefficient optoacoustic transducers, the noble metals appear to be quite useful for the acoustic detection. We will show below that the hybrid metal-ferromagnet bilayer structures can be used to generate and detect ultrashort acoustic pulses in different configurations.

## Optical measurements of picosecond acoustic pulses

It was shown recently that optical pump-probe experiments can be used to measure the ultrashort acoustic strain pulses in hybrid gold-cobalt-sapphire multilayer structures. In the first proof-of-principle experiment (see Fig. 1a) the ultrashort acoustic pulses were excited by optical pumping (pump wavelength 400 nm) of the cobalt layer through the sapphire substrate and detected via the acoustic modulation of the wave vector of time-delayed femtosecond surface plasmon polariton pulses (probe wavelength 800 nm) propagating at the gold-air interface. Measurements were performed by the time-resolved surface plasmon interferometry with tilted slit-groove plasmonic microinterferometers. Acoustic pulses with spatial extent of 10 nm and duration of 3ps were observed in these experiments. Remarkably large strain amplitudes exceeding 1% of lattice constant could be directly measured.

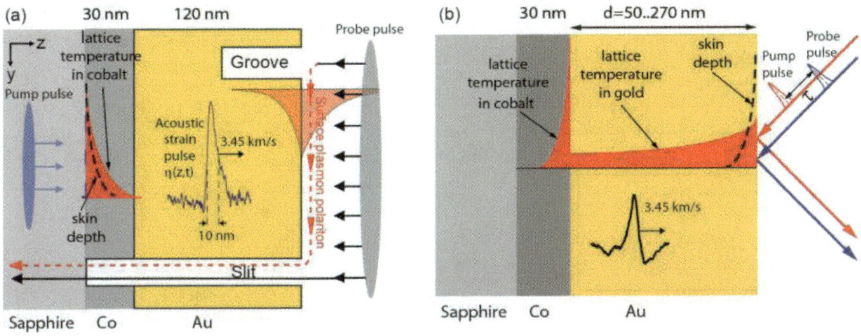


**Fig. 1.** (a) Ultrafast acousto-plasmonics in gold-cobalt bilayer structures is used to measure an ultrashort acoustic strain pulse η(z,t) generated in cobalt (see the inset). (b) An unusual generation of ultrashort acoustic pulses in cobalt via super-diffusive hot electrons through gold is revealed in simple pump-probe reflectivity measurements. The time derivative of probe reflectivity (see the inset) gives the profile of a 2ps-long acoustic pulse generated in cobalt.

The detailed analysis revealed that the shape of ultrashort acoustic pulses can be extracted not only from the sophisticated plasmonic measurements, but also from much simpler reflectivity measurements [7]. For the acoustic pulses shorter than the acoustic passage time through the skin depth of probe light, $\tau_{ac} < \delta_{skin}/c_s = 4.8$ ps, the time derivative of probe reflectivity gives the acoustic pulse shape. We have used this technique to study the acoustic generation in gold-cobalt-sapphire samples when exciting and probing the gold surface (see Fig. 1b). A surprising observation is that a very efficient super-diffusive transport by hot electrons through gold leads to the overheating of electrons in cobalt resulting in the efficient heating of cobalt lattice. Due to the strong electron-phonon coupling the lattice temperature in cobalt becomes much larger as compared to the adjacent gold. The 2ps duration of the acoustic pulse emitted by thermal expansion of cobalt is determined by diffusion length $\xi_{heat} = 12$ nm of hot electrons in cobalt. These pulses are even shorter than 3ps long pulses generated by the direct optical pumping of cobalt transducer. The exponential decay of the acoustic pulse amplitude as a function of gold thickness (data not shown) reveals the heat penetration depth in gold of $\xi_{heat} = 110$ nm, in agreement with previous theoretical estimates [8].

Besides an exotic mechanism to generate ultrashort acoustic pulses by transient overheating of cobalt through the cold layer of gold, our measurements provide a different perspective on the phenomenon of transient demagnetization in nickel mediated by hot electron diffusion through gold [9].

**Acknowledgments** Funding by 'Nouvelle équipe, nouvelle thématique de la Region Pays de la Loire' and the french 'Agence Nationale de la Recherche' (project Equipex UNION, ANR-10-EQPX-52-01), and discussions with Pascal Ruello and Jean-Yves Bigot are acknowledged.

## References

[1] A.V. Scherbakov et al., "Coherent magnetization precession in ferromagnetic (Ga,Mn)As induced by picosecond acoustic pulses", Phys. Rev. Lett. **105**, 117204 (2010).

[2] J. Kim, M. Vomir, J.-Y. Bigot, "Ultrafast magnetoacoustics in nickel films", Phys. Rev. Lett. **109**, 166601 (2012).

[3] O. Kovalenko, T. Pezeril, V.V. Temnov, "New concept for magnetization switching by ultrafast acoustic pulses", Phys. Rev. Lett. **110,** 266602 (2013).

[4] V.V. Temnov, "Ultrafast acousto-magneto-plasmonics", Nature Photonics **6**, 728 (2012).

[5] V.V. Temnov et al., "Femtosecond nonlinear ultrasonics in gold probed with ultrashort surface plasmons", Nature Communications **4**, 1468 (2013).

[6] C. Thompsen, H.T. Grahn, H.J. Maris, J. Tauc, "Surface generation and detection of phonons by picosecond light pulses", Phys. Rev. B **34**, 4129 (1986).

[7] K.J. Manke et al., "Detection of shorter-than-skin-depth acoustic pulses in a metal film via transient reflectivity", arXiv:1306.0641 (2013).

[8] V. Gusev and O. B. Wright, "Ultrafast nonequilibrium dynamics of electrons in metals", Phys. Rev. **B** 57, 2878 (1998).

[9] A. Eschenlohr et al., "Ultrafast spin transport as key to femtosecond demagnetization", Nature Materials **12**, 332-336 (2013).

# Ultrafast Demagnetization Rates in Two-Component Magnetic Materials

**O. Chubykalo-Fesenko[1,§], U. Atxitia[2,3], P. Nieves[2], J. Barker[1] and R.W. Chantrell[1]**

[1]Instituto de Ciencia de Materiales de Madrid, CSIC, Cantoblanco, c. Sor Juana Ines de la Cruz, 3, Madrid, 28440, Spain

[2]University of York, UK

[3] Departamento de Física de Materiales, UPV/EHU, San Sebastian,

[§] oksana@icmm.csic.es

**Abstract** The recently derived Landau-Lifshitz-Bloch equation of motion for two-component magnetic systems, including ferri- ferro- and antiferromagnets, provides a framework to analyze species dependent ultra fast demagnetization rates. In agreement with reported experimental observations, we show that Gd sublattice demagnetizes slower then FeCo in GdFeCo. However, we predict that at high initial temperatures Gd should be faster than Fe, providing the possibility to control the polarity of the transient antiferromagnetic state. Our results indicate that typically Fe should be slower than Ni in FeNi. The situation is opposite if Fe is 4 times stronger coupled to the electronic system than Ni.

The element-specific synchrotron measurements, using X-ray circular magnetic dichroism, have opened the possibility to evaluate separately the ultra fast demagnetization rates in multi-component magnetic alloys. Prominent examples are measurements in GdFeCo [1], where it was established that Gd sub-lattice is much slower than Fe. Differently, in NiFe the two sub-lattices have similar demagnetization rates and the debate on which is faster is still going on [2-4]. It has been proposed that the ratio between the magnetic moment and the Curie temperature, $\mu$/TC, is a good figure of merit to establish the ultrafast demagnetization rates of materials [5]. Following this ratio, any rare earth material is clearly at least 3-5 times slower than any transition metal. At the same time, recently it has been established that a typical "fast" transition metal Ni can become "slow" approaching the Curie temperature [6]. Thus, the question what is the relative demagnetization speed in each metal comprising a multi-component ferromagnet is not trivial.

The Landau-Lifshitz-Bloch (LLB) equation, recently derived for a two-component system [7], can help to elucidate this question. For pure longitudinal processes (when the sublattices magnetizations remain parallel) we write the LLB

 
J.-Y. Bigot et al. (eds), *Ultrafast Magnetism I*, Springer Proceedings in Physics 159
DOI 10.1007/978-3-319-07743-7_78

equation for each sublattice ν and for the reduced magnetization $\vec{m}_\nu=\vec{M}_\nu/M(T=0)$ in the form:

$$\frac{d\vec{m}_\nu}{dt}=-\gamma_\nu\alpha_\nu^{||}\left[\frac{\Gamma_{\nu\nu}}{2}\left(\frac{m_\nu^2}{m_{e,\nu}^2}-1\right)-\frac{\Gamma_{\nu\kappa}}{2}\left(\frac{m_\kappa^2}{m_{e,\kappa}^2}-1\right)\right]\vec{m}_\nu$$

Here $\gamma_\nu$ is the gyromagnetic ratio and $m_{e,\kappa}$are the equilibrium magnetization values. The longitudinal damping and the rate parameters are evaluated in the mean-field approximation. Namely,

$$\alpha_\nu^{||}\cong\frac{2\lambda_\nu k_B T m_{e,\nu}}{J_{0,\nu}m_{e,\nu}+|J_{0,\nu\kappa}|m_{e,\kappa}},$$

where $\lambda_\nu$ is the parameter representing the atomistic coupling to the heat bath (proportional to the scattering rate), $k_B$ is the Boltzmann constant, $T$ is the temperature, $J_{0,\nu}=z_\nu J_\nu$, $J_{0,\nu\kappa}=z_{\nu\kappa}J_{\nu\kappa}$, $J_\nu$ is the intra-sublattice exchange parameter and $z_\nu$ is the average number of the neighbors of the same species, $J_{\nu\kappa}$ is the inter-sublattice exchange parameters and $z_{\nu\kappa}$ is the average number of neighbors of the opposite species. The temperature-dependent rate parameters are expressed in terms of the longitudinal susceptibilities $\chi_\nu^{||}=(\partial m_\nu/\partial H)_{H\to 0}$ as:

$$\Gamma_{\nu\nu}=\frac{1}{\chi_\nu^{||}}\left(1+\frac{|J_{0,\nu\kappa}|}{\mu_\nu}\chi_\kappa^{||}\right),\qquad \Gamma_{\nu\kappa}=\frac{|J_{0,\nu\kappa}|}{\mu_\nu}\frac{m_{e,\kappa}}{m_{e,\nu}}$$

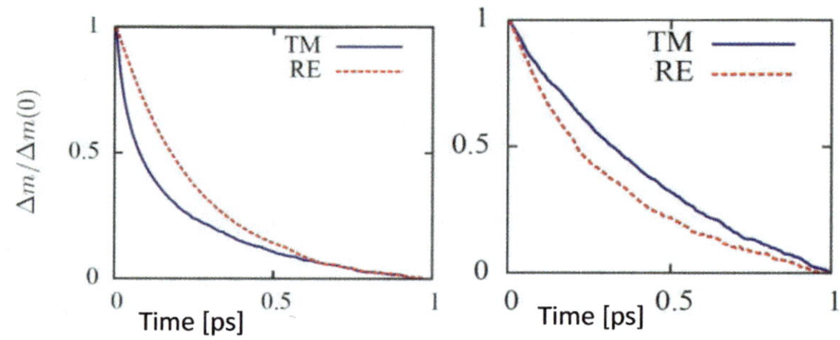


**Fig. 1.** Atomistic simulations of longitudinal relaxation in disordered GdFeCo ferrimagnet with 25% of Gd under a heat pulse with initial temperature 300 K and final temperature $0.75T_C$=600 K (left) and $0.975T_C$=780 K (right). The demagnetization curves are normalized to the value at 2ps. The parameters of GdFeCo are taken from Ref.[9].

The longitudinal relaxation rates can be evaluated as the eigenvalues of the two coupled linearized LLB equations. This indicates that one cannot attribute a one-exponential decay to each of the sublattices. Nevertheless, for FeCoGd the analysis of the demagnetization rates shows that for temperatures not close to $T_C$,

$\Gamma_{FeCo,Gd} \sim \Gamma_{Gd,FeCo} < \Gamma_{Gd,Gd} \ll \Gamma_{FeCo,FeCo}$ and FeCo and Gd sublattices can be practically described by one eigenvalue. In this temperature region Gd is much slower than FeCo. Close to $T_C$, however, the situation surprisingly changes and our model predicts that Gd becomes faster than FeCo. To confirm these predictions, we present the corresponding atomistic simulations in Fig.1. One of the consequences of the change of relaxation rates is the possibility to observe the "inverted" transient ferromagnetic state at high temperatures [8], i.e. directed towards Fe instead of Gd.

As for the relaxation in NiFe, our results show that since the two-sublattices Fe and Ni are strongly coupled, one cannot characterize their individual ultra fast dynamics in terms of one-exponential behavior. Nevertheless, following the common experimental practice, we fit the magnetization relaxation curves to one exponential decay. Fig.2 presents the ratio between Ni and Fe demagnetization times as a function of their coupling to the bath parameters and the pulse temperature. Because Fe and Ni are similar transition metals, one can expect similar coupling to the bath parameters for both in their alloy. This indicates that in normal situation Ni is faster than Fe. However if one assumes that Fe is 4 times stronger coupled to the electronic system, as also discussed in Ref.[4], the situation is opposite.

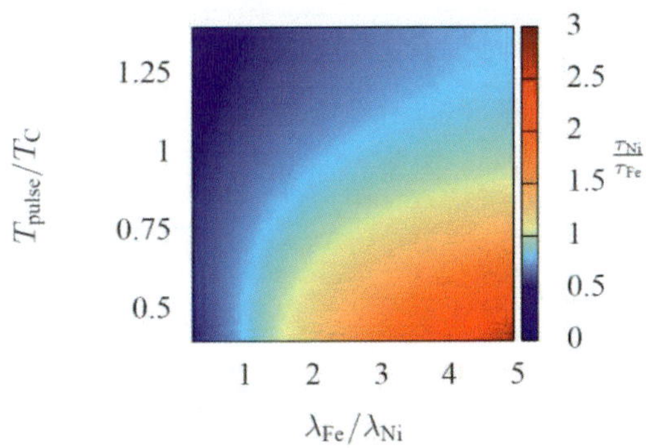


**Fig. 2.** The ration between Ni and Fe relaxation times in FeNi alloy, obtained from the LLB-based simulations, as a function of the ratio of the coupling to the bath parameters and the maximum pulse temperature.

**Acknowledgments** The authors would like to acknowledge the Femtospin project from European Community and the Spanish grant FIS2010-20979-C02-02 from Ministry of Science and Innovation. U.A. gratefully acknowledges support from Basque Country Government.

## References


[1] I. Radu et. al., "Transient ferromagnetic-like state mediating ultrafast reversal of antiferromagnetically coupled spins" Nature **472**, 205 (2011).

[2] I. Radu *et. al,* submitted.

[3] S. Mathias et. al., “Probing the timescale of the exchange interaction in a ferromagnetic alloy” Proc. Nat. Acad. Sci. USA **109**, 4792 (2012).
[4] A. J.Schellekens and B.Koopmans, “Microscopic model for ultrafast magnetization dynamics of multisublattice magnets” Phys. Rev. B **87**, 020407 (R) (2013).
[5] B. Koopmans, G. Malinowski, F. Dalla Longa, D. Steiauf, M. Fähnle, T. Roth, M. Cinchetti & M. Aeschlimann, “Explaining the paradoxical diversity of ultrafast laserinduced demagnetization” Nature Mat. **9**, 259 (2010).
[6] T. Roth, A. J. Schellekens, S. Alebrand, O. Schmitt, D. Steil, B. Koopmans, M. Cinchetti, and M. Aeschlimann, “Temperature Dependence of Laser-Induced Demagnetization in Ni: A Key for Identifying the Underlying Mechanism” Phys. Rev. X, **2**, 021006 (2012).
[7] U.Atxitia, P.Nieves and O.Chubykalo-Fesenko, “Landau-Lifshitz-Bloch equation for ferrimagnetic materials” Phys. Rev. B **86**, 104414 (2012).
[8] U.Atxitia et. al, submitted.
[9] T. A. Ostler et. al., “Crystallographically amorphous ferrimagnetic alloys: Comparing a localized atomistic spin model with experiments” Phys. Rev.B **84**, 024407 (2011).

# Lattice-Mediated Optical Control of Magnetic Anisotropy in FeBO3

**D. Afanasiev[1], I. Razdolski[1], D. Bolotin[2], S.V. Yagupov[2], M.B. Strugatsky[2], A. Kirilyuk[1], Th. Rasing[1] and A.V. Kimel[1]**

[1]IMM, Radboud University Nijmegen, 6525 AJ Nijmegen, the Netherlands

[2]Solid-State Physics Chair, Department of Physics, Taurida National University, 95036 Simferopol, Yaltinskaya St. 4, Ukraine

d.afanasiev@science.ru.nl

**Abstract.** Ultrafast optical control of magnetization dynamics based on interference of coherent spin-wave resonances and limited to media with long spin coherence times. Here we report that femtosecond laser pulse may trigger oscillations of the magneto-crystalline anisotropy and thus induce long-living magnetization oscillations.

## Introduction

The optical control of the magneto-crystalline anisotropy by intense femtosecond laser pulses opens up an intriguing new opportunity to control magnetization by light.

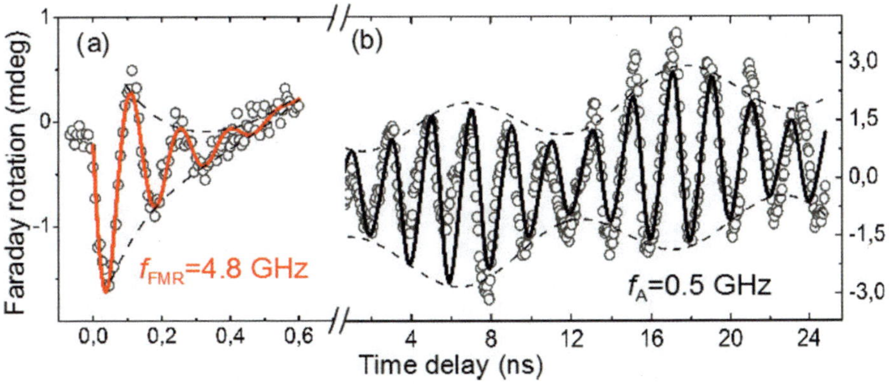


**Fig. 1.** Femtosecond laser excitation of magnetization dynamics in $FeBO_3$ measured via the Faraday effect in the external magnetic field 40 Oe: (a) short-living high frequency FMR mode at $f_{FMR}$=4.8 GHz and (b) long-living magneto-acoustic mode at the frequency of size acoustic resonance along $C_3$ axis in $f_A$=0.5 GHz



J.-Y. Bigot et al. (eds), *Ultrafast Magnetism I*, Springer Proceedings in Physics 159
DOI 10.1007/978-3-319-07743-7_79

It has been demonstrated that ultrafast laser excitation can suddenly change the orientation of the easy-axis of magnetite- anisotropy and trigger spin-precession around the new equilibrium with the frequency of ferromagnetic resonance (FMR) see Ref. 1 and Fig. 1a. However, the lifetime of these effects is limited by the time of the spin-lattice interaction, which is in the range of 10 ps-1 ns. Optically excited lattice vibrational modes [2] could also drive magnetization dynamics at the acoustic frequencies by inducing coherent oscillations of magneto-crystalline anisotropy value. These oscillations affect magnetization equilibrium and lead to the magnetization tracking of the new equilibrium Fig. 1b. Unlike pure spin resonances, these coupled modes are characterized by small damping and able to maintain coherence of the magnetic mode up to microseconds [3].

## Results and discussion

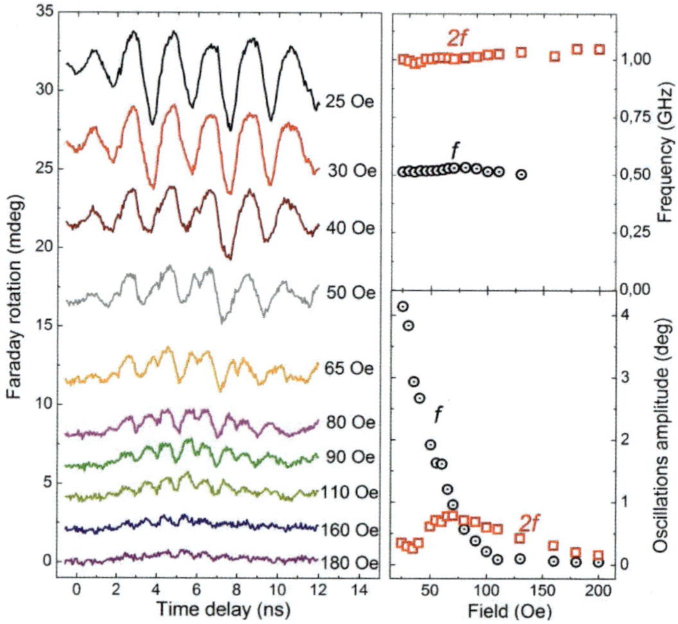


**Fig. 2.** Femtosecond laser excitation of magnetization oscillations in FeBO3 at the frequencies f=0.5 GHz and 2f, measured at room temperature for different magnetic fields.

Here we demonstrate that coherent oscillations of laser-induced magnetic anisotropy can be realized in single crystalline $FeBO_3$, an easy-plane weak ferromagnet [4]. It has extremely low value of the intrinsic in-plane magnetic anisotropy, what makes it almost isotropic in the easy-plane and very sensitive to external perturbations including stress and optical excitation.

Laser-induced magnetization dynamics are characterized by relatively big oscillations amplitudes $\leq 5^0$ and very small damping. Unlike FMR the frequency of the oscillations is hardly sensitive to the external magnetic field while their amplitude drops significantly upon a slight increase of the field (Fig.2). We argue that these oscillations can be explained in terms of light-induced longitudinal mechanical strain which via modulation of the in-plane anisotropy value acts like a driving force on the magnetization dynamics. Strong anharmonicity of the observed magnetization dynamics includes single and doubled frequencies and originates from nonlinear coupling between the excited longitudinal sound and magnetization. Such coupling becomes effective because big values of laser-induced dynamical stain $\sim 10^{-4}$, which is one order of magnitude higher than static magnetostrictive deformation in $FeBO_3$[5].

## References


[1] F. Hansteen, A. Kimel, A. Kirilyuk and Th. Rasing, "Femtosecond photomagnetic switching of spins in ferrimagnetic garnet films" Phys. Rev. Lett. **95,** 047402 (2005)

[2] A. Kobyakov, M. Sauer and D. Chowdhury,"Stimulated Brillouin scattering in optical fibers" Advances in Optics and Photonics **2,** 1-59 (2010)

[3] B. Ya. Kotyuzhanskii, L. A. Prozorova, "Parametric excitation of phonons in antiferromagnetic $FeBO_3$" JETP Lett. **35,** 244-246 (1982).

[4] A. J. Kurtzig, R.Wolfe, R. C. LeCraw, and J. W. Nielsen, "Magneto-optical properties of a green room temperature ferromagnet: $FeBO_3$" Appl. Phys. Lett. **14,** 350 (1969).

[5] Yu. N. Mitsay *et al.*, "Gakel'-Turov oscillations in iron borate" JMMM **219,** 340-348 (2000).

# Dual-Pump Manipulation of Ultrafast Demagnetization in TbFeCo

**T.Y. Cheng[1,3], J. Wu[1*], R.W. Chantrell[1], X. Zou[1], T. Liu[2], J.W. Cai[2], Y.B. Xu[3]**

1. Department of Physics, University of York, YO10 5DD, UK
2. Institute of Physics, Chinese Academy of Sciences, Beijing 100190, China
3. Department of Electronics, University of York, YO10 5DD, UK

* jing.wu@york.ac.uk

**Abstract** Laser-induced ultrafast demagnetization in TbFeCo has been studied with a dual-pumping system. Five different laser power combinations were performed at three different time intervals between the two pump pulses. The experimental results are also compared with computational simulations using the atomistic model. The results demonstrate that the ultrafast demagnetization can be controllably manipulated in both its magnitude and temporal response.

## Introduction

Laser-induced ultrafast demagnetization has been investigated for many years[1-3] since Beaurepaire et al first demonstrated it in ferromagnetic Ni film[1], and it is still one of the most important issues in the application of high density and fast write-read speed magnetic recording techniques.

This paper is devoted to a dual-pump manipulation of ultrafast demagnetization of the TbFeCo sample using two laser pump pulses in sequence, with variable laser power and time intervals. The intensity of Pump1 or Pump 2 can be varied by using a separate attenuator, and the time interval between them can be tuned via an optical delay line.

A group of dual-pump induced ultrafast demagnetization measurements have been presented and compared with computational simulations based on the atomistic model. As shown in Fig 1, the additional reduction in magnetization excited by the 2nd pump pulse demonstrates a controllable manipulation of the magnitude and temporal response of demagnetization dynamics, by tuning the intensity and time interval of the 2nd pump. In Fig 2, it also proves that with the same total pump intensity, the total demagnetization rate is independent of the intensity combination or the time interval between two pumps. And one may have more freedom to manipulate the additional demagnetization induced by Pump 2 if

 
J.-Y. Bigot et al. (eds), *Ultrafast Magnetism I*, Springer Proceedings in Physics 159
DOI 10.1007/978-3-319-07743-7_80

Pump1 hasn't fully demagnetized the spin system. Moreover, the temporal response of the ultrafast demagnetization can be manipulated by setting different time intervals between Pump1 and Pump2, from 0 to 0.97 ps, as shown in Fig 3. This dual-pumping scheme provides an effective technique to manipulate both the magnitude and temporal response of the ultrafast demagnetization, which would play a significant role in inscribing information to magnetic materials, at a desired time scale, for future magnetic recording technologies.

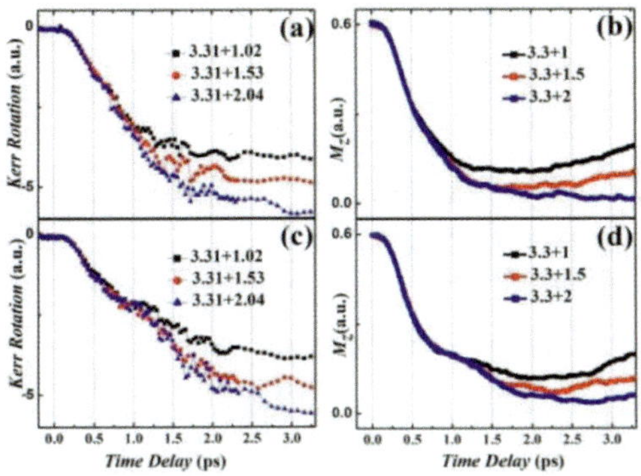


**Fig. 1.** Dual-pump induced demagnetization with an increasing Pump2 (a): Fixed Pump1 and variable Pump2 at a time interval of 0.47 ps(experiment); (b): Fixed Pump1 and variable Pump2 at a time interval of 0.5 ps(simulation); (c): Fixed Pump1 and variable Pump2 at a time interval of 0.97 ps(experiment); (d): Fixed Pump1 and variable Pump2 at a time interval of 1ps (simulation)

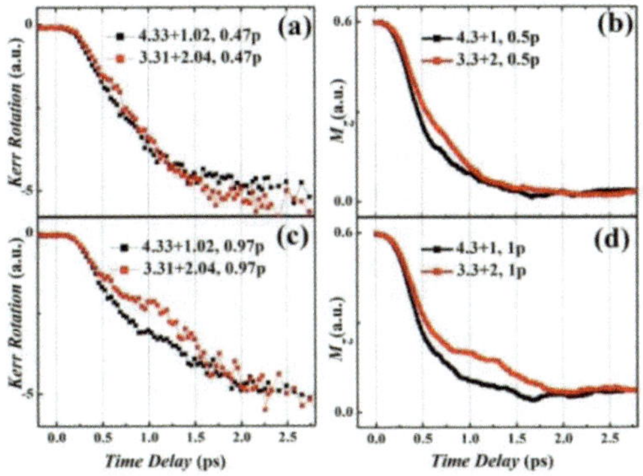


**Fig. 2.** Dual-pump induced demagnetization with a total pump intensity of 5.35/5.3 J/m$^2$ (a): Demagnetization induced by dual-pump (4.33+1.02 J/m$^2$ ) and (3.31+2.04 J/m$^2$) at a time interval of 0.47 ps (experiment); (b): Demagnetization induced by dual-pump (4.3+1 J/m$^2$) and (3.3+2 J/m$^2$) at a time interval of 0.5 ps (simulation); (c): Demagnetization induced by dual-pump (4.33+1.02 J/m$^2$ ) and (3.31+2.04 J/m$^2$) at a time interval of 0.97 ps (experiment); (d): Demagnetization induced by dual-pump (4.3+1 J/m$^2$) and (3.3+2 J/m$^2$) at a time interval of 1 ps (simulation)

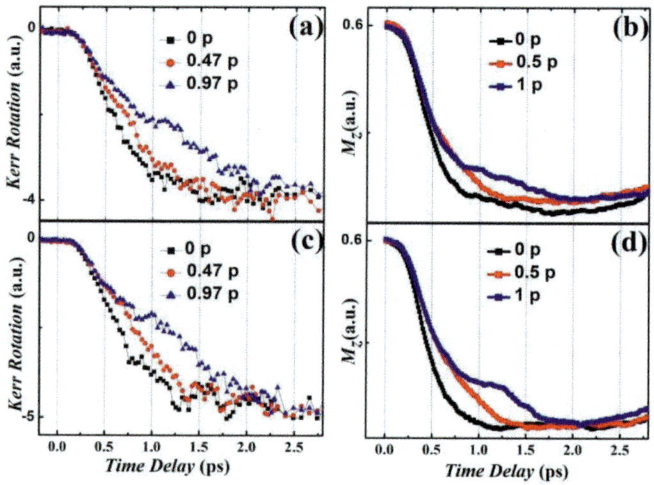


**Fig. 3.** Dual-pump induced demagnetization with different time intervals (a): Demagnetization induced by dual-pumping system (3.31+1.02 J/m$^2$)(experiment); (b): Demagnetization induced by dual-pumping system (3.3+1 J/m$^2$)(simulation); (c): Demagnetization induced by dual-pumping system (3.31+1.53 J/m$^2$) (experiment); (d): Demagnetization induced by dual-pumping system (3.3+1.5 J/m$^2$)(simulation)

# References


[1] E. Beaurepaire, J.-C. Merle, A. Daunois, and J.-Y. Bigot, "Ultrafast Spin Dynamics in Ferromagnetic Nickel", Phys. Rev. Lett. 76, 4250 (1996).

[2] G. P. Zhang and W. Hübner, "Laser-induced Ultrafast Demagnetization in Ferromagnetic Metals", Phys. Rev. Lett. **85**, 3025 (2000).

[3] K. Vahaplar, A. M. Kalashnikova, A. V. Kimel, D. Hinzke, U. Nowak, R. Chantrell, A. Tsukamoto, A. Itoh, A. Kirilyuk, and Th. Rasing, "Ultrafast Path for Optical Magnetization Reversal via a Strongly Nonequilibrium State", Phys. Rev. Lett. **103**, 117201 (2009).

# Terahertz response and ultrafast laser-induced dynamics of spins and charges in $CoFe/Al_2O_3$ multilayers

**J.D. Costa[1,2*], T. Huisman[3], R. Mikhaylovskiy[3], J. Ventura[1], J.M. Teixeira[1], D. Schmool[1],G. Kakazei[1], S. Cardoso[4], P. Freitas[4,2], Th. Rasing[3] and A.V. Kimel[3]**

[1]IN-IFIMUP and Dep. Física, Rua do Campo Alegre 687, 4169-007 Porto, Portugal

[2] International Iberian Nanotechnology Laboratory, INL, Braga, Portugal

[3]Radboud University Nijmegen, Institute for Molecules and Materials, Nijmegen, The Netherlands

[4]INESC-MN and IN-Institute of Nanoscience and Nanotechnology, Rua Alves Redol, 9-1, 1000-029 Lisbon, Portugal

zediogo.costa@gmail.com

Spintronics is a recently emerged research field in both fundamental science and technology, a successful development which promises a breakthrough in energy efficient information processing. The ever increasing demand for faster information processing raises a fundamental question about the feasibility to realize the concepts of spintronics at THz frequencies. To answer this question we investigate THz response of magnetoresistive materials using time-resolved pump-probe technique [1], THz absorption and THz emission spectroscopies [2].

The studied sample consisted of a metal-insulator multilayer of $[Co_{80}Fe_{20}$ (1.8 nm) / $Al_2O_3$ (3 nm)$]_{x10}$ prepared by ion beam sputtering on glass substrate. The corresponding magnetic and electrical properties were reported previously [3].

We first measured the THz transmission through the CoFe/Al2O3 multilayer as a function of temperature (T) and found a similar dependence with that of the electrical resistance [R; Fig. 1(a)]. This behavior clearly demonstrates that a correlation between the two exists, which should enable the ultrafast probing of the magnetoresistance with the help of THz spectroscopy.

 
J.-Y. Bigot et al. (eds), *Ultrafast Magnetism I*, Springer Proceedings in Physics 159
DOI 10.1007/978-3-319-07743-7_81

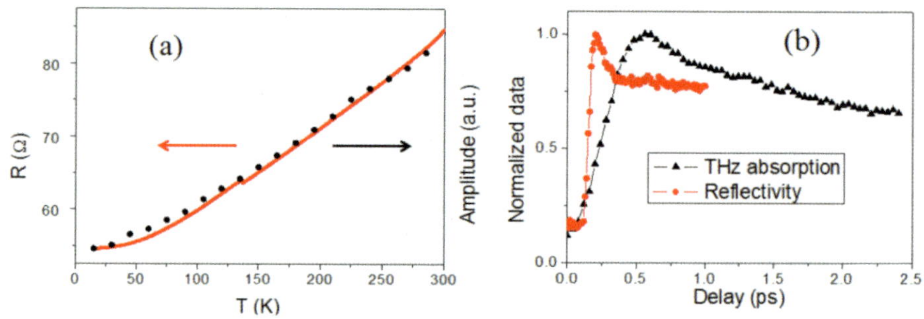


**Fig. 1.** (a) THz transmission and resistance as a function of the temperature. The black dots represent the THz transmission and the red line the resistance behavior for the same sample as reported elsewhere [3]. (b) Comparison between the photo-induced near-infrared reflectivity and the THz transmission.

To obtain more information on electronic transitions and thus the conductivity of the sample, we performed optical pump - THz probe measurements. The resulting data reveals a step like behavior of the transmission with three different regimes: an initial sharp increase, followed by two relaxation processes with different time scales until a new equilibrium is reached [Fig. 1(b)]. The steep increase (~500 fs) is caused by the incident optical pump that generates electronic transitions that change the samples resistance and therefore the THz transmission. This is followed by a fast decay with a time scale of ~0.8 ps which is the characteristic time of electron-phonon relaxation. After a few ps the dynamics is determined by a slow thermal relaxation (~140 ps). The change in THz transmission is related with a change in the resistivity, which will affect the materials reflectivity over a broad spectral range. Using a near-infrared pump-probe setup we measured the ultrafast variation of the reflectivity at optical frequencies. Figure 1(b) shows the comparison between the THz transmission and the optical reflectivity measurements. The reflectivity measured with the pump-probe setup has a similar behavior but a shorter time scale which can be explained by some dispersion in the ZnTe crystal used for the THz detection.

It is also shown that an ultrafast laser excitation leads to a THz emission from the samples such that the polarization of the THz wave is defined by the magnetization of the sample before the excitation. The phenomenon is explained in terms of magnetic dipole emission of THz due to ultrafast demagnetization of the magnetic material leading to an emission proportional to the second derivative of the magnetization [4]:

$$E_y(t) \sim \frac{\partial^2 M_x}{\partial t^2} \tag{1}$$

Both the angular and magnetic dependence were consistent with this explanation. Moreover, the magnetization dynamics induced by the pump was also studied by all optical spectroscopy, thus measuring the optically induced demagnetization. To further verify the validity of Eq. (1) we calculated the second derivative of the demagnetization curve and compared it with the THz emitted from the sample in the same conditions (Fig. 2). Qualitatively similar behaviors are observed. Nevertheless, the values of the frequencies obtained using the two approaches are different [as in Fig. 1(b)]. Again this is ascribed to dispersion occurring in the THz detector, broadening the detected signal. Also, Eq. (1) is just an approximation itself assuming the emitter to be a point dipole.

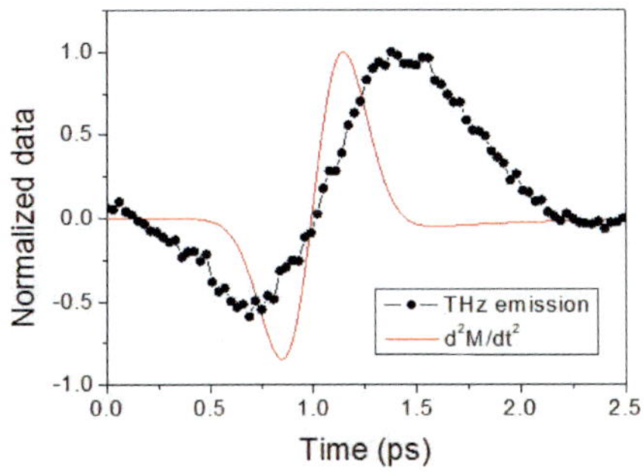


**Fig. 2.** Comparison between the THz emission and the second derivative of the demagnetization.

This thorough study of the THz transmission and emission in CoFe/Al2O3 allowed us to verify that the THz transmission has the same temperature dependence as the resistance, opening the possibility of achieving spintronic devices in the THz spectral range. Moreover, the effect of a femtosecond laser excitation on the THz transmission, and thus the electrical resistivity, was studied. The dependence of the THz emission on the applied magnetic field, in conjugation with the all optical pump probe measurements, enabled a deeper comprehension of the magnetic origin of the THz emission in ferromagnetic materials.

## References

[1] A. Kirilyuk, A.V. Kimel, and T. Rasing, “Ultrafast optical manipulation of magnetic order“ Rev. Mod. Phys. **82**, 2731 (2010).

[2] R. Ulbricht, E. Hendry, J. Shan, T.F. Heinz, M. Bonn, “Carrier dynamics in semiconductors studied with time-resolved terahertz spectroscopy” Rev. Mod. Phys. **83**, 543 (2011).

[3] W. Kleemann, O. Petracic, C. Binek, G.N. Kakazei, Y.G. Pogorelov, J. B. Sousa, S. Cardoso, and P. P. Freitas, “Interacting ferromagnetic nanoparticles in discontinuous $Co_{80}Fe_{20}/Al_2O_3$ multilayers: From superspin glass to reentrant superferromagnetism” Phys. Rev. B. **63**, 134423 (2001).

[4] E. Beaurepaire, G. M. Turner, S. M. Harrel, M. C. Beard, J.-Y. Bigot, and C. A. Schmuttenmaer, “Coherent terahertz emission from ferromagnetic films excited by femtosecond laser pulses” Appl. Phys. Lett. **84**, 3465 (2004).

# Nonthermal magnetization switching by ultrashort acoustic pulses

**O. Kovalenko, T. Pezeril and V.V. Temnov**

Institut des Molécules et Matériaux du Mans, UMR CNRS 6283, Université du Maine,

72085 Le Mans cedex, France

oleksandr.kovalenko.etu@univ-lemans.fr

**Abstract** Magnetoelastic interactions in a thin layer of magnetostrictive material Terfenol-D are investigated theoretically. The phenomenon of ultrafast magnetoacoustic switching by a single picosecond strain pulse is predicted.

## Introduction

The knowledge of the fundamental speed limits of information transfer and recording is crucial for the development of new ultrafast data recording technologies. So far the ultrafast magnetization switching has been demonstrated by applying pulses of magnetic field and spin-polarized current, and, in the most spectacular way, by single circularly polarized femtosecond laser pulses [1]. Trying to take advantage of the recently reported generation of giant ultrashort acoustic pulses in metal-ferromagnet structures [2-4] here we present a new concept for the *nonthermal* magneto-acoustic switching in Terfenol-D [5].

## Magnetoelastic interactions

This rear-earth-based compound is well suited for magnetoacoustic investigations due to the large value of magnetoelastic (magnetostrictive) coupling [6]. The realistic design of a hybrid multilayer structure for magnetoacoustics is presented in Fig.1a. An optoacoustic cobalt transducer is excited through the dielectric substrate by a femtosecond laser pulse and generates a giant ultrashort acoustic pulse, which is injected in a thin (110) layer of Terfenol-D after propagation through gold. If the Terfenol-D film is sufficiently thin and is covered by an acoustically matched dielectric layer, the action of the acoustic pulse can be adequately described by applying an ultrafast but spatially homogeneous time-dependent strain.

 
J.-Y. Bigot et al. (eds), *Ultrafast Magnetism I*, Springer Proceedings in Physics 159
DOI 10.1007/978-3-319-07743-7_82

The dynamics of magnetization induced by an ultrashort acoustic strain pulse η(t) are governed by the Landau-Lifshitz-Gilbert (LLG) equation

$$\frac{d\vec{M}}{dt} = -\gamma_L \vec{M} \times \vec{H}_{eff} - \frac{\alpha\gamma_L}{M_s}[\vec{M} \times (\vec{M} \times \vec{H}_{eff})], \quad 1$$

which describes the damped precession of the magnetization vector around the effective time-dependent magnetic field $\vec{H}_{eff}(t) = -\frac{1}{\mu_0}\frac{dF(\eta(t))}{d\vec{M}}$. The phenomenological expression for the free energy density $F = F_k + F_{MEL} + F_d + F_z$ in Terfenol-D [5] is dominated by the competition of the magnetocrystalline anisotropy

$$F_k = K_1(\alpha_x^2\alpha_y^2 + \alpha_x^2\alpha_z^2 + \alpha_y^2\alpha_z^2) + K_2(\alpha_x^2\alpha_y^2\alpha_z^2) \quad 2$$

and the magnetoelastic energy density

$$F_{MEL} = b_1(\alpha_x^2\varepsilon_{xx} + \alpha_y^2\varepsilon_{yy} + \alpha_z^2\varepsilon_{zz}) + b_2(\alpha_x\alpha_y\varepsilon_{xy} + \alpha_x\alpha_z\varepsilon_{xz} + \alpha_z\alpha_y\varepsilon_{zy}) \quad 3$$

The minima of free energy define four metastable in-plane magnetization directions 1,2,3 and 4 (see Fig. 1a). The explicit dependence of the magnetoelastic term on the components $\varepsilon_{ij}$ of the strain tensor and the directional cosines $\alpha_i$ of the magnetization $\vec{M} = M_s(\alpha_x, \alpha_y, \alpha_z)$ vector suggests that the application of a time-dependent strain η(t) will modify the total free energy density, F=F(η(t)), and induce magnetization precession. Whereas for arbitrary acoustic pulses only numerical solutions of Eq. (1) can be obtained, an ultrashort acoustic pulse with a duration $\tau_{ac} << T_{prec}$ significantly shorter than the precession period $T_{prec} \approx$ 25 ps results in the deflection of the magnetization vector out of the sample plane by an angle

$$\Delta\theta_{ac} \cong \gamma \frac{(b_2 + 2b_1)}{2M_s} \alpha_y^{'} \alpha_x^{'} \int \eta(t)dt, \quad 4$$

which is proportional to the acoustic pulse area $\int \eta(t)dt$ [4].

## Magnetization switching

Numerical simulations show that sufficiently strong acoustic pulses can switch the magnetization vector back and forth between two different metastable states 1 and 2 (Fig. 1b). The magnetoacoustic switching diagram in Fig. 1c shows that depending on the acoustic pulse duration $\tau_{ac}$ and amplitude $\eta_{ac}$, switching is possible between all four metastable magnetization directions. Whereas at short pulse durations below 10 ps the switching strain is inversely proportional to pulse duration, for longer pulses it saturates and reaches the value of 0.3%. Therefore, according to our simulations, the conditions for ultrafast magnetoacoustic switching in Terfenol-D are experimentally accessible.

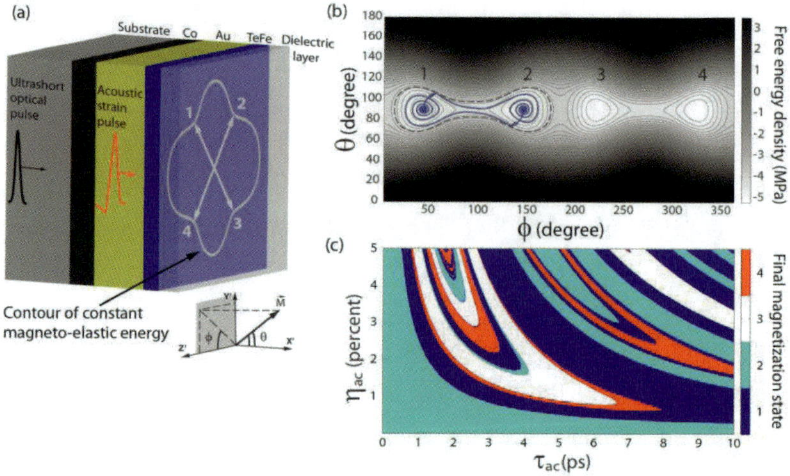


**Fig. 1.** (a) Design of a multilayer structure for the acoustic magnetization switching. An opto-acoustic cobalt transducer is excited by an ultrashort optical pulse and generates a giant acoustic pulse with 1% amplitude and 4 ps duration (see Ref. [3] for details). An acoustic pulse propagates through the gold buffer layer and interacts with magnetization in Terfenol-D. (b) The magnetization vector in a thin (110) Terfenol-D film possesses four metastable in-plane magnetization directions: 1,2,3 and 4. An ultrashort tensile strain pulse with an amplitude of 0.9% and 3 ps duration deflects the magnetization vector out of the sample plane, resulting in its precession and decay into another minimum (magnetization switching between energy minima 1 and 2 occurs). (c) Depending on the duration and amplitude of rectangular acoustic pulses, the magnetization can be switched between its all four metastable states. The initial magnetization state is 2.

**Acknowledgments** The authors thank the "Nouvelle équipe, nouvelle thématique de la Région Pays de la Loire" for funding.

## References

[1] C. D. Stanciu, F. Hansteen, A. V. Kimel, A. Kirilyuk, A. Tsukamoto, A. Itoh, and Th. Rasing, "All-optical magnetic recording with circularly polarized light" Phys. Rev. Lett. **99,** 047601 (2007).
[2] J.-W. Kim, M. Vomir, and J.-Y. Bigot, "Ultrafast Magnetoacoustics in Nickel Films" Phys. Rev. Lett. **109**, 166601 (2012).
[3] V. V. Temnov, "Ultrafast acousto-magneto-plasmonics" Nature Photonics **6,** 728 (2012).
[4] V. V. Temnov et al., "Femtosecond nonlinear ultrasonics in gold probed with ultrashort surface plasmons", Nature Communications **4,** 1468 (2013).
[5] O. Kovalenko, T. Pezeril and V.V. Temnov, "New concept for magnetization switching by ultrafast acoustic pulses" Phys. Rev. Lett. **110**, 266602 (2013).
[6] C. de la Fuente et al., "Magnetocrystalline anisotropy in a (110) $Tb_{0.27}Dy_{0.73}Fe_2$ thin-film" J. Phys. Condens. Matter **16**, 2959 (2004).

# Improving the efficiency of ultrafast optical control of magnetism in GdFeCo continuous films and submicron structures

**R. Medapalli[1], M. Savoini[1], I. Razdolski[1], S. Khorsand[1], A.M. Kalshnikova[2], A. Tsukamoto[3], A. Itoh[3], A. Kirilyuk[1], Th. Rasing[1], and A.V. Kimel[1]**

[1]Radboud University Nijmegen, Institute for Molecules and Materials, Heyendaalseweg 135, Nijmegen, The Netherlands

[2]Ioffe Physical-Technical Institute, Russian Academy of Sciences, 194021 St. Petersburg, Russia

[3]College of Science and Technology, Nihon University, 7-24-1 Funabashi, Chiba, Japan

r.medapalli@science.ru.nl

**Abstract** The goal of this work is to define conditions for the most efficient ultrafast optical control of magnetism. Results show that tuning the composition of the GdFeCo alloys towards the magnetization compensation point as well as reducing the sizes of structures allowing one to reduce the energy of the laser pulse required for ultrafast demagnetization or magnetization reversal.

## Introduction

The possibility to manipulate magnetic order with the help of a femtosecond laser pulse has been a subject of intense discussions in modern magnetism since the seminal observation of subpicosecond laser-induced demagnetization [1]. Ten years later it was also demonstrated that if a rare earth (RE)–transition metal (TM) ferrimagnet having antiferromagnetically coupled nonequivalent RE and TM magnetic sublattices, is brought into a transient state with no net magnetization on a subpicosecond time scale, the subsequent relaxation from this state leads to a deterministic reversal of the initial net magnetization of the medium [2,3]. Here we report about our studies aimed to improve the efficiency of the ultrafast laser-induced demagnetization and magnetization reversal. In particular, we show that varying the ratio between Gd and FeCo in the alloy or the sizes of structures fabricated from GdFeCo thin films we can reduce the energy of the femtosecond laser pulse required for the complete ultrafast demagnetization or magnetization reversal.

 
J.-Y. Bigot et al. (eds), *Ultrafast Magnetism I*, Springer Proceedings in Physics 159
DOI 10.1007/978-3-319-07743-7_83

## Results

For our experiments, we chose continuous films of four different compositions of GdxFe100−x−yCoy (x=18, 22, 24, and 30 %). With the help of these samples, it is possible to create three important cases at a given temperature of the sample (see Fig. 1a): 1) the magnetization of Gd is smaller than that of Fe i.e., MGd < MFe, 2) the magnetizations are approximately equal i.e., MGd ~ MFe, and 3) the magnetization of Gd is larger than that of Fe i.e., MGd > MFe. For the experiments we used a time-resolved stroboscopic magneto-optical pump-probe setup. We investigated the degree of ultrafast demagnetization as a function of sample temperature, laser pump fluence, and concentration of Gd [4]. The degree of demagnetization as a function of laser pulse fluence shows a similar trend in all studied samples. However, results clearly show that the degree of the ultrafast laser-induced demagnetization strongly depends on Gd-concentration and, in particular, on the relative temperature of the ferrimagnet compared to its magnetization compensation point (TM-T). The findings reveal that the ratio between the magnetizations of the Gd and Fe sub-lattices plays a crucial role in the process of the laser-induced demagnetization.

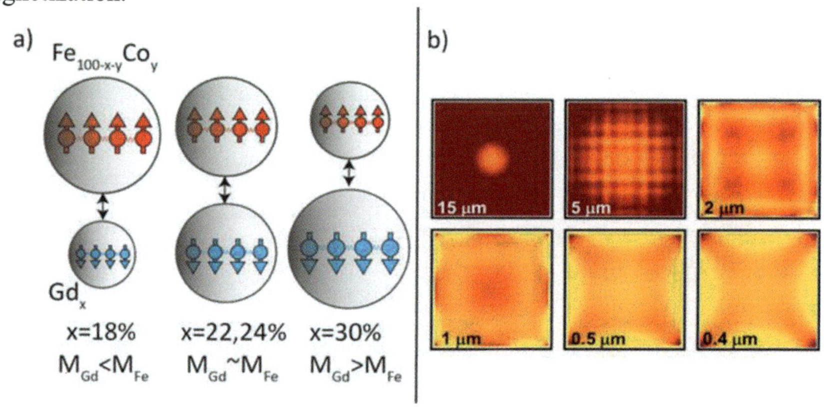


**Fig. 1.** a) Illustration of the three different cases ( MFe>MGd, MFe~MGd, and MFe<MGd) realized in four different compositions of GdFeCo (18%, 22%, 24% , and 30% of Gd) at room temperature, b) shows the electric field profiles for the different structure sizess. The brighter the central spot, the higher the intensity of the radiation is. It is noted that when the structure size is below 1 x 1 μm2, the intensity reaches it maximum.

We also studied ultrafast optical control of magnetism in structures fabricated from continuous films of $Gd_{26}Fe_{64.5}$Co9.5 alloy. For this, we employed a magnetization sensitive microscopy technique [4]. We studied ultrafast demagnetization in structure of various sizes ranging from 15 x 15 $\mu m^2$ to 1 x 1 $\mu m^2$ [5]. We find that smaller structures require a lower energy density to undergo light-induced ultrafast demagnetization and magnetization reversal. In order to understand this, we implemented Finite Difference Time Domain (FDTD) simulations. From the FDTD simulations we can explain this reduction in terms of an enhanced light absorption by interference of the incident and reemitted light-waves. Figure 1b shows the profile of the electric field density in all the structures. The brighter the area, the more it absorbs. Taking the line profiles of the absorbed electric field intensity through the center of the structures, we calculated the absorbed energy. The comparison of the calculations with and without interference effects clearly shows that the smaller structures require up to 40% less energy in order to achieve the full demagnetization.

In conclusion, our experiments clearly show that the ultrafast laser-induced demagnetization and all-optical magnetization reversal are the most effective and require less energy per laser pulse when the sample temperature is in the vicinity of its magnetization compensation temperature and the size of the structures are less than 1 x 1 $\mu m^2$.

**Acknowledgments** The authors thank A. van Etteger and T. Toonen for their technical support. This research was supported by the European Research Council ERC Grant Agreement No. 257280 (Femtomagnetism), the European Union's Seventh Framework Program (FP7/2007-2013) FEMTOSPIN, de Nederlandse Organisatie voor WetenschappelijkOnderzoek (NWO) and de Stichting voor Fundamenteel Onderzoek der Materie (FOM) and the program "Invited Scientist" funded by the Russian Ministry of Education and Science (Grant Agreement 14.B37.21.0899).

## References

[1] E. Beaurepaire, J.-C. Merle, A. Daunois, and J.-Y. Bigot, "Ultrafast spin dynamics in ferromagnetic nickel" Phys. Rev. Lett. 76, 4250 (1996).

[2] C. D. Stanciu, A. V. Kimel, F. Hansteen, A. Tsukamoto, A. Itoh, A. Kirilyuk, and Th. Rasing, "Ultrafast spin dynamics across compensation points in ferrimagnetic GdFeCo: The role of angular momentum compensation" Phys.Rev.B 73, 220402 (2006).

[3] I. Radu, K. Vahaplar, C. Stamm, T. Kachel, N. Pontius, H. A. D¨ urr, T. A. Ostler, J. Barker, R. F. L. Evans, R. W. Chantrell, A. Tsukamoto, A. Itoh, A. Kirilyuk, Th. Rasing, and A. V. Kimel, "Transient ferromagnetic-like state mediating ultrafast reversal of antiferromagnetically coupled spins " Nature 472, 205 (2011).

[4] R. Medapalli, I. Razdolski, M. Savoini, A. R. Khorsand, A. Kirilyuk, A. V. Kimel, Th. Rasng, A. M. Kalashnikova, A. Tsukamoto, A. Itoh, "Efficiency of ultrafast laser-induced demagnetization in GdxFe100-x-yCoy alloys" Phys. Rev. B 86 054442 (2012).

[5] M. Savoini, R. Medapalli, B. Koene, A. R. Khorsand, L. Le Guyader, L. Duo, M. Finazzi, A. Tsukamoto, A. Itoh, F. Nolting, A. Kirilyuk, A. V. Kimel, Th. Rasing, "Highly efficient all-optical switching of magnetization in GdFeCo microstructures by interference-enhanced absorption of light" Phys. Rev. B 86 140404 (2012).

# Magneto-optical resistance induced and controlled by laser pulses

**Michèle Albrecht, Mircea Vomir, Jean-Yves Bigot**

Institut de Physique et Chimie des Matériaux de Strasbourg, UMR7504, CNRS, Université de Strasbourg, BP 43, 23 rue du Lœss, 67034 Strasbourg, France

vomir@ipcms.unistra.fr

**Abstract** We demonstrate that the electrical resistance of metallic magnetic microstructures can be controlled by optically inducing magnetic domains. These domains are patterned with femtosecond laser pulses, in a completely reversible manner. In the case of adjacent parallel stripes, it is shown that the change of the electrical resistance has a linear dependence with the number of domain walls separating the stripes.

## Introduction

In spintronics, the control and transport of spins is a major challenge to improve the speed and density of recording. Regarding the transport properties of spins, an important aspect is the control of the magnetization of layers or of domains. For example, the use of electric currents has proven an efficient way to control domain walls in magnetic wires [1, 2]. Alternatively, in Spin-photonics one can think of using laser beams to manipulate the magnetization of micro and nanostructures. The purpose of this article is to show that the electrical resistance of metallic magnetic microstructures can be controlled by optically writing patterns consisting of adjacent magnetic domains. These induced domains can be erased and rewritten at will. The corresponding resistivity change is related to the spin scattering within the domain walls.

## Results and discussion

We use a time-resolved magneto-optical confocal microscope with femtosecond laser pulses to write and read the magnetic domains [3]. First, the writing sequence is done by locally switching the magnetization, a process which is known to occur via an initial ultrafast thermal change of the spins [4]. The written pattern is then read using a Magneto Optical Pump-Probe Imaging (MOPPI) technique.

 
J.-Y. Bigot et al. (eds), *Ultrafast Magnetism I*, Springer Proceedings in Physics 159
DOI 10.1007/978-3-319-07743-7_84

The sample is a 15 nm thick CoPt multilayer wire with lateral dimensions of $l$=10 μm and length L= 40 μm. The resistance measurement is done using the four points method, with gold electrodes realized by photolithography. Various patterns can be written with a typical size determined by the focus of the pump beam (600 nm diffraction limited).

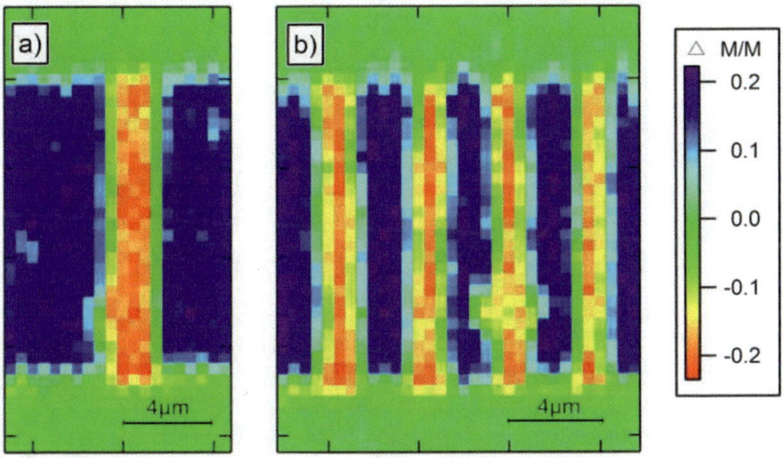


**Fig. 1.** Magnetic domains optically written on a CoPt multilayer wire. (a) single transverse stripe (b) four adjacent stripes.

Figure 1a) shows a typical stripe with a width of 1.5 μm, written transversally to the magnetic wire. Depending on the desired transport properties of the device, several such adjacent stripes can be written as shown in figure 1b) (4 stripes).

The resistance of the wire without stripes, measured for a current applied along the wire, is 104 Ω. As displayed in figure 2, the change of the electric resistance ΔR varies linearly as a function of the number of written magnetic domains. ΔR does not depend on the period of the stripes. Our interpretation of this resistance change is the mechanism of spin scattering within the domain walls. It can be described by a diffusive spin model, taking into account the size of the domain walls (here optically induced).

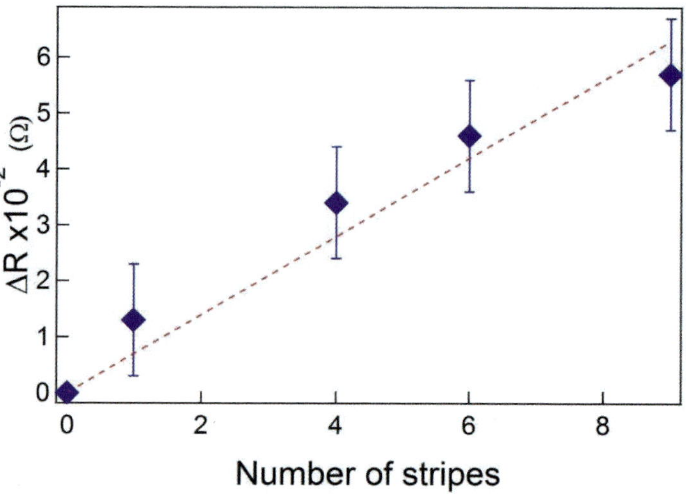


**Fig. 2.** Resistance change as a function of the number of stripes in a 10μm X 15nm X 40μm CoPt wire. The resistance of the wire without stripes is 104Ω. The dashed line is a linear fit.

In conclusion, it is shown that magneto-optics can be used for manipulating the transport properties of magnetic nanostructures. Our approach presents several advantages. The domains are erasable and can be reproduced in a controlled manner, in contrast with magnetic structures having natural stripes which are generally difficult to manipulate with an external control. Such optical devices are elementary bricks for more elaborated Spin-photonic circuits.

**Acknowledgments** We thank the European Research Council for financial support (Project ATOMAG: ERC-2009-AdG-20090325 247452). We are grateful to G. Versini, M. Acosta and S. Barre for their technical support. The microstructures have been made in the STnano platform facility at IPCMS.

## References

[1] C. H. Marrows, "Spin-polarised currents and magnetic domain walls" Adv. Phys. **54**, 585 (2005).
[2] S. S. P. Parkin, M. Hayashi, and L. Thomas, "Magnetic Domain-Wall Racetrack Memory" Science **320**, 190 (2008).
[3] A. Laraoui, M. Albrecht, J.-Y. Bigot, "Femtosecond magneto-optical Kerr microscopy" Opt. Lett., **32**, 936 (2007).
[4] E. Beaurepaire, J. C.Merle, A.Daunois, and J.-Y. Bigot, "Ultrafast spin dynamics in ferromagnetic nickel", Phys. Rev. Lett. **76**, 4250 (1996).

# Part VII
# Spin Photo-Emission Dynamics

# The valence band structure of Gadolinium studied with time-resolved photoemission

**B. Frietsch, J. Bowlan, R. Carley, M. Teichmann,J. Wolter, and M. Weinelt**

Fachbereich Physik, Freie Universität Berlin, Arnimallee 14, 14195 Berlin, Germany and Max-Born-Institut, Max-Born-Strasse 2A, 12489 Berlin, Germany

weinelt@physik.fu-berlin.de

**Abstract** We have studied the response of the exchange split valence bands of ferromagnetic gadolinium tofemtosecond laser excitation. We observe a drop of the exchange splitting with a time constant of 0.9 ps but different response times of minority and majority spin bands. Furthermore, even above the Curie temperature there is a finite exchange splitting, which also decreases with laser excitation.

## Introduction

It is well established that optical excitation of thin ferromagnetic films leads to demagnetization within a few hundred femtoseconds [1] and even ultrafast optical magnetic switching of ferrimagnets has been demonstrated [2]. However, it remains controversial which microscopic processes are fast enough to provoke femtomagnetism: direct interaction with the laser field [3], scattering among electrons [4], phonons [5], and magnons [6], and/or spin-transport [7].To establish microscopic models it is necessary to unravel the transient electronic structure [8] and clarify on which timescale the exchange splitting $\Delta E_{ex}$ and spin polarization collapses.

To approach these problems we perform time- and angle-resolved photoemission (TR-ARPES). We use laser-driven high-order harmonic generation as an extreme ultraviolet tight source [9]. A Ti:Sapphire CPA laser systemdelivers1.5mJ, 40 fs IR pulses at 10 kHz repetition rate, which are focused into a gas cell containing argon at 100 mbar. Individual harmonics are selected by a toroidal grating monochromator (TGM) with a single grating. The time resolution $\tau \sim \lambda / cN$ is primarily determined by the transmitted bandwidth ($N$ is the number of illuminated lines). While the temporal spread is nearly independent of the TGM slit-size, the latter allows tuning of the energy resolution $\Delta E$. For 10 μm slits we obtain $\tau = 100$ fs, $\Delta E$ =150 meV, and $2 \cdot 10^4$ photons/pulse at $h\nu$ = 35.6 eV.

 
J.-Y. Bigot et al. (eds), *Ultrafast Magnetism I*, Springer Proceedings in Physics 159
DOI 10.1007/978-3-319-07743-7_85

## Gadolinium valence-band dynamics and exchange splitting

We have studied ultrafast demagnetization of the local-moment ferromagnet gadolinium [10], prepared as epitaxial film of 10 nm thickness on a W(110) substrate. Photoelectrons are detected after a hemispherical energy analyzer. Tuning the TGM to 35.6 eV allows us to map the valence bands of Gd at the Γ-point of the 4thBrillouin zone. Binding energy and exchange splitting of the $\Delta_2$-like Σbands are shown in Fig.1a. The initial binding energies of 1.39 and 2.25 eV are in agreement with synchrotron measurements conducted at a comparable sample temperature of 100 K [11]. Upon laser excitation (hν = 1.6eV, s-pol., 300 fs, fluence1.2 mJ/cm$^2$) the exchange splitting of the valence bands drops with a time constant of 0.9 ±0.1 ps. This value coincides with the values observed in MOKE [12] and XMCD[13] measurements. The latter experiment probes the *4f* spin subsystem but was conducted at higher pump fluences (3-5 mJ/cm$^2$). Comparing ultrafast de- and thermal re-magnetization [10] we conclude that $\Delta E_{ex}$ maps the true magnetization. From Fig. 1a it is evident that the minority valence band reacts immediately after laser excitation while the response of its majority counterpart is delayed and is only half as fast. Delay zero was independently determined from the first occurrence of pump-induced hot electrons above the Fermi level $E_F$. We attribute the instantaneous response of the minority spin band to superdiffusive spin transport [7]. Photoemission probes the first few surface layers and is thus very sensitive to spin currents into the bulk. The delayed response of the majority band can be explained by electron-phonon scattering which increases with increasing lattice temperature [5]. Our data suggest that both spin transport and electron-phonon scattering contribute to ultrafast demagnetization. In addition, the molecular field acting on the spins may delay the response of the majority spin band [14].

It has been a long standing question whether the exchange splitting vanishes at the Curie temperature $T_C$ = 293 K [15]. For the Gd surface state scanning tunnel-

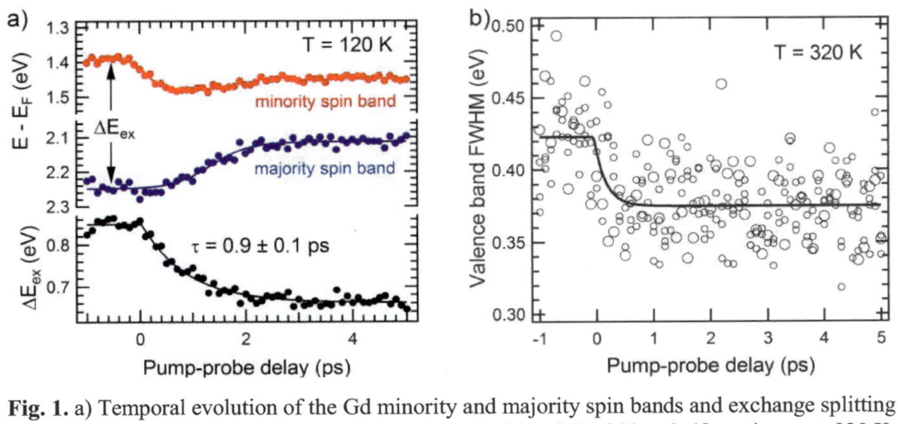


**Fig. 1.** a) Temporal evolution of the Gd minority and majority spin bands and exchange splitting upon optical excitation. b) Decrease of the valence band's full width at half maximum at 320 K.

Ing spectroscopy revealed a sizeable exchange splitting at $T_C$ [16]. However, photoemission measurements of the bulk valence bands remain ambiguous since the minority and majority spin bands coalesce into a single peak due to thermal broadening [11,15]. In Fig. 1b we show the linewidth of this peak as a function of pump-probe delay at 320 K. Upon laser excitation we observe a small but significant decrease of the linewidth, which can only be explained by a further reduction of $\Delta E_{ex}$. Above $T_C$there exists in fact a finite exchange splitting, which implies local magnetic order.

**Acknowledgments** We gratefully acknowledge funding by the Deutsche Forschungsgemeinschaft, the Leibniz graduate school *D in L,* and the Helmholtz Virtual Institute *Dynamic Pathways in Multidimensional Landscapes*. J.B. is indebted to the Humboldt Foundation for a scholarship.

## References

[1] E. Beaurepaire, J.-C.Merle, A. Daunois, and J.-Y. Bigot, “Ultrafast Spin Dynamics in Ferromagnetic Nickel” Phys. Rev. Lett. **76**, 4250 (1996).
[2] I. Radu *et al.*, “Transient ferromagnetic-like state mediating ultrafast reversal of antiferromagnetically coupled spins” Nature **472**, 205 (2011).
[3] J.-Y. Bigot, M. Vomir, and E. Beaurepaire, “Coherent ultrafast magnetism induced by femtosecond laser pulses” Nature Physics **5**, 515 (2009).
[4] M. Krauß *et al.*, “Ultrafast demagnetization of ferromagnetic transition metals: The role of the Coulomb interaction” Phys. Rev. B **80**, 180407 (2009).
[5] B. Koopmans *et al.*, “Explaining the paradoxical diversity of ultrafast laser-induced demagnetization” Nature Mater. **9**, 259–265 (2010).
[6] A.B. Schmidt et al., „Ultrafast Magnon Generation in an Fe Film on Cu(100)“ Phys. Rev. Lett. **105**, 197401 (2010).
[7] M. Battiato, K. Carva, and P. M. Oppeneer, “Superdiffusive Spin Transport as a Mechanism of Ultrafast Demagnetization” Phys. Rev. Lett.**105**, 027203 (2010).
[8] A. J. Schellekens and B. Koopmans, “Comparing Ultrafast Demagnetization Rates Between Competing Models for Finite Temperature Magnetism” Phys. Rev. Lett. **110**, 217204 (2013); K. Carva, M. Battiato, D. Legut, and P. M. Oppeneer, “Ab initio theory of electron-phonon mediated ultrafast spin relaxation of laser-excited hot electrons in transition-metal ferromagnets” Phys. Rev. B **87**, 184425 (2013).
[9] B. Frietsch et al., “A high-order harmonic generation apparatus for time- and angle-resolved photoelectron spectroscopy” Rev. Sci. Instrum.**84** (2013), in press.
[10] R. Carley et al., “Femtosecond Laser Excitation Drives Ferromagnetic Gadolinium out of magnetic Equilibrium” Phys. Rev. Lett. **109**, 057401 (2012).
[11] C. Schüßler-Langeheine, “Magnetic properties of thin films of heavy lanthanide metals studied by magnetic x-ray diffraction and high-resolution photoemission,” Ph.D. thesis, Fachbereich Physik, Freie Universität Berlin, 1999.
[12] M. Sultan et al., “Electron- and phonon-mediated ultrafast magnetization dynamics of Gd(0001)”, Phys. Rev. B **85**, 184407 (2012).
[13] M. Wietsruk et al., “Hot-electron-driven enhancement of spin-lattice coupling in Gd and Tb 4f ferromagnets observed by Femtosecond X-ray magnetic circular dichroism” Phys. Rev. Lett. **106**, 127401 (2011).
[14] M.Avignon and K.-H. Bennemann, http://arxiv.org/pdf/1305.5698.pdf.

[15] K. Maiti, M. C. Malagoli, A. Dallmeyer, and C. Carbone, “Finite Temperature Magnetism in Gd: Evidence against a Stoner Behavior” Phys Rev. Lett. **88**, 167205 (2002).
[16] M. Bode, M. Getzlaff, A. Kubetzka, R. Pascal, O. Pietzsch, and R. Wiesendanger, “Temperature-Dependent Exchange Splitting of a Surface Stateon a Local-Moment Magnet: Tb(0001)”, Phys. Rev. Lett. **83**, 3017 (1999).

# Mechanisms of Multiphoton Photoemission from Metal Surfaces

**Xuefeng Cui, Cong Wang, Adam Argondizzo, and Hrvoje Petek**

Department of Physics and Astronomy, University of Pittsburgh, Pittsburgh PA 15260 USA

petek@pitt.edu

**Abstract** Multiphoton photoemission (MPP) from metal surfaces with energy, momentum, time, and resolution enables probing the electronic structure, optical response, and electron dynamics at metal surfaces. We discuss recent progress in understanding of the MPP excitation process for Ag(111) and Cu(111) surfaces using tunable femtosecond laser excitation and photoelectron *E(k)* imaging.

## Introduction

Interferometric time-resolved two-photon photoemission (ITR-2PP) has proven capable of measuring quasiparticle correlations in noble metals.[1] For instance, the dephasing of Shockley surface state and top of the *d*-band holes in copper has been studied in a nonresonant 2PP process with 3.1 eV, 10 fs excitation.[2,3] Modeling of the coherent 2PP process by optical Bloch equations proved inadequate to reproduce experimental data, suggesting that the coherent polarization dynamics in a metal are more complex than could be reproduced by a simple excitation scheme treating the optically coupled levels at the momentum of observation. We reexamine MPP processes from Shockley surface (SS) states of Ag(111) and Cu(111) surfaces with a broadly tunable <15 fsexcitation NOPA source for various resonance conditions using 3D (energy, momentum, and time resolved) MPP.

## Results and Discussion

Three and four photon photoemission processes are excited with a 1-2 MHz repetition rate fiber-pumped NOPA system (Clark MXR). Photoemitted electrons are detected with an energy and momentum imaging photoelectron spectrometer (SPECS). The degenerate interferometric pump-probe measurements are taken by splitting the excitation pulses into two replicas in a Mach-Zehnder interferometer; acquiring photoelectron images in ~100 as delay intervals obtains movies of the coherent polarization dynamics excited in the single crystal metal samples.



J.-Y. Bigot et al. (eds), *Ultrafast Magnetism I*, Springer Proceedings in Physics 159
DOI 10.1007/978-3-319-07743-7_86

Figure 1 shows the MPP spectra of Ag(111) surface excited with and 1.81 and 2.20 photon excitation showing dispersive bands due to the initial SS and intermediate image potential (IP) states at -0.063 and 3.79 eV relative to the Fermi level ($E_F$). The range of SS is limited to $|k_{||}|$=0.07 Å$^{-1}$by its intersection with $E_F$. The band dispersions are consistent with the known band structure of Ag(111). Tuning the laser into two-photon resonance between the SS and IP state causes dramatic changes in the MPP spectra, which can be interpreted through strong surface electron-hole correlation. 3D photoelectron measurements show evidence for this correlation in the time domain. In short, the MPP process is found to be driven by the local field of the correlated state.

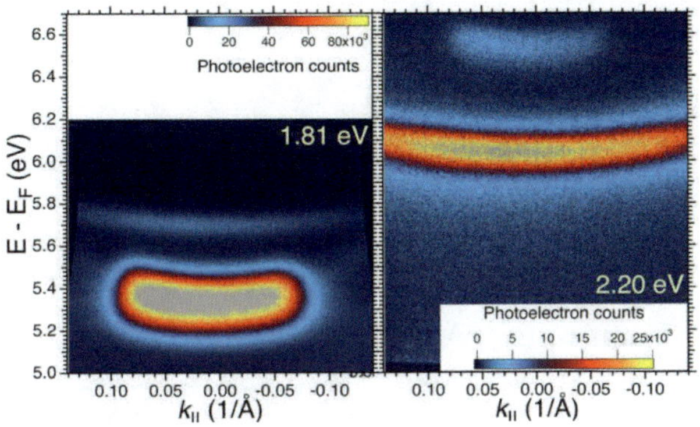


**Fig. 1.**E(k) images of MPP from Ag(111) surface excited with 1.81 and 2.20 eV light.

Similar experiments have suggested that the intermediate "virtual" state in nonresonant 2PP from SS of Cu(111) surface has a finite "lifetime", which is inconsistent with its detuning from real states.[2] Measurements over a broad excitation photon energy range suggest that the apparent slow dephasing of the linear polarization is associated with the coherent polarization induced in the sample at *E(k)* regions that are distant from that of the observation. Apparently, the coherent polarization dynamics detected at the SS feature reflect the optical response of bulk Cu, and coupling between the surface and bulk polarizations.

The detailed 3D measurements on Ag and Cu provide unprecedented view into optically induced correlation in noble metals, which is hidden in the linear optical response. The lessons learned from noble metals are applicable to more complicated materials where novel properties emerge from quasiparticle correlations.

**Acknowledgments** This research was supported by Division of Chemical Sciences, Geosciences, and biosciences, office of Basic Energy Science of the U.S. Department of Energy through grant DE-FG02-09ER16056.

## References

[1] H. Petek, S. Ogawa,"Femtosecond Time-Resolved Two-Photon Photoemission Studies of Electron Dynamics in Metals"Prog. Surf. Sci. **56**, 239-310 (1997).
[2] S. Ogawa, H. Nagano, H. Petek, A. P. Heberle, "Optical Dephasing in Cu(111) Measured by Interferometric Two-photon Time-resolved Photoemission"Phys. Rev. Lett.**78**, 1339-1342 (1997).
[3] H. Petek, H. Nagano, S. Ogawa, "Hole Decoherence of d-Bands in Copper"Phys. Rev. Lett.**83**, 832-835 (1999).

# Time-Resolved Photo-Emission Electron Microscopy of Nanomagnetic Logic Chains

**Z. Gu[1], R. Storz[2], M. Marcus[3], A. Doran[3], A. Young[3], A. Scholl[3], W. Chao[3], D. Carlton[3], B. Lambson[1], M. Nowakowski[1] and J. Bokor[1]**

[1]Department of Electrical Engineering and Computer Sciences, University of California, Berkeley, CA 94720. [2]Thorlabs, Inc., 56 Sparta Avenue, Newton, New Jersey 07860. [3]Lawrence Berkeley National Laboratory, Berkeley, CA 94720

zgu@eecs.berkeley.edu

**Abstract** We report a time-resolved study of precessional timescale nanomagnetic logic dynamics. We find both the desired cascade-like signal transmission behavior as well as various logical defects. For cascade-like behavior, we observe an average settling time of ~100 picoseconds per nanomagnet.

## Introduction

Nanomagnetic Logic is a post-CMOS logic device technology based on stray field-coupled single-domain ferromagnetic islands[1] with a promising path towards ultra-low power operation[2]. One critical aspect of the operation is the reset and clocking of the magnetic logic elements in the computation path. Much of the past work on nanomagnetic logic used external magnetic fields to perform this reset operation. There have been some experiments that used integrated clock lines for this purpose[3], though the dynamics were not studied and the speed of operation was slow (~usec time scale). In practical applications, the reset operation should be fast (sub-nsec). For the same reason, the circuit dynamics should also be fast. We report the first time-resolved study of sub-nsec precessional timescale nanomagnetic logic dynamics following fast clock pulses. We further suggest an exciting possibility for the future that ultrafast (psec timescale) magnetization dynamical processes to be discussed in this conference could be applied to dramatically speed up magnetic logic.

## Experimental results

A fundamental logic structure in nanomagnetic logic is a linear chain. The stray fields of neighboring nanomagnets couple such that the lowest energy state for a

 
J.-Y. Bigot et al. (eds), *Ultrafast Magnetism I*, Springer Proceedings in Physics 159
DOI 10.1007/978-3-319-07743-7_87

chain of magnets with easy axes orthogonal to the chain axis is antiparallel magnetization. Such a chain is therefore a series of logical inverters and can thus be used to transmit information. Logic gates have also been demonstrated[1]. One requirement for this process is that it must be causal; a nanomagnet in the chain must not flip until information has been transmitted to it through the preceding nanomagnets in the chain. This results in nanomagnets flipping one by one along a chain in a cascade-like behavior. Nanomagnets that have been reset through the clock process must remain reset until a neighbor flips. To implement this experimentally, we used configurational anisotropy to engineer a secondary easy axis orthogonal to the primary easy axis[4]. The concave nanomagnets with indentations along their short edges as shown in Fig. 1a give rise to this secondary easy axis. We fabricated chains of 7 concave nanomagnets each from 10 nm thick evaporated Permalloy. Each nanomagnet is 150 nm wide and is separated from its neighbors by a 30 nm gap.

This pattern was created by electron beam lithography and liftoff on top of 160 nm thick rectangular copper lines. When current passes through these clock lines, they produce magnetic fields along the direction of the secondary easy axes to reset the chains. We perform a pump-probe x-ray magnetic circular dichroism experiment using the electron yield technique at the PEEM-3 beamline at the Advanced Light Source at Lawrence Berkeley National Lab by passing current pulses synchronized to the x-ray pulses through these clock lines as the pump. Fig. 1b shows an image intensity spectrum that qualitatively represents the pulse shape. A pulse width of roughly 2 ns, a trailing edge transition time of roughly 500 ps, and an overshoot are all readily apparent. We calculate the peak near field produced to be 135 mT based on the measured voltage and resistance of the clock lines.

We capture magnetic contrast images by averaging exposure at a particular time delay between pump and probe with both left-circular and right-circular x-ray polarizations. Images taken with the two polarizations are then divided to produce an intensity contrast that shows black and white for opposite magnetizations. Fig. 2 shows a time series captured in this manner for a chain of nanomagnets 410 nm in length. We first verify that our clock pulses are successfully resetting the nanomagnets by observing in the +1.0 ns image that they all appear gray and have no net magnetization along the horizontal axis. We then examine the series of images given by times +2.0, +2.2, and +2.4 ns. The nanomagnet in the middle of the chain first flips to black. After 200 ps, the two following nanomagnets flip to white and black respectively. After another 200 ps, the final nanomagnet has flipped to white. This is cascade-like behavior though it did not occur throughout the chain. This series allows us to estimate that in cascade-like behavior, each nanomagnet takes approximately 100 ps to flip.

Fig. 2 also shows two defects: a cluster of two neighboring black nanomagnets at the beginning of the chain, and the nanomagnet that flipped black at +2.0 ns without the influence of either neighbor. We attribute defects to lithographic roughness as well high temperature effects arising from poor thermal transfer in vacuum. Further optimization to reduce defects is planned for future work.

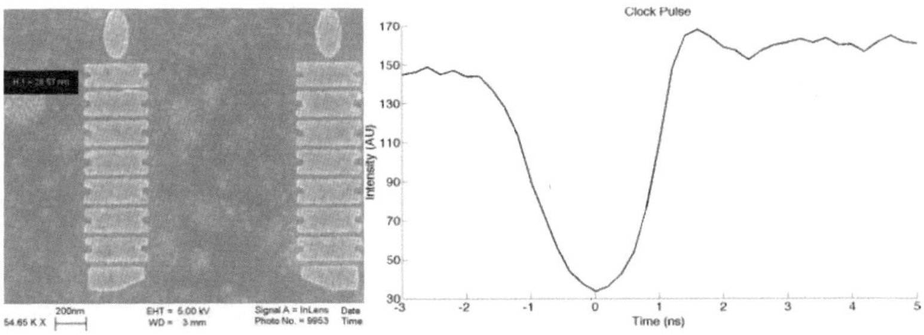


**Fig. 1.** a (left): Scanning Electron Micrograph of a pair of chains on top of a clock line. b (right): Image intensity spectrum representing the average shape of the clock pulses.

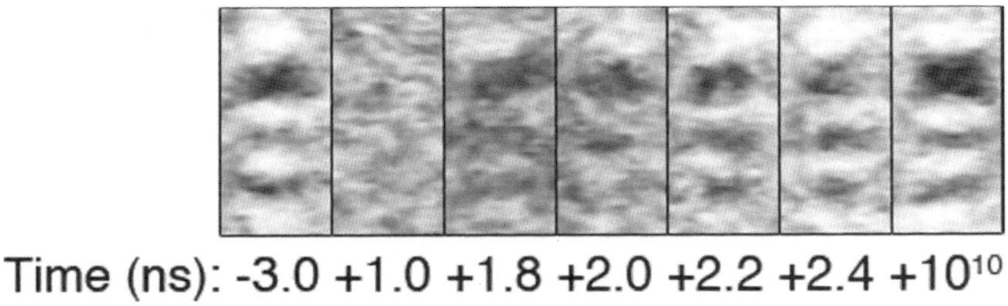


**Fig. 2.** Magnetic contrast images of a chain of 410 nm by 150 nm nanomagnets at different times relative to the pulse peak. Orientation of the chain is upside down relative to Fig. 1. Black and white regions represent opposite magnetizations along the horizontal direction. The $10^{10}$ image is a static image taken minutes after pulsing has been turned off.

**Acknowledgments** The authors thank the Advanced Light Source and Center for X-Ray Optics Nanofabrication Lab at Lawrence Berkeley National Lab, and the Marvell Nanofabrication Lab at the University of California, Berkeley for use of their facilities. This work was supported by the Western Institute of Nanoelectronics, the National Science Foundation, and the Department of Energy.

# References

[1] A. Imre et al., "Majority Logic Gate for Magnetic Quantum-Dot Cellular Automata" Science **311**, 205–208 (2006).

[2] B. Lambson, D. Carlton, and J. Bokor, "Exploring the Thermodynamic Limits of Computation in Integrated Systems: Magnetic Memory, Nanomagnetic Logic, and the Landauer Limit" Phys. Rev. Lett. **107**, 10604 (2011).

[3] M. T. Alam et al., "On-chip clocking of nanomagnet logic lines and gates" IEEE T. Nano. **11**, 273–286 (2012).

[4] B. Lambson, Z. Gu et al., "Cascade-like signal propagation in chains of concave nanomagnets" Appl. Phys. Lett. **100**, 152406 (2012).

# Spin-selective excitation pathways in nonlinear photoemission from metal surfaces

**A. Winkelmann, C.-T. Chiang, M. Pazgan, T.R.F. Peixoto, J. Kirschner**

Max-Planck-Institut für Mikrostrukturphysik, Weinberg 2, 06120 Halle(Saale), Germany

winkelm@mpi-halle.mpg.de

**Abstract** We observe an enhanced spin-polarization and a reduction of surface sensitivity due to intermediate unoccupied electronic states in spin-resolved two-photon photoemission (2PPE) from ferromagnetic cobalt films. We discriminate 2PPE signals coming from initial and intermediate electronic states. This allows us to determine the influence of unoccupied quantum-well states in the two-photon transitions.

## Introduction

Spin-polarized photoemission is a widely applicable method to analyze the electronic structure of materials with spin resolution. This method has evolved into a powerful technique especially in combination with femtosecond lasers, allowing the manipulation of spin-polarized excited electrons at surfaces and interfaces using multi-photon optical transitions in nonlinear photoemission.

## Spin-dependent two-photon photoemission via quantum-well states

Because of the relatively abundant possibilities for non-linear photoexcitation pathways between the continua of electronic bands, it is challenging to discriminate spin-polarized electrons optically injected into specific bulk electronic states. Suitable conditions for the study of selective optical transitions via bulk-derived electronic states can be offered by ultrathin films. Within a few atomic layers, the electronic states confined within the interior of the film form discrete quantum-well (QW) states, providing an opportunity to study bulk-derived unoccupied states by 2PPE.

 
J.-Y. Bigot et al. (eds), *Ultrafast Magnetism I*, Springer Proceedings in Physics 159
DOI 10.1007/978-3-319-07743-7_88

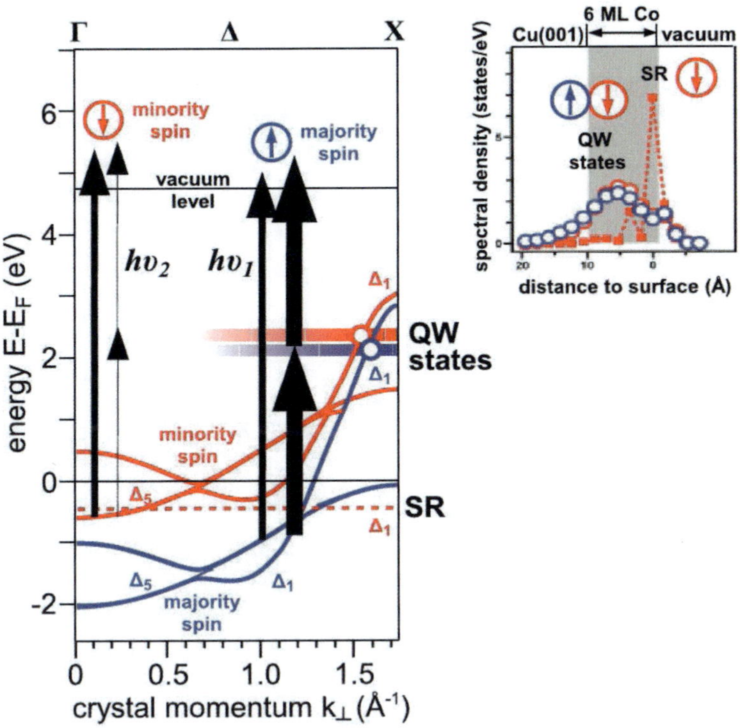


**Fig. 1.** Left: Schematic 1PPE and 2PPE excitation pathways in the fcc Co band structure relevant for 6ML Co films grown on Cu(001). Right: depth-resolved spectral density of the quantum-well states compared to the minority surface-resonance on Co(001) (from [2]).

In our specific case, we observe an increased spin polarization in 2PPE via intermediate quantum-well states in Co/Cu(001) films [1]. We observe 2PPE spin polarizations near 40% compared to about 20% in one-photon photoemission (1PPE) [2]. This effect is attributed to the spin, energy, and momentum selection in the two-photon transitions and leads to the preferred excitation of majority quantum well states (see Fig.1, left). For these states, we observed magnetic dichroism due to spin-orbit coupling [1] and we obtain significant lifetime effects in time-resolved measurements. Our results also show that the selection of specific spin-dependent excitation pathways in nonlinear photoemission can change the sensitivity to the spatial distribution of the states involved (see Fig.1, right). In our case, we go from more surface-specific excitations in 1PPE toward bulk-derived transitions in 2PPE, which is important for mapping *bulk* unoccupied states using angle-resolved nonlinear photoemission.

## References

[1] C.-T. Chiang, A. Winkelmann, P. Yu, and J. Kirschner, "Magnetic Dichroism from Optically Excited Quantum Well States" Phys. Rev. Lett. **103**, 077601 (2009)
[2] C.-T. Chiang, A. Winkelmann, J. Henk, F. Bisio, J. Kirschner, "Spin-selective pathways in linear and nonlinear photoemission from ferromagnets" Phys. Rev. B **85**, 165137 (2012)

# Part VIII
# X-Ray and Far UV-Spin Dynamics

# Ultrafast demagnetization dynamics in the presence of nanometer sized magnetic domains

**Jan Lüning**

Laboratoire de Chimie Physique – Matière et Rayonnement (UMR CNRS 7614)
Université Pierre et Marie Curie, 11 Rue Pierre et Marie Curie, 75005 Paris, France

Jan.Luning@upmc.fr

**Abstract** Experiments probing ultrafast demagnetization dynamics in the presence of a nanometer sized magnetic domain structure provide evidence that angular momentum transport by the excited spin polarized electrons may play an important role in those systems. These results are discussed also in regard of more refined follow-up X-ray based experiments.

## Introduction

Since the discovery of the ultrafast demagnetization phenomenon by E. Beaurepaire and his colleagues in 1996 [1], the field of femtomagnetism has developed to an active research area. Initial experiments relied mostly on all-optical pump-probe techniques, which raised concerns about optical artifacts affecting the measurement, a subject which was controversially discussed, see for example Ref. [2] and references therein. Since X-ray based techniques can overcome these limitations, the advent of sources providing femtosecond short X-ray pulses was waited for by the interested community. Indeed, first experiments realized at the BESSY femtoslicing source by C. Stamm and colleagues [3] fulfilled these expectations and yielded novel insight on details of the underlying mechanism.

In addition to giving access to the entire electronic structure of the valence band, X-ray techniques offer additional key advantages. First of all, this is their shorter wavelength, which matches naturally the nanometer length scales expected to be of relevance in ultrafast magnetization dynamics. Furthermore, X-ray techniques provide via the accessible core electron absorption resonances element sensitivity and a wide variety of magnetic dichroism effects that can, for example, be exploited in scattering experiments as contrast mechanism. This allows individual probing of components of complex, heterogeneous materials with a combined femtosecond temporal and nanometer spatial resolution.



J.-Y. Bigot et al. (eds), *Ultrafast Magnetism I*, Springer Proceedings in Physics 159
DOI 10.1007/978-3-319-07743-7_89

## Results and Discussion

In two recent experiments we employed resonant magnetic small angle X-ray scattering at femtosecond pulsed XUV sources to probe ultrafast magnetization dynamics in the presence of nanometer sized magnetic domains. Magnetic multilayer films with an out-of-plane magnetic anisotropy were used as samples, which can be prepared to exhibit a magnetic domain structure of randomly oriented or linearly aligned worm domains. In both experiments scattering contrast was obtained by tuning of the photon energy to match the magnetically dichroic Co $M_{2,3}$ absorption edges.

The first experiment [4] was realized at the higher harmonic generation (HHG) source of the Laboratoire d'Optique Appliquée (Palaiseau, France), which is driven by a 1 kHz Ti:sapphire laser system providing 800 nm pulses of 35 fs duration and 20 mJ pulse energy. Since the duration of the harmonics is short in comparison to the pulse length of the driving IR laser, and since pump and probe pulse originate from the same laser (jitter-free), an excellent overall time resolution can be obtained in such experiments. This has allowed us to reveal in Co/Pd multilayer and alloy films a demagnetization dynamic which is about a factor of 2 faster than what has been observed before on saturated, single domain films of similar or other Co containing materials. In addition, we find that the time constant describing the initial demagnetization dynamics does not depend on the intensity of the pump pulse [4].

Both these observations can be understood by postulating the presence of a significant angular momentum transfer by superdiffusion of hot spin polarized electrons as recently proposed by Battiato and coworkers [5]. The presence of the nanometer sized magnetic domain structure offers an additional channel for angular momentum transfer between neighboring magnetic domains exhibiting antiparallel spin polarization. Furthermore, we note that the velocity of the excited electrons does uniquely depend on the energy of the pump photons, but not on the intensity of the pump pulse. It is thus understandable that the time constant of the initial demagnetization dynamics is not influenced by the pump intensity.

The idea of a significant direct angular momentum transfer by the excited valence band electrons is further supported by our second experiment [6], which has been realized at the XUV Free Electron Laser FLASH (Hamburg, Germany). The high peak brilliance of these sources enables single x-ray pulse based snapshot characterization of the transient state following an ultrafast excitation. We have exploited this to probe the demagnetization dynamics stimulated by very intense pump pulses leading to a nearly complete quenching of the magnetization. Under these conditions, we have observed a modification in the magnetic small angle scattering pattern, which can be interpreted as the signature of a spin accumulation in the domain wall region, thus broadening the magnetization profile between neighboring domains [6].

We note that both these experiments provide only indirect evidence for the relative importance of angular momentum transport to the overall demagnetization process. More refined X-ray based techniques, on the other hand, allow further scrutinizing the proposed model and some of these approaches, as well as preliminary results, will be discussed.

**Acknowledgments** The author acknowledges financial support from the French Agence Nationale de la Recherche (ANR) under grant NT09491966 (FEMTO-X-MAG – 'Nanoscale imaging of femtosecond magnetization dynamics by scattering of coherent soft X-ray laser harmonics'), from Ile-de-France via the CNano program (contracts DYNAVO and DYNAVO-PostDoc), from the Triangle de la physique (contracts IMMAGE and POL-IMMAGE), and from the CNRS through the PEPS project SASELEX.

## References

[1] E. Beaurepaire, J.-C. Merle, A. Daunois, J.-Y. Bigot, "Ultrafast Spin Dynamics in Ferromagnetic Nickel" Phys. Rev. Lett. **76**, 4250–4253 (1996).
[2] B. Koopmans, M. van Kampen, J. T. Kohlhepp, W. J. M. de Jonge, "Ultrafast Magneto-Optics in Nickel: Magnetism or Optics?" Phys. Rev. Lett. **85**, 844–847 (2000).
[3] C. Stamm, T. Kachel, N. Pontius, R. Mitzner, T. Quast, K. Holldack, S. Khan, C. Lupulescu, E. F. Aziz, M. Wietstruk, H. A. Dürr, W. Eberhardt, "Femtosecond modification of electron localization and transfer of angular momentum in nickel" Nat Mater **6**, 740–743 (2007).
[4] B. Vodungbo, J. Gautier, G. Lambert, A. B. Sardinha, M. Lozano, S. Sebban, M. Ducousso, W. Boutu, K. Li, B. Tudu, M. Tortarolo, R. Hawaldar, R. Delaunay, V. López-Flores, J. Arabski, C. Boeglin, H. Merdji, P. Zeitoun, J. Lüning, "Laser-induced ultrafast demagnetization in the presence of a nanoscale magnetic domain network" Nature Communications **3**, 999 (2012).
[5] M. Battiato, K. Carva, P. M. Oppeneer, "Superdiffusive Spin Transport as a Mechanism of Ultrafast Demagnetization" Phys. Rev. Lett. **105**, 027203:1-4 (2010).
[6] B. Pfau, S. Schaffert, L. Müller, C. Gutt, A. Al-Shemmary, F. Büttner, R. Delaunay, S. Düsterer, S. Flewett, R. Frömter, J. Geilhufe, E. Guehrs, C. M. Günther, R. Hawaldar, M. Hille, N. Jaouen, A. Kobs, K. Li, J. Mohanty, H. Redlin, W. F. Schlotter, D. Stickler, R. Treusch, B. Vodungbo, M. Kläui, H. P. Oepen, J. Lüning, G. Grübel, and S. Eisebitt, "Ultrafast optical demagnetization manipulates nanoscale spin structure in domain walls" Nature Communications **3**, 1100 (2012).

# Catching the moment – magnetization dynamics studied with X-ray Photoemission Electron Microscopy

**L. Le Guyader[1,2], S. El. Moussaoui[1], M. Buzzi[1] and F. Nolting[1]**

[1]Swiss Light Source, Paul Scherrer Institut, 5232 PSI-Villigen, Switzerland

[2]Helmholtz-Zentrum Berlin für Materialien und Energie GmbH, Albert-Einstein-Strasse 15, 12489 Berlin, Germany

Frithjof.nolting@psi.ch

**Abstract** Time resolved Photoemission Electron Microscopy (PEEM) using a stroboscopic pump probe mode in combination with X-ray magnetic circular dichroism (XMCD) has proven to be a versatile tool to access the magnetization dynamics in microstructures and thin films. Here, PEEM studies of the dynamics of Co/$SmFeO_3$ and of GdFeCo will be presented.

## Introduction

Ultrafast control of the magnetization of thin films using femtosecond laser pulses has attracted remarkable interest in the last fifteen years. Beginning with the report of an unexpected ultrafast demagnetization in a nickel thin film in 1996 [1], the research on ultrafast magnetization dynamics quickly developed. While the proper microscopic description of ultrafast demagnetization is still intensely debated, novel laser-induced magnetic phenomena have been discovered during these years in a large variety of materials ranging from ferromagnets (FMs) to antiferromagnets (AFMs) and from metals to insulators [2]. Using synchrotron-based X-ray techniques, the different elements composing a heterostructure can be selectively investigated due to the element specificity and sensitivity to different magnetic order, e.g. ferromagnetic and antiferromagnetic, provided by widely tunable polarized X-rays and allowing investigations in a wide range of materials and experimental conditions. Here we present the application of Photoemission Electron Microscopy (PEEM) for the study of laser-induced magnetization dynamics.

At the Surface/Interface: Microscopy (SIM) beamline of the Swiss Light Source (SLS) at the Paul Scherrer Institut (PSI) a femtosecond (fs) laser is used as the pump-pulse (Femtosource Scientific XL 500, 500 nJ per pulse, 50 fs pulsewidth, 800 nm wavelength, 5.2 MHz repetition rate) [3]. Here, the laser is used to directly excite the measured sample area, in contrary to other setups where

 
J.-Y. Bigot et al. (eds), *Ultrafast Magnetism I*, Springer Proceedings in Physics 159
DOI 10.1007/978-3-319-07743-7_90

it is used to close a circuit by a photodiode to create for example a magnetic field pulse or a current pulse. The technique and its potential will be presented by two examples. The first one is about fast heating through a spin reorientation phase transition and the second about the investigation of all-optical switching.Dynamics of laser-induced spin reorientation in Co/$SmFeO_3$ heterostructure

## Dynamics of laser-induced spin reorientation in Co/$SmFeO_3$ heterostructure

Ultrafast control of a ferromagnet (FM) via exchange coupling with an antiferromagnet (AFM) is demonstrated in a Co/$SmFeO_3$ heterostructure [4]. Employing time-resolved photoemission electron microscopy combined with x-ray magnetic circular dichroism, a sub-100-ps change of the Co spins orientation by up to 10° driven by the ultrafast heating of the $SmFeO_3$ orthoferrite substrate through its spin reorientation phase transition is revealed. Numerical modeling of the ultrafast-laser-induced heat profile in the heterostructure, and the subsequent coupled spins dynamics and equilibration of the spin systems suggest that the localized laser-induced spin reorientation is hindered compared with the static case. Moreover, numerical simulations show that a relatively small Co/$SmFeO_3$ exchange interaction could be sufficient to induce a complete and fast spin reorientation transition (SRT).

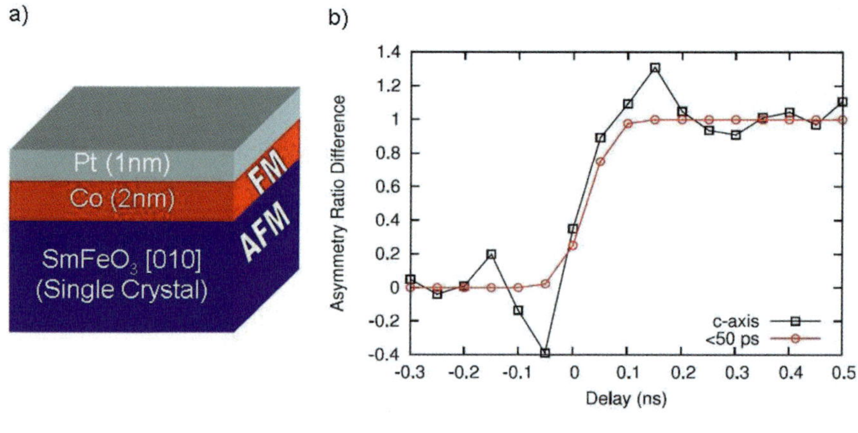


**Fig. 1.** a) sketch of the sample heterostructure. b) Time dependence of the XMCD signal after heating with a 50fs laser pulse. Black squares are the measurements and red circles simulations which give an upper limit of the time needed of the orientation change, which is given by the time resolution of the experiment.

## All-optical magnetization switching in GdFeCo based ferrimagnetic alloys

Recently it was discovered that an ultrafast laser pulse alone, without any magnetic field, is enough to deterministically reverse the magnetization: the magnetization reversal in a GdFeCo ferrimagnet can be triggered solely by a single, linearly polarized laser pulse [5, 6]. Here, we investigate the influence of the magnetization compensation point on the all-optical magnetization switching in GdFeCo samples. Switching is possible below and above the magnetization compensation temperature $T_M$ but in its immediate vicinity.

**Acknowledgments** This work was partially supported by the European Union's Seventh Framework Programme (FP7/2007-2013) Grants No. NMP3-SL-2008-214469 (UltraMagnetron) and No. 214810 (FANTOMAS).

## References

[1] E. Beaurepaire, J.-C.Merle, A. Daunois, and J.-Y. Bigot, "Ultrafast Spin Dynamics in Ferromagnetic Nickel" Phys. Rev. Lett. **76**, 4250 (1996).
[2] A. Kirilyuk, A. V. Kimel, and Th. Rasing, "Ultrafast optical manipulation of magnetic order" Rev. Mod. Phys. **82**, 2731 (2010).
[3] L. Le Guyader, A. Kleibert, A. Fraile Rodríguez, S. El Moussaoui, A. Balan, M. Buzzi, J. Raabe, and F. Nolting, "Studying nanomagnets and magnetic heterostructures with X-ray PEEM at the Swiss Light Source" J. Electron. Spectrosc. Relat. Phenom. **185**, 371 (2012).
[4] L. Le Guyader, A. Kleibert, F. Nolting, L. Joly, P. M. Derlet, R. V. Pisarev, A. Kirilyuk, Th. Rasing, and A. V. Kimel, "Dynamics of Laser induced spin reorientation in Co/$SmFeO_3$ heterostructure" Phys. Rev. B **87**, 054437 (2013).
[5] T. A. Ostler, J. Barker, R. F. L. Evans, R. Chantrell, U. Atxitia, O. Chubykalo-Fesenko, S. El Moussaoui, L. Le Guyader, E. Mengotti, L. J. Heyderman, F. Nolting, A. Tsukamoto, A. Itoh, D. Afanasiev, B. A. Ivanov, A. M. Kalashnikova, K. Vahaplar, J. Mentink, A. Kirilyuk,Th. Rasing and A. V. Kimel, "Ultrafast Heating as a Sufficient Stimulus for Magnetization Reversal in a Ferrimagnet" Nat. Commun. **3**, 666 (2012).
[6] L. Le Guyader, S. El Moussaoui, M. Buzzi, R. V. Chopdekar, L. J. Heyderman, A. Tsukamoto, A. Itoh, A. Kirilyuk, Th. Rasing, A. V. Kimel, and F. Nolting, "Demonstration of laser induced magnetization reversal in GdFeCo nanostructures" Appl. Phys. Lett. **101**, 022410 (2012).

# Accessing the magnetic susceptibility of FeRh on a sub-nanosecond time scale

**Federico Pressacco [1,2], E. Mancini[2], V. Uhlir[3], E.E . Fullerton[3] and C.H. Back[2]**

[1]Synchrotron SOLEIL L'Orme des Merisiers Saint-Aubin - BP 48 91192 GIF-sur-YVETTE CEDEX France

[2]Universitaet Regensburg, Universitaetsstr. 31, D-93040 Regensburg, Germany

[3]University of California, San Diego, La Jolla, CA 92093-0401, USA

federico.pressacco@synchrotron-soleil.fr

**Abstract** We measure the magnetization curves as a function of time delay in FeRh thin films. Due to a first order magnetic phase transition, an anhysteretic curve is measured giving access to the magnetic susceptibility on a sub-nanosecond timescale.

## Introduction

FeRh is a metallic compound which undergoes a first order phase transition from an Anti-Ferromagnetic Phase (AFP) to a Ferromagnetic Phase (FP) when heated above a critical temperature which is usually around 400 °K. The change in magnetic order results in a change in the net magnetization of the system from zero up to 2.2 Bohr magnetons after increasing the system temperature together with a volume expansion of about 1% [1]. This is an uncommon characteristic for a magnetic material since usually one observes a decrease of the magnetization upon heating. This discloses the possibility to use FeRh to Heat-Assisted Magnetic Recording (HAMR) devices.

The magnetization dynamics has been investigated using the Magneto Optical Kerr Effect (MOKE) in an all-optical pump-probe experimental scheme. An infrared (800 nm), ultrafast (200 fs) laser pulse has been used to impulsively heat the system above the critical temperature while a second pulse (200 fs 800 nm) has been used as probe. At different time delays we swept the external field H in a range of ± 2 kOe and measured the induced Kerr rotation. The system relaxes back to the antiferromagnetic phase in about 3 nanoseconds as can be seen in Fig.1. We show how the measured curves are equivalent to the first magnetization curve measured statically [2,3]. From these curves we were able to extract the susceptibility of the system during the ferromagnetic phase expansion on a picose-

 
J.-Y. Bigot et al. (eds), *Ultrafast Magnetism I*, Springer Proceedings in Physics 159
DOI 10.1007/978-3-319-07743-7_91

cond time scale Fig.2. This measurement is possible due to the presence of the peculiar phase transition present in the FeRh system.

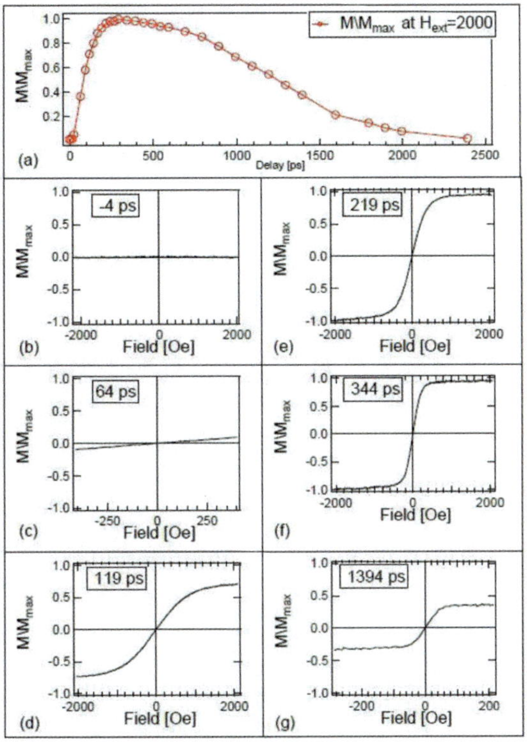


**Fig. 1.** (a) Kerr rotation measured on a 30 nm thick film of FeRh as a function of time delay. The values have been extracted from the magnetization curves for Hext = 2 kOe. (b)-(e) Magnetization curves for selected time delays.

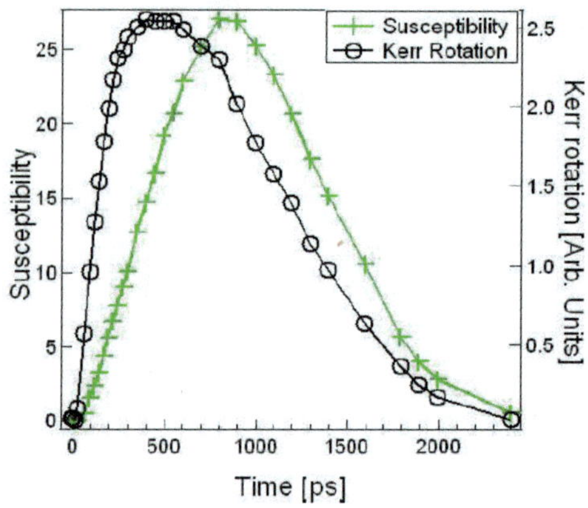


**Fig. 2.** The susceptibility and the Kerr rotation at Hext=2 kOe extracted from the magnetization curves are plotted as a function of time delay.

**Acknowledgments** aa The authors thank the Marie Curie Actions and FANTOMAS project which founded the project.

# References

[1] S. Mariager, F. Pressacco, et al., "Structural and magnetic dynamics of a laser induced phase transition in FeRh" Phys. Rev. Lett. **108**, 08721 (2012).
[2] J. Pearson, P. T. Squire, et al. "Which anhysteretic magnetization curve?" IEEE transaction on magnetics **33,** 0018-9464 (1997).
[3] G. Bertotti, "Hysteresis in Magnetism", (Academic Press, 1998).

# Engineering Ultrafast Magnetism

**I. Radu[1], C. Stamm[1], A. Eschenlohr[1], F. Radu[1], R. Abrudan[2], K. Vahaplar[3], T. Kachel[1], N. Pontius[1], R. Mitzner[1], K. Holldack[1], A. Föhlisch[1], R.F.L. Evans[4], T.A. Ostler[4], J. Mentink[3], R.W. Chantrell[4], A. Tsukamoto[5], A. Itoh[5], A. Kirilyuk[3], A.V. Kimel[3], Th. Rasing[3]**

[1]Helmholtz-Zentrum Berlin für Materialien und Energie GmbH, Berlin, Germany

[2]Institut für Experimentalphysik/Festkörperphysik, Ruhr-University Bochum, Germany

[3]Institute for Molecules and Materials, Radboud University Nijmegen, The Netherlands

[4]Department of Physics, University of York, United Kingdom

[5]College of Science and Technology, Nihon University, Chiba, Japan

ilie.radu@helmholtz-berlin.de

**Abstract** We employ time-resolved X-ray magnetic circular dichroism with fs time resolution to investigate the ultrafast, laser-driven dynamics of multi-sublattice materials, with both ferromagnetic and antiferromagnetic coupling. These measurements provide evidence for a demagnetization time that scales with the elemental magnetic moment and varies with the exchange interaction symmetry.

## Introduction

Controlling magnetic states of matter is crucial for understanding the underlying fundamental physical phenomena and for engineering the next-generation magnetic devices combining ultrafast data processing with ultrahigh-density data storage. An appealing notion in this context is the employment of femtosecond (fs) laser pulses to fully set the orientation and the magnetization magnitude of a spin ensemble. Achieving this on ultrashort timescales, *i.e.* comparable to the excitation event itself, remains however a challenge mainly due to the lack of understanding the dynamical behavior of the elementary building blocks forming such a spin ensamble, *i.e.* the magnetic moments and the exchange interaction.

Here, we report on the latest developments in our studies on ultrafast magnetism, which reveal highly unexpected and sometime counterintuitive results [1] by employing a novel experimental approach combining the fs laser excitation with an ultrafast, element-specific X-ray probing of spins. In particular, by investigating the magnetization dynamics in a variety of ferromagnetic and ferrimagnetic alloys, we demonstrate a simple but powerful way of controlling the demagnetization/switching dynamics in a large class of magnetic materials.

 
J.-Y. Bigot et al. (eds), *Ultrafast Magnetism I*, Springer Proceedings in Physics 159
DOI 10.1007/978-3-319-07743-7_92

## Results

Most of the ultrafast magnetization studies performed so far were done on materials consisting of several non-equivalent magnetic sublattices – see e.g. Reference [2] for a review work. This fact has been however largely ignored, assuming that the exchange-coupled magnetic sublattices are at any time in equilibrium with each other being thus undistinguishble and loosing their magnetization on the very same time-scale. However, recent experiments, using fs X-ray Magnetic Circular Dichroism (XMCD), have revealed distinctly different magnetization dynamics of the constituent magnetic moments in a Gd(FeCo) alloy excited by a femtosecond optical pulse [1]. Moreover, it was subsequently shown that this distinct dynamics in combination with the exchange interaction between the sublattices, can lead to a deterministic switching of this ferrimagnetic Gd(FeCo) by an ultrafast heat pulse alone [3]. These results lead to the intriguing question whether such element-specific spin dynamics and heat-induced deterministic switching are general phenomena or something strictly related to a narrow class of GdFe-like ferrimagnetic materials?

In the present contribution we address this question by investigating the laser-driven dynamics of several multi-sublattice magnetic materials, with both ferromagnetic (NiFe alloys) and antiferromagnetic (GdFe and DyCo alloys) coupling between sublattices, using element-specific, femtosecond time-resolved XMCD. These measurements, fully supported by phenomenological modeling and atomistic spin simulations, provide evidence for a demagnetization time that scales with the elemental magnetic moment and varies with the sign of the exchange interaction. As such, one can tune the speed of magnetization processes in multi-sublattices alloys, being either switching or demagnetization, by properly choosing the magnitude of the constituent magnetic moments and the sign of the exchange interaction that couples them. These findings allow designing new synthetic ferrimagnetic materials with tailor-made optimized properties, resulting in faster, more energy-effective switching strategies, as exemplified for the case of a synthetic Fe/FePt ferrimagnet.

**Acknowledgments** Funding from the European Community's Seventh Framework Program (FP7/2007-2013 Grants No. NMP3-SL-2008-214469 (UltraMagnetron) and No. 214810 (FANTOMAS) and ERC Grant No. 257280 (Femtomagnetism) as well as Grant No. 226716), the German Federal Ministry of Education and Research (BMBF) Grant No. 05K10PG2 (FEMTOSPEX), the Foundation for Fundamental Research on Matter (FOM) and the Netherlands Organization for Scientific Research (NWO) is gratefully acknowledged.

## References

[1] I. Radu, K. Vahaplar, C. Stamm, T. Kachel, N. Pontius, H. A. Dürr, T. A. Ostler, J. Barker, R. F. L. Evans, R. W. Chantrell, A. Tsukamoto, A. Itoh, A. Kirilyuk, Th. Rasing, A. V.

Kimel, “Transient ferromagnetic-like state mediating ultrafast reversal of antiferromagnetically coupled spins” Nature **472**, 205–208 (2001).
[2] A. Kirilyuk, A.V. Kimel, Th. Rasing, “Ultrafast optical manipulation of magnetic order” Rev. Mod. Phys. **82,** 2731 (2010).
[3] T. A. Ostler, J. Barker, R. F. L. Evans, R. Chantrell, U. Atxitia, O. Chubykalo-Fesenko, S. El Moussaoui, L. Le Guyader, E. Mengotti, L. J. Heyderman, F. Nolting, A. Tsukamoto, A. Itoh, D. Afanasiev, B. A. Ivanov, A. M. Kalashnikova, K. Vahaplar, J. Mentink, A. Kirilyuk,Th. Rasing and A. V. Kimel, “Ultrafast Heating as a Sufficient Stimulus for Magnetization Reversal in a Ferrimagnet” Nat. Commun. **3**, 666 (2012).

# Ultrafast, element-specific magnetization dynamics of multi-constituent magnetic materials by use of high-harmonic generation

**T.J. Silva[a], E. Turgut[b], S. Mathias[b,c], C. La-o-vorakiat[b], P. Grychtol[b,d], R. Adam[d], D. Rudolf[d], H.T. Nembach[a], M. Aeschlimann[c], C.M. Schneider[d], H.C. Kapteyn[b], M.M. Murnane[b], and J.M. Shaw[a]**

[a]Electromagnetics Division, NIST, Boulder, CO, 80305 USA

[b]Department of Physics/JILA, University of Colorado and NIST, Boulder, CO, 80309 USA

[c]University of Kaiserslautern/Research Center OPTIMAS, 67663 Kaiserslautern, Germany

[d]Peter Grünberg Institut PGI-6 & JARA-FIT, Research Centre Jülich, 52425 Jülich, Germany

silva@boulder.nist.gov

**Abstract** We have studied femtosecond magnetization dynamics probed by extreme ultraviolet pulses from high-harmonic generation, with element-selectivity and ultrafast time resolution. By use of this technique, we identify the microscopic processes that drive magnetization dynamics on femtosecond timescales. Here, we concentrate on controlling superdiffusive spin-currents in magnetic multilayers.

We review recent progress in the study of ultrafast magnetization dynamics by use of high harmonic generation (HHG) light sources [1] [2] [3]. Since HHG results in broadband, coherent, short-wavelength beams, it is ideal for the simultaneous capture of magnetic dynamics with elemental specificity. To date, bright HHG light sources have been demonstrated from the VUV to the keV photon energy region (> 1.5 keV), while retaining the polarization, coherence, and perfect synchronization properties of the driving laser under phase-matched generation conditions [4].

Here we show that tabletop HHG sources are ideal probes of ultrafast magnetization dynamics because of their femtosecond-to-attosecond pulse duration, in combination with simultaneous element-specific access to multiple atomic species. These unique experimental capabilities make it possible to solve long-standing problems in far-from-equilibrium magnetism, and also enable the study ultrafast processes in more complex and technologically relevant magnetic materials. In recent work, HHG was used to capture the fastest magnetization dynamics in elemental materials, complex magnetic alloys, and multilayer systems, thereby elucidating the role of exchange interaction [2], superdiffusive spin-currents [3], and spin scattering in ultrafast magnetization dynamics [1].

 
J.-Y. Bigot et al. (eds), *Ultrafast Magnetism I*, Springer Proceedings in Physics 159
DOI 10.1007/978-3-319-07743-7_93

By use of ultrashort (<5 fs), extreme ultraviolet (EUV) pulses from a tabletop HHG source tuned to the *M*-edges of Ni and Fe [1] [3], we observed (via the transverse magneto-optic Kerr effect) layer-specific spin-dynamics in substrate/Ta (3 nm)/Ni (5 nm)/X (t nm)/Fe (4 nm)/$Si_3N_4$ (6 nm) magnetic multilayers, where the non-magnetic spacer X is Ru, Ta, W, or $Si_3N_4$ [1]. For the case of X = Ru and parallel magnetic orientation of the two layers, we observed that the Fe moment is *enhanced* by ~ 20% after ultrafast optical pumping [3], as shown in Fig. 1a.

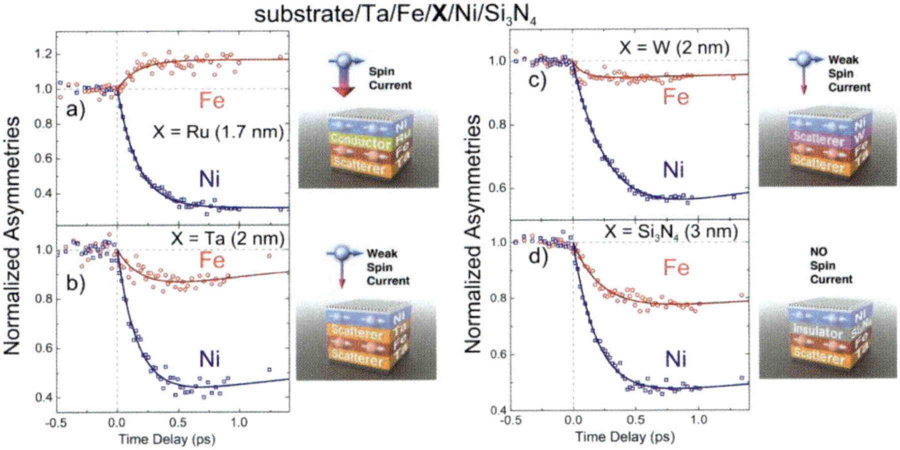


**Fig. 1.** Magnetization dynamics in substrate/Ta(3 nm)/Fe(4 nm)/X/Ni(5 nm)/ $Si_3N_4$(6 nm, top) multilayers with varying X. In a), enhancement of the Fe magnetization is observed in the presence of good spin transport across the 1.7 nm Ru spacer layer. In b) and c), varying degrees of demagnetization for the Fe layer are observed when Ta (2 nm), W (2 nm), or $Si_3N_4$ (3 nm) are used as the nonmagnetic spacer between the Ni and Fe layers.

However, for the cases of X = Ta, W, or $Si_3N_4$, both the Ni and Fe moments always *demagnetize* on the timescale of approximately 100 fs [1]. We interpret the data in terms of a competition between ultrafast, local spin-flip processes and laser-generated, superdiffusive, non-local spin-currents that flow between the layers, whereby a majority spin-current injected into the Fe layer acts to counter the local spin-flip process. Such spin-currents are massive in scale: the observation of increased magnetization for X = Ru when the moments are parallel allows us to quantify the resultant spin-current, with a peak value equivalent to a 100% polarized charge current of $10^{10}$ A/cm$^2$. The theoretical expectation, based on the calculated spin asymmetry of the quasiparticles lifetimes in Ni and Fe, is that an initially ballistic excited spin-current flows preferentially from the Ni to Fe layer [5]. After 10 fs, the spin-current becomes diffusive in character as the optically excited electrons thermalize; hence the superdiffusive character of the spin-current.

While the preferential flow from Ni to Fe with X = Ru is not due to strong differences in optical absorption for the top and bottom layers (see Table 1), we also find that the majority spin-current continues to flow from the top to bottom layer even when the order of the Ni and Fe layers is inverted (not shown), suggesting that the Ta seed layer adjacent to the Ni layer plays an important role in breaking the stack symmetry [1].

For antiparallel orientation of the layer magnetizations, the Fe layer always demagnetizes [3], as expected for a spin-current-driven process. Likewise, the absence of any Fe magnetization enhancement for X = Ta, W, or $Si_3N_4$ is expected due to the poor spin conductivity of these layers [6]. However, comparison of the data for all samples with different spacer layer materials indicates that interlayer spin-current is not the only mechanism that drives ultrafast magnetization dynamics. Given that we observe demagnetization of both Ni and Fe for all of these cases, local spin-flip scattering must contribute to the dynamics [1].

**Table 1.** a Calculated absorption ratios of the pump beam by each layer in terms of the percentage of the incident pump beam [1].

| Substrate/…/spacer/…/capping | Fe | Spacer | Ni |
|---|---|---|---|
| Fe(4 nm)/Ru(1.7 nm)/Ni(5 nm) | 15.9 | 11.7 | 19.3 |
| Fe(4 nm)/Ta(2 nm)/Ni(5 nm) | 19 | 3.2 | 23.5 |
| Fe(4 nm)/W(2 nm)/Ni(5 nm) | 18.9 | 5.4 | 23.1 |
| Fe(4 nm)/$Si_3N_4$(3 nm)/Ni(5 nm) | 21.5 | 0 | 26.4 |
| Ni(5 nm)/Ru(1.7 nm)/Fe(4 nm) | 18.5 | 12 | 17.1 |

**Acknowledgments** The authors gratefully acknowledge funding from the U.S. Department of Energy Office of Basic Energy Sciences, Award #DE-FG02-09ER46652 and from the Deutsche Forschungsgemeinschaft, #Schn-353/17, #AE-19/20 and #GR 4234/1-1.

## References

[1] E. Turgut, C. La-o-vorakiat, J. M. Shaw, et al., "Controlling the Competition between Optically Induced Ultrafast Spin-Flip Scattering and Spin Transport in Magnetic Multilayers" Phys. Rev. Lett. **110**, 197201 (2013).

[2] S. Mathias, C. La-O-Vorakiat, P. Grychtol, et al., "Probing the timescale of the exchange interaction in a ferromagnetic alloy" Proceedings of the National Academy of Sciences **109**, 4792-4797 (2012).

[3] D. Rudolf, C. La-O-Vorakiat, M. Battiato, et al., "Ultrafast magnetization enhancement in metallic multilayers driven by superdiffusive spin current" Nat. Commun. 3, 1037 (2012).

[4] T. Popmintchev, M.-C. Chen, D. Popmintchev, et al., "Bright Coherent Ultrahigh Harmonics in the keV X-ray Regime from Mid-Infrared Femtosecond Lasers" Science **336**, 1287-1291 (2012).

[5] M. Battiato, K. Carva, and P. M. Oppeneer, "Theory of laser-induced ultrafast superdiffusive spin transport in layered heterostructures" Phys. Rev. B **86**, 024404 (2012).

[6] C. T. Boone, H. T. Nembach, J. M. Shaw, et al., "Spin transport parameters in metallic multilayers determined by ferromagnetic resonance measurements of spin-pumping" J. Appl. Phys. **113**, 153906 (2013).

# Ultrafast Spin Dynamics on the Nanoscale

**C.E. Graves, A.H. Reid, H.A. Dürr**

Stanford Institute for Materials and Energy Sciences, SLAC National Accelerator Laboratory, 2575 Sand Hill Road, Menlo Park CA 94025, USA

hdurr@slac.stanford.edu

**Abstract** We summarize the first x-ray free electron laser experiment that determines the extraordinary spin dynamics in the ferrimagnetic alloy GdFeCo on the femtosecond time- and nanometer length-scale. The results show that ultrafast nanoscale spin transport plays an important role in the all-optical switching observed in this material.

## Introduction

One of the mechanisms central to the fs laser induced quenching of magnetic order [1] is angular momentum transfer to the lattice [2]. However, laser excited hot electrons may transport angular momentum away from the excitation area [3,4]. The latter process could open up the possibility to manipulate magnetic order at a distance with the Fermi velocity (~1nm/fs) rather than with the speed of sound (~1nm/ps) [4]. The fundamental question is how and when long-range magnetic order emerges from complete disorder. In equilibrium, magnetic order forms due to spontaneous magnetization mediated by the short-range exchange interaction.

Here we show that the non-equilibrium flow of angular momentum via spin currents can also achieve long-range magnetic order. To reveal this mechanism, we employ intense fs x-ray pulses to probe magnetic ordering on the nm length and fs timescale after fs optical laser excitation has brought a metallic system into a highly non-equilibrium spin state [5].

## Results

The model system for all-optical magnetic switching is the ferrimagnetic alloy Co9.5Fe66.5Gd24, in which the interplay between itinerant Fe, Co spins and localized Gd 4f magnetic moments lead to a laser-induced magnetization reversal on picosecond timescales [6]. So far the nanoscale aspects of this process have not been investigated. However, as Fig. 1 illustrates, even the chemical composition



J.-Y. Bigot et al. (eds), *Ultrafast Magnetism I*, Springer Proceedings in Physics 159
DOI 10.1007/978-3-319-07743-7_94

of Co9.5Fe66.5Gd24 displays pronounced variations on the nm lengthscale. The sample segregates into Fe-rich (blue) and Gd-rich (red) regions as determined by scanning transmission electron microscopy with x-ray fluorescence analysis (STEM-EDX). X-ray diffraction probes these regions when the x-ray energy is tuned to the Gd 3d - 4f and the Fe 2p - 3d resonance. Using circularly polarized x-rays, the magnetic component, Sq, for Fe and Gd can be obtained as a function of wavevector, q [5]. The wavevector q is inversely proportional to the probed lengthscale, d, i.e. $q = 2\pi/d$.

$S_q$ measures magnetization dynamics following laser excitation at low-q values (white) and high-q values (gray) (Fig. 1, bottom left). High-q $S_q$ ($\sim 2\pi/10$nm) probes the dynamics related to the chemical enrichment-depletion regions while low-q $S_q$ probes the average sample magnetization. The average sample magnetization displays the expected ultrafast demagnetization, as our laser excitation was higher than in ref. [6]. Therefore, the sample was heated above the Curie point without switching magnetization direction. However, the temporal behavior of high-q $S_q$ is noticeably different. High-q $S_q$ directly probes the magnetization difference between enriched and depleted regions (blue and red areas in Fig. 1,top). Critically, the high-q Gd signal increases while the corresponding Fe signal decreases. This demonstrates that there is an increasing difference in Gd magnetization between enriched and depleted regions on a ~10nm length scale while the Fe magnetization averages out. The reconstructed total magnetization demonstrates that the Gd magnetization reverses direction in the Gd-rich regions after 1ps (Fig. 1, bottom right). The complete analysis of these data identifies the driving force for this nanoscale spin reversal as a net transfer of Fe spin polarization from the Fe-rich into the Gd-rich nanoregions.

These results clearly demonstrate how x-ray free electron laser (XFEL) radiation can access the smallest space-time dimensions in magnetic solids. XFELs enable the study of emerging phenomena such as the disappearance and appearance of order in highly non-equilibrium systems. The demonstrated far-from-equilibrium spin transfer proceeds against the magnetic disorder at elevated temperatures and could provide an effective tool for spin manipulation in future applications.

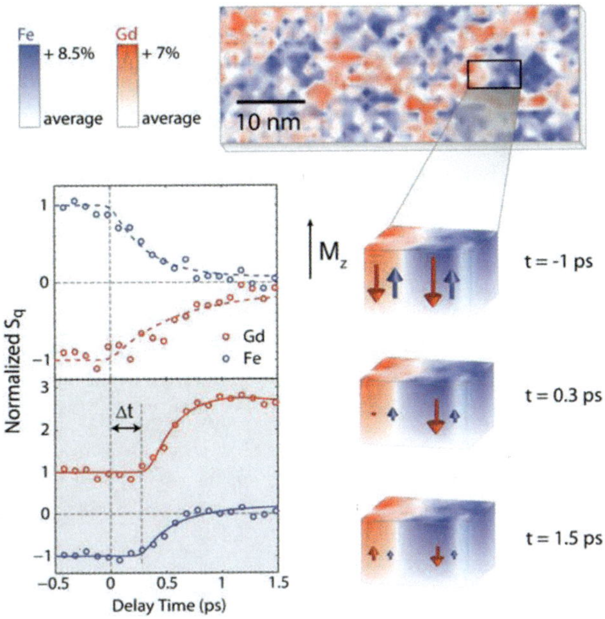


**Fig. 1.** (top panel) Chemical composition map of $Co_{9.5}Fe_{66.5}Gd_{24}$ . (Bottom left panel) Fe 3d and Gd 4f magnetization dynamics, $S_q$, determined by x-ray diffraction. For low q values ($q< 0.2$ nm$^{-1}$), the measured delay scans probe the average magnetization (white background) while at high q ($q> 0.2$ nm$^{-1}$) the difference in magnetization between Gd and Fe rich regions is probed (gray background). (Bottom right panels) Reconstructed Fe and Gd magnetization in the Fe-rich (blue) and Gd-rich (red) nanoscale regions. (from ref. [5])

**Acknowledgments** This article summarizes the first experiment of its kind performed at an x-ray free electron laser and would not have been possible without the contributions of T. Wang, B. Wu, S. de Jong, K. Vahaplar, I. Radu, D.P. Bernstein, M. Messerschmidt, L. Müller, R. Coffee, M. Bionta, S.W. Epp, R. Hartmann, N. Kimmel, G. Hauser, A. Hartmann, P. Holl, H. Gorke, J. H. Mentink, A. Tsukamoto, A. Fognini, J.J. Turner, W.F. Schlotter, D. Rolles, H. Soltau, L. Strüder, Y. Acremann, A.V. Kimel, A. Kirilyuk, Th. Rasing, J. Stöhr and A.O. Scherz. Research at Stanford was supported through the Stanford Institute for Materials and Energy Science (SIMES) under contract DE-AC02-76SF00515 by the US Department of Energy, Office of Basic Energy Sciences

## References

[1] E. Beaurepaire, J.-C. Merle, A. Daunois, J.-Y. Bigot, "Ultrafast spin dynamics in ferromagnetic nickel" Phys. Rev. Lett. **76**, 4250 (1996).

[2] B. Koopmans, et al. "Explaining the paradoxical diversity of ultrafast laser-induced demagnetization" Nature Mater. **9**, 259 (2010).

[3] M. Battiato, M. Carva, P.M. Oppeneer, "Superdiffusive Spin Transport as a Mechanism of Ultrafast Demagnetization" Phys. Rev. Lett. **105**, 027203 (2010).

[4] G. Malinowski, et al., "Control of speed and efficiency of ultrafast demagnetization by direct transfer of spin angular momentum" Nature Phys. **4**, 855 (2008).
[5] C.E. Graves, et al. "Nanoscale spin reversal by non-local angular momentum transfer following ultrafast laser excitation in ferrimagnetic GdFeCo" Nature Mater. **12**, 293 (2013).
[6] I. Radu, et al., "Transient ferromagnetic-like state mediating ultrafast reversal of antiferromagnetically coupled spins" Nature **472**, 205 (2011).

# Element Selective Investigation of Spin Dynamics in Magnetic Multilayers

**Dennis Rudolf[1], Chan La-O-Vorakiat[2], Marco Battiato,[3] Roman Adam[1], Patrik Grychtol[2], Justin M. Shaw[4], Emrah Turgut[2], Pablo Maldonado[3], Stefan Mathias[5], Hans T. Nembach[4], Thomas J. Silva[4], Martin Aeschlimann[5], Henry C. Kapteyn[2], Margaret M. Murnane[2], Peter M. Oppeneer[3] and Claus M. Schneider[1]**

[1]Peter Grünberg Institut PGI-6, Research Centre Jülich, 52425, Jülich, Germany

[2]Department of Physics and JILA, University of Colorado, Boulder, CO 80309-0440, USA

[3]Department of Physics and Astronomy, Uppsala University, SE-75120 Uppsala, Sweden

[4]Electromagnetics Division, NIST, Boulder, CO 80305-3328, USA

[5]University of Kaiserslautern and Research Center OPTIMAS, 67663, Kaiserslautern, Germany

r.adam@fz-juelich.de

**Abstract** Our understanding of ultrafast switching processes in novel spin-based electronics depends on our detailed knowledge of interactions between spin, charge and phonons in magnetic structures. We present element-selective studies, using extreme ultraviolet (XUV) light, to gain insight into spin dynamics in exchange coupled magnetic multilayers on the femtosecond time scale.

In recent years, intensive studies of spin dynamics have been driven largely by a demand for high-speed- and high-density magnetic storage. Encoding data in the electron spin promises a new route for the magnetic storage technology with the potential to reduce energy requirements and heat dissipation. At the same time an increasing effort has been directed towards mastering the control of the spin alignment by laser light. A number of distinct models have been proposed to describe how laser excitation can couple to the spins. Most of these models are based on phonon-, electron-, or magnon mediated spin-flip processes [1], direct laser-induced spin-flips [2] or relativistic spin-light interaction [3]. Recently, a model based on superdiffusive spin transport has been proposed [4].

The variety of these distinct interpretations suggests that several physical mechanisms may govern femtosecond spin-dynamics even in a simple elemental magnet. The analysis becomes even more complex for heterogeneous magnetic systems, such as magnetic alloys [5,6] and exchange-coupled layered structures [7]. The detailed understanding of spin dynamics in such complex magnetic alloys



J.-Y. Bigot et al. (eds), *Ultrafast Magnetism I*, Springer Proceedings in Physics 159
DOI 10.1007/978-3-319-07743-7_95

and multilayers requires testing of the electronic and magnetic properties element-selectively with femtosecond time resolution, at the same time. Element selective studies of complex alloys and materials have been addressed for some time using synchrotron radiation by tuning the light to the characteristic resonances of a particular material. In our studies we used the synchrotron-generated XUV light to explore *M*-edge resonances in magnetic materials to gain insight into exchange coupling between magnetic thin films in magnetic multilayers. By exploiting the transversal magneto-optical Kerr effect of Co(5nm)/Si(1.5-4nm)/Ni(10nm) near the *M* absorption edges of cobalt (59.5eV) and nickel (66.5eV) we obtained magnetic contrast as large as 80% for Co and 25% for Ni near the Brewster angle of about 45° [8]. By tuning the angle and energy we recorded magnetic asymmetries as well as element-selective magneto-optical hysteresis loops reflecting switching of the individual ferromagnetic layers as a function of the interlayer coupling (Fig.1).

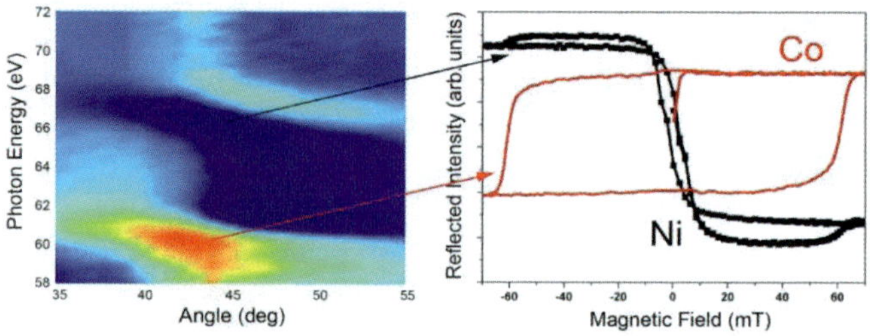


**Fig. 1.** Angle- and energy dependent magnetic asymmetry (left) and layer selective hysteresis loops (right) of Co(5nm)/Si(4nm)/Fe(10nm) trilayers recorded near *M* absorption edges of cobalt (59.5eV) and nickel (66.5eV) at the Brewster incidence angle.

Requirements for femtosecond element-selective measurements were recently met by strong progress in laser-based XUV light sources generating femtosecond pulses with photon energies reaching up to 72 eV. Such high-energy photons are produced when an intense, ultrafast laser pulse is focused into a collection of gas atoms, generating higher-order harmonics of the fundamental light [9]. Similar to synchrotron experiments, tuning the probing beam to a chosen absorption edge results in a resonant increase of the signal for the corresponding element.

Combining element selectivity with femtosecond time resolution we studied the magnetic response of ferromagnetic Ni(5nm)/Ru(1.5nm)/ Fe(4nm) multilayers. By exciting the multilayer with infrared laser light we observed the evolution of magnetization response in the Ni and Fe layers simultaneously but separately using synchronized XUV probe pulses detecting spin response within 300 fs after laser excitation (Fig.2). Following the excitation, we detected an unexpected fluence-dependent magnetization quenching, as well as magnetization enhancement in the buried Fe layer for parallel alignment of Fe and Ni magnetization. We ascribe the observed response to the optically generated superdiffusive spin currents between the layers observed experimentally for the first time [7].

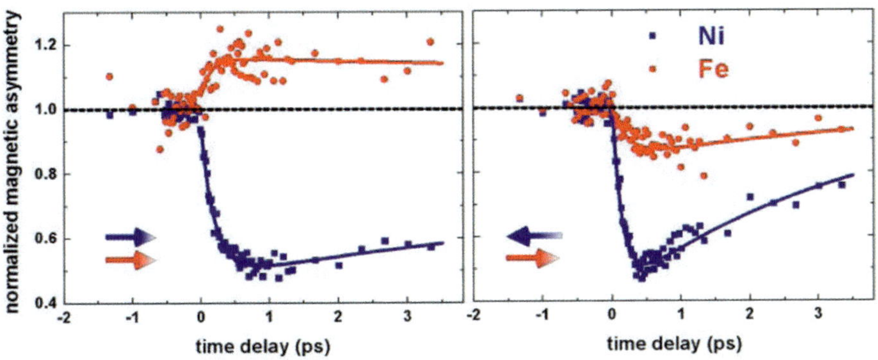


**Fig. 2.** Experimentally measured time- and layer resolved magnetization for Ni(5nm)/Ru(1.5nm)/ Fe(4nm) trilayers, after femtosecond laser excitation. The signal at the Fe *M* absorption edge (52 eV) anomalously increases for the parallel and decreases for antiparallel magnetic orientation of Ni and Fe layers. The arrows in the panel show the relative magnetization of Ni and Fe layers.

**Acknowledgments** The authors gratefully acknowledge financial support from BMBF (Project 05KS7UK1), from the DFG (SFB 491) and from the EU (grant No. 281043, "FemtoSpin").

## References

[1] B. Koopmans et al., "Explaining the paradoxical diversity of ultrafast laser-induced demagnetization" Nature Materials **9**, 259-265 (2010).
[2] G.P. Zhang, W. Hübner, G. Lefkidis, Y. Bai, T.F. George, "Paradigm of the time-resolved magneto-optical Kerr effect for femtosecond magnetism" Nature Phys, **5**(7), 499-502 (2009).
[3] J.-Y. Bigot, M. Vomir, E. Beaurepaire, "Coherent ultrafast magnetism induced by femtosecond laser pulses" Nature Phys, **5**(7), 515-520 (2009).
[4] M. Battiato, K. Carva and P. M. Oppeneer "Superdiffusive Spin Transport as a Mechanism of Ultrafast Demagnetization" Phys. Rev. Lett. **105**, 027203 (2010).
[5] S. Mathias et al. "Probing the timescale of the exchange interaction in a ferromagnetic alloy" PNAS, 1201371109 v1-6 (2012).
[6] C. La-O-Vorakiat, M. Siemens, M. Murnane, H. Kapteyn, S. Mathias, M. Aeschlimann, P. Grychtol, R. Adam, C. M. Schneider, J. Shaw, H. Nembach and T. Silva, "Ultrafast Soft X-Ray Magneto-Optics at the M-edge Using a Tabletop High-Harmonic Source" Phys. Rev. Lett. **103**, 257402 (2009).
[7] D. Rudolf et al., " Ultrafast magnetization enhancement in metallic multilayers driven by superdiffusive spin current" Nature Commun. **3**, 1037 (2012).
[8] P. Grychtol, R. Adam, S. Cramm, D. Bürgler, C. M. Schneider and S. Valencia, "Interlayer coupling in a Co/Si/NiFe multilayer wedge system studied by resonant magnetic reflectivity measurements in the extreme ultraviolet range" Phys. Rev. B. **82**, 054433 (2010).
[9] H.C. Kapteyn, M.M. Murnane and I.P. Christov, " Extreme Nonlinear Optics: Coherent X Rays from Lasers" Physics Today **58**(3), 39-46 (2005).

# Element- and time-resolved dynamics in rare-earth/transition metals alloys

**N. Bergeard[1], V. López-Flores[1*], V. Halté[1], M. Hehn[2], C. Stamm[3], N. Pontius[4], E. Beaurepaire[1], and C. Boeglin[1]**

[1] Institut de Physique et de Chimie des Matériaux de Strasbourg, UMR7504, CNRS et Université de Strasbourg, 67034 Strasbourg, FRANCE

[2] Institut Jean Lamour, Université de Lorraine, Nancy, FRANCE

[3] Department of Materials, ETH Zurich, CH-8093 Zurich, SWITZERLAND

[4] Institut für Methoden und Instrumentierung der Forschung mit Synchrotronstrahlung Helmholtz-Zentrum Berlin für Materialien und Energie GmbH, Albert-Einstein-Str. 15, 12489 Berlin, GERMANY

bergeard@ipcms.unistra.fr

**Abstract:** Ultrashort laser pulses are used to induce changes of the magnetization in ferrimagnetic $Co_xTb_{1-x}$ alloys. Ultrafast magnetization dynamics has been probed by tr-XMCD at the $CoL_3$ and $TbM_5$ edges. We demonstrated that demagnetization of the 4f magnetic moment is much faster than in pure Tb when the excited-state-temperature is below the compensation temperature.

## Introduction

Ultrafast magnetization dynamics is an important issue for both fundamental science and for applications in order to optimize spin manipulation on a microscopic level. Application of ultrashort laser pulses allows ultimately the manipulation of the magnetization of magnetic films down to the femtosecond time scale. Therefore it is essential to understand the different fundamental processes taking place during the first hundred femtoseconds. Since the first observation of laser induced spin dynamics [1] the mechanisms of femtosecond demagnetization have been widely debated. The standard technique used in these studies is time-resolved magneto-optic Kerr effect. Nowadays time-resolved X-ray magnetic circular dichroism (TR-XMCD) using synchrotron light with femtosecond time resolution is accessible as a complementary and element-specific tool to study ultrafast magnetization dynamics [2-5]. It is now possible to measure absolute values of the magnetization with a high temporal resolution (130 fs) using the f-slicing source at the Helmholtz-Zentrum Berlin [2-3]. One of the most striking riddle deals with the intrinsic time scale of dynamics, faster for 3d transition metals (TM) (~100fs) than

 
J.-Y. Bigot et al. (eds), *Ultrafast Magnetism I*, Springer Proceedings in Physics 159
DOI 10.1007/978-3-319-07743-7_96

for 4f rare earths (RE) (~700 fs) [4]. In RE/TM alloys, the indirect 4f-4f coupling is enhanced due to the presence of 3d electrons in the conduction band. It allows investigating the effect of $3d_{TM}$-$5d_{RE}$ exchange coupling. Recently, in ferrimagnetic $Fe_{.75}Co_{.34}Gd_{.22}$ it was shown that Gd shows a long demagnetization time of 430 fs whereas Fe is much faster despite the strong exchange interaction between both elements [5]. The numerical values of the thermalization times have not been explicitly addressed.

In this context, we have used laser pulses to induce changes of the magnetization in ferrimagnetic $Co_xTb_{1-x}$ at different working temperatures. The magnetization dynamic has been probed at the $CoL_3$ and $TbM_5$ edges [6]. We show that the demagnetization time of the 4f magnetic moment can be reduced to values down to $\tau$ = 280 fs compared to pure Tb ($\tau$=740 fs). Our observations also suggest that in case of multisublattice ferrimagnets the 4f demagnetization is faster when the excited-state temperature is below the compensation temperature ($T_{comp}$) whereas it is slower when the excited-state temperature approaches the Curie temperature ($T_C$) [6].

The pump fluences used during our experiment were adjusted to 12 mJ/cm$^2$ for $Co_{0.74}Tb_{0.26}$ and 21 mJ/cm$^2$ for $Co_{0.86}Tb_{0.14}$ in order to reach similar amplitudes of demagnetization. The 15 nm Co-RE alloys have been grown on $Si_3N_4$ membranes. The films are characterized by $T_{comp}$, where the magnetic moments of the Co and the rare-earth sublattices compensate and by a specific $T_C$ where the magnetic order is lost. A temperature dependent XMCD study at the $CoL_3$ and rare-earth $M_5$ edges has been performed to characterize the different films ( $T_{comp}$ and $T_C$) [6]. The pump-probe experiment was performed at film temperatures close to room temperature or slightly above so that the laser excited state temperature is either in the vicinity of $T_C$ ($Co_{0.86}Tb_{0.14}$ ) or $T_{comp}$.($Co_{0.74}Tb_{0.26}$). In Figure 1 a, and b we show the ultrafast dynamics obtained at the Co $L_3$ and Tb $M_5$ edges for the $Co_{0.74}Tb_{0.26}$ alloy. The XMCD amplitudes are normalized to the quantitative static XMCD measurements [6]. A model detailed elsewhere [6] suggests that for localized 4f electrons an increase of the spin relaxation time leads to larger thermalization times in the vicinity of $T_C$. This model also explains why only minor changes are observed for 3d electrons in Co. This is confirmed by comparing the ultrafast dynamics at the $TbM_5$ edge in $Co_{0.74}Tb_{0.26}$ (~280fs) and $Co_{0.86}Tb_{0.14}$ (~500fs) where the excited-state temperature is below $T_{comp}$ resp. $T_C$ (Figure 1 b and c). Finally, we wish to point out that although the 4f elements studied so far are demagnetization rate limited close to $T_C$ by the divergence of 4f spin fluctuations, this can be overcome using $T_{comp}$ in the ferrimagnetic alloys. In the future, more studies on specific ferrimagnetic materials, consisting of two antiparallel-coupled sublattices, are requested to fully describe this effect.

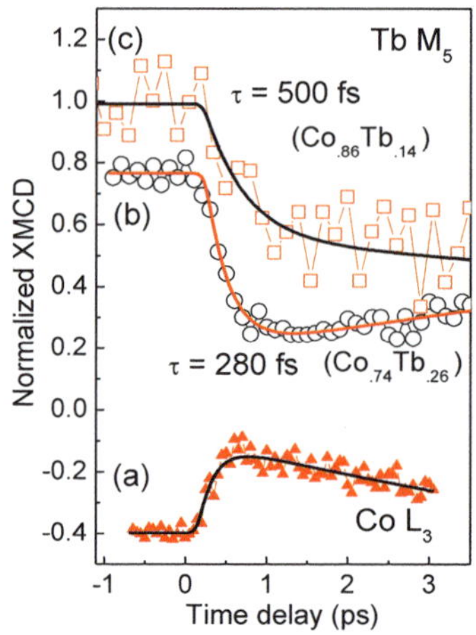


**Fig. 1. (a, b)** Ultrafast dynamics of Co $L_3$ (triangles) and Tb $M_5$ (open circles) edges in $Co_{0.74}Tb_{0.26}$. **(c)** Ultrafast dynamics of Tb $M_5$ edge (open squares) in $Co_{0.86}Tb_{0.14}$.

**Acknowledgments** This work was supported by the CNRS – PICS, by Université de Strasbourg and the E.U. Contract Integrated Infrastructure Initiative I3 in FP6-Project No. R II 3 CT-2004-5060008, BESSY IA-SFS Access Program, by the "Agence Nationale de la Recherche" in France via the project EQUIPEX UNION: # ANR-10-EQPX-52" and by the German Ministry of Education and Research BMBF Grant 05K10PG2 (FEMTOSPEX). V.L.-F. acknowledges Ministry of Education of Spain (Programa Nacional de Movilidad de Recursos Humanos del Plan Nacional de I-D+i 2008-2011) for financial support.

# References

* Now at Synchrotron SOLEIL, L'Orme des Merisiers, Saint-Aubin, FRANCE

[1] E. Beaurepaire, J.C. Merle, A. Daunois, and J.-Y. Bigot, "Ultrafast Spin Dynamics in Ferromagnetic Nickel" Phys. Rev. Lett. **76**, 4250-4253 (1996).
[2] C. Stamm et al., " Femtosecond modification of electron localization and transfer of angular momentum in nickel" Nature Mater. **6**, 740-743 (2007).
[3] C. Boeglin, E. Beaurepaire, V. Halté, V. Lopez-Flores, C. Stamm, N. Pontius, H. A. Dürr, and J.-Y. Bigot, " Distinguishing the ultrafast dynamics of spin and orbital moments in solids" *Nature* **465**, 458-461 (2010).
[4] M. Wietstruk et al., "Hot-Electron-Driven Enhancement of Spin-Lattice Coupling in Gd and Tb 4f Ferromagnets Observed by Femtosecond X-Ray Magnetic Circular Dichroism" Phys. Rev. Lett. **106**, 127401 (2011).
[5] I. Radu et al., "Transient ferromagnetic-like state mediating ultrafast reversal of antiferromagnetically coupled spins" *Nature* **472**, 205-209 (2011).
[6] V. López-Flores et al. " Role of critical spin fluctuations in ultrafast demagnetization of transition-metal rare-earth alloys" Phys. Rev. B **87**, 214412 (2013).

# Space charge effects occurring during fast demagnetization processes

**Nathan Beaulieu[a], Gregory Malinowski[b], Azzedine Bendounan[a], Mathieu G. Silly[a], Christian Chauvet[a], Damjan Krizmancic[c] and Fausto Sirotti[a]**

[a] Synchrotron SOLEIL, L'Orme des Merisiers, Saint-Aubin BP 48, 91192 Gif-sur-Yvette Cedex (France)

[b] Laboratoire de Physique des Solides, Université Paris Sud, Orsay (France)

[c] Instituto Officina dei Materiali (IOM)-CNR Laboratorio TASC, in Area Science Park S.S.14, Km 163.5, I-34149 Trieste (Italy)

nathan.beaulieu@synchrotron-soleil.fr

**Abstract** Fast demagnetization processes are observed with 50-100 femtosecond pulsed laser excitations and energy densities of some mJ/cm$^2$. In these experimental conditions, space charge effects are usually observed in photoemission experiments. We present the study of a laser induced demagnetization of an epitaxial Gd layer with a quantitative measurement of the space charge.

Since the first experiments showing a fast demagnetization processes induced by short laser pulses [1,2], the role played by the hot electrons excited by the pulsed laser on the observed magnetization dynamics is becoming more and more important [3,4]. The hot electrons create a large amount of unbalanced negative charge which is self-accelerated by a coulomb repulsion [5]. This phenomenon is also well known in pump probe photoemission experiments [6]. We present the magnetization dynamics of an epitaxial Gd layer on W(110) studied with a fs laser pump synchrotron radiation probe experiment performed using partial yield soft-X-ray absorption at the Gd $M_5$ edge [7]. The effect of the laser is directly measured on the photoelectron created by the synchrotron before and after the laser pulse.

The experiments were performed on the UHV-Photoemission experimental station of the TEMPO beamline at the SOLEIL synchrotron radiation facility [8]. The experimental geometry is summarized in Fig. 1. It allows us to perform magnetic circular dichroism experiments in absorption and photoemission under laser excitation. The time resolved measuring principle is presented in Fig. 2. The 50 fs, 6 μJ laser pulses from the Coherent REGA 9050 laser system are emitted at a frequency of about 282 KHz and synchronized to the Soleil synchrotron radiation time structure which has a base frequency of about 846 KHz. Soleil was operated



J.-Y. Bigot et al. (eds), *Ultrafast Magnetism I*, Springer Proceedings in Physics 159
DOI 10.1007/978-3-319-07743-7_97

in Hybrid mode where an isolated bunch is centered in ¼ of the ring while the other three quarters are filled with smaller bunches separated of 2.8 ns.

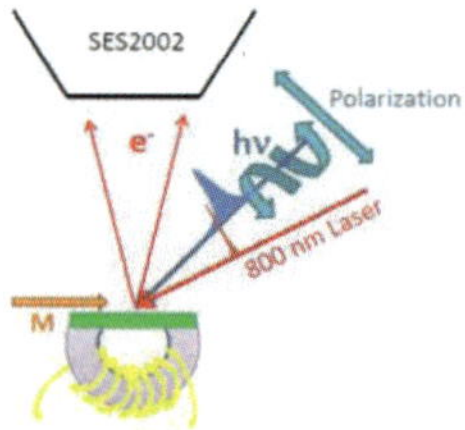


**Fig. 1.** Schematic representation of the sample environment and the experimental geometry in the UHV photoemission chamber

Our time resolved delay line detector allow us to separate in different channels the photoelectrons originated from the isolated bunch: several spectra are measured at the same time, each corresponding to a defined time interval of the laser time period.
The photoemission intensity measured in each time channel is shown in the top of Fig. 2 where the photoemission intensity from the isolated bunch is presented as a function of the time channel (x axis) and the kinetic energy (vertical axis). The isolated bunch intensity is well separated from the generated by the ¾ machine filling.

The integrated intensity corresponding to the isolated bunch in four adjacent periods indicated in the top panel is presented in the bottom one. The photoelectrons synchronized with the laser are shifted at higher kinetic energy.

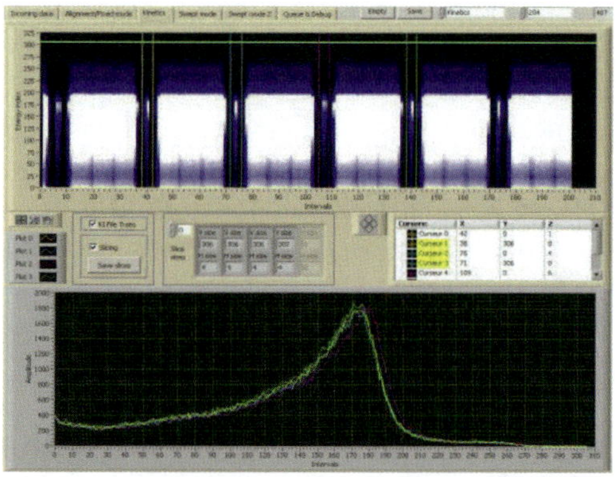

**Fig. 2.** Time resolved photoelectron spectroscopy data acquisition. Top: photoemission intensity in a color scale as a function of the electron kinetic energy (vertical axis) and time (horizontal axis). The acquisition is synchronized with the synchrotron radiation periodicity. Bottom: Photoemission spectra obtained for the isolated bunch during four adjacent periods obtained grating the four sections indicated in the top panel.

In order to observe phenomena at ps timescale, we vary the delay between the laser and the synchrotron radiofrequency signal. The kinetic energy shift of the Gd

4f photoemission peak measured during the laser excitation is presented in Fig. 3 for two kinetic energies: 100 eV and 1100 eV.

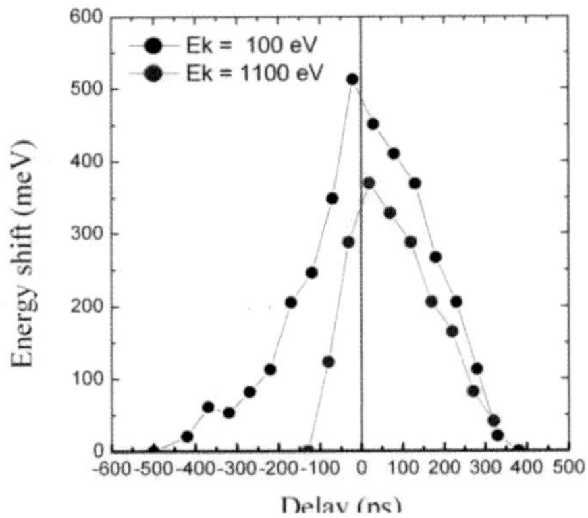


**Fig. 3.** kinetic energy shift observed on Gd-4f core level photoemission at 1100 et and 1100 eV kinetic energy as a function of the delay between the synchrotron and laser pulses.

The two light pulses are superimposed at time t=0 where the kinetic energy shift is maximum. At positive delays, where the synchrotron arrives after the laser pulses, we probe the duration of the space charge effect. It extends to a time interval of about 400 ps and is dependent from the laser focusing and pulse energy density. An energy shift is also observed at negative delays, when the laser arrives after the synchrotron bunch. The time interval on which the shift is observed depends on the kinetic energy of the electrons. It is due to the acceleration induced by the space charge on the photoelectrons which are travelling towards the detector. It is compatible with an interaction region of about 2.4 mm from the sample surface.

In conclusion, the determination of detection thresholds in pump/probe experiments is an essential point for the design of new experiments. This is particularly true when other thresholds are imposed by the physical problems as it is the case for the fast demagnetization.

**Acknowledgments** The research described here has been supported by Triangle de la physique contract 2010-005T-FEMTOMAG.

# References

[1] A. Vaterlaus, T. Beutler, and F. Meier, "Spin-lattice relaxation time of ferromagnetic gadolinium determined with time-resolved spin-polarized photoemission" Phys. Rev. Lett. **67**, 3314 (1991).

[2] E. Beaurepaire, J.-C. Merle, A. Daunois, and J.-Y. Bigot, "Ultrafast Spin Dynamics in Ferromagnetic Nickel", Phys. Rev. Lett. 76, 4250 (1996).

[3] G. Malinowski, F. Dalla Longa, J. H. H. Rietjens, P. V. Paluskar, R. Huijink, H. J. M. Swagten & B. Koopmans "Control of speed and efficiency of ultrafast demagnetization by direct transfer of spin angular momentum" Nat. Phys. **4**, 855 (2008).

[4] M. Battiato, K. Carva, and P. M. Oppeneer “Superdiffusive Spin Transport as a Mechanism of Ultrafast Demagnetization” Phys. Rev. Lett. **105**, 027203 (2010).
[5] C. Cirelli, M. Hengsberger, A. Dolocan, H. Over, J. Osterwalder and T. Greber “Direct observation of space charge dynamics by picosecond low-energy electron scattering” EPL **85**, 17010 (2009).
[6 S. Hellmann, K. Rossnagel, M. Marczynski-Bühlow, and L. Kipp “Vacuum space-charge effects in solid-state photoemission” Phys. Rev. B, **79**, 035402 (2009).
[7] N. Beaulieu, et al. Journal of Electron Spectroscopy and Related Phenomena (2013).
[8] N. Bergeard, et al. “Time-resolved photoelectron spectroscopy using synchrotron radiation time structure” Journal of Synchrotron Radiation, **18**(2), 245-250 (2011).

# Ultrafast spectroscopy with spin polarization

**V. Lollobrigida[a,b], R. Ciprian[c], F. Offi[b] and G. Panaccione[c]**

[a]Mathematics and Physics Department, University of Roma Tre, Via della Vasca Navale 84, I-00146 Rome, Italy

[b]Science Department, University of Roma Tre, Via della Vasca Navale 84, I-00146 Rome, Italy

[c]CNR - IOM, Materials Institute, Lab. TASC, Area Science Park, S.S. 14 – km 163.5, I-34149 Trieste, Italy

lollobrigida@fis.uniroma3.it

**Abstract** A project for the design and realization of an experimental station dedicated to ultrafast spin polarization dynamics allowing for spin polarization measurements of photoelectron yield as excited by free electron laser pulses is presented.

## Introduction

Research on magnetic materials in the last years has been focusing on the determination of the speed with which is possible to modify the magnetic state of a material and of the spatial length over which this occurs. To investigate these problems it is necessary to examine the dynamical magnetic properties of the materials, which could be done by taking snapshots of the spin dynamics on femtosecond timescales and with nanometer-scale spatial resolution as well. For this kind of measurements state-of-the-art probes are required. Regarding the time domain, advances were made over the years by using femtosecond laser pulses in a pump-probe configuration.

The dynamics of the magnetization can be uncovered by measuring the spin polarization (SP) of the photoelectron yield of a photoemission experiment [1]. One of the most reliable methods to measure SP is the elastic Mott scattering of electrons of 50 keV-1 MeV diffusing onto the heavy ion cores of a target. The scattering angles must be chosen in order to optimize the spin-orbit scattering (which yields the SP asymmetry) ratio over the dominant charge scattering, which determines an overall reduction of 4 orders of magnitude of the unscattered photoelectron beam intensity. As a consequence, experiments which perform SP analysis downstream of high resolution k (angular) and E (dispersion) analysis are affected by poor statistics.

 
J.-Y. Bigot et al. (eds), *Ultrafast Magnetism I*, Springer Proceedings in Physics 159
DOI 10.1007/978-3-319-07743-7_98

Novel linac-based free electron laser (FEL) sources could represent a new possibility for SP analysis due to the high brilliance and total intensity in a single photon pulse. By combining this to the time resolution (typical duration of a FEL pulse is 50 fs) and to SP measurement by Mott scattering it should be possible to address fundamental questions on magnetization dynamics and ultrafast demagnetization in nanostructured-solids and surfaces.

## State of the art

The need to design and build an experimental station dedicated to these studies is addressed by the ULTRASPIN (ULTRAfast spectroscopy with SPIN polarization) project. Within ULTRASPIN, spin analysis by Mott scattering of the photoelectron yield as excited by individual FEL radiation pulses will be implemented. The core instrument will be a four-axis Mott scattering device allowing for full vectorial reconstruction of the SP both in the static and dynamic (pump and probe) modes [2]. The acquisition chain usually present in traditional Mott scattering detection will be reviewed and eventually modified to achieve linearity, in order to preserve the information about the number of electrons scattered in each direction. At the same time the detection system should compatible to the time distribution of the photoelectrons, in order to obtain information in a time-of-flight (TOF) mode analysis.

We plan to investigate the dynamics of correlated systems displaying orbital and spin ordering transitions, especially for nanostructured materials, atomically clean surfaces and interfaces. The SP of the photoelectrons will be measured, at energies above photoemission threshold, as well as at characteristics energies of the systems under consideration. After each FEL pulse the photoelectrons will be collected, accelerated and focused into a periaxial beam aimed at a couple of four-axis Mott scattering devices equipped with fast "total energy" detectors providing asymmetry data of Mott scattering with respect to four scattering planes, i.e. four vectorial projections on the three independent axis of the SP vector. The energy range exploitable by FEL Mott scattering is from threshold photoemission up to hard X-rays, while the 50 fs duration of FEL pulses translates in an 80 meV energy broadening. This energy resolution is exploitable for SP measurements at threshold photoemission and adequate for SP measurements at core level thresholds (Cr, Mn, Fe, Co, Ni, Cu) and valence band energy density of states analysis with spin selectivity (without k-selection).

The project includes the design and implementation of a compact UHV multi-functional station for sample preparation and analysis. The system is composed of two minor-chambers [Fig.1 (a)], the first dedicated to sample preparation and growth, and the second one to sample analysis using different complementary techniques. Both chambers are equipped with a load lock for sample fast entry.

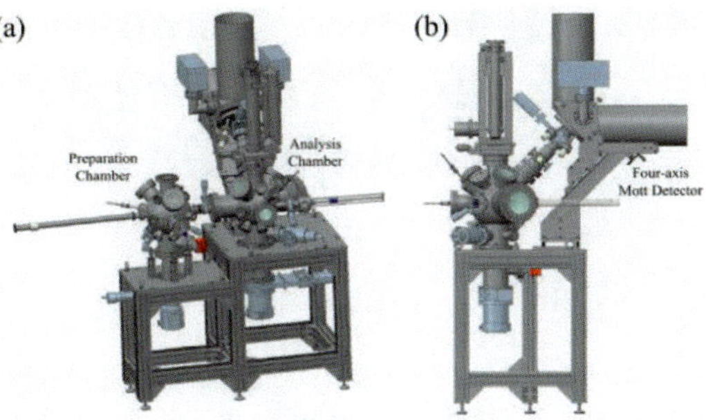


**Fig. 1.** (a) Design of the UHV multifunctional station. (b) View of the analysis chamber equipped with a four-axis Mott detector for spin polarization measurements.

The first chamber is equipped with a sputtering ion gun, a heating stage (temperatures up to 1000°C), two/three electron-gun evaporators and a LEED system. The sample can be positioned on a rotatable three-axial manipulator onto a holder equipped with a small electromagnet for magneto-optical Kerr effect measurements.

After the preparation, the sample can be easily translated in the second chamber manufactured in mu-metal in order to avoid perturbations due to static or low frequency stray-magnetic fields. A rotatable three-axial manipulator has been mounted, equipped with an open cycle cryostat and a sample environment which can hold up to three samples simultaneously. Two of the three stages are dedicated to sample cooling and heating.

Our first goal has been the design [Fig. 1(b)] and setup of the electrostatic extraction optics of the Mott detectors accelerators for spin electron polarization measurements. The first in progress experiments will be carried out in static conditions at threshold energy using tabletop lasers in order to precisely evaluate the effect and the evolution of space charge on both total intensity and spin polarization.

The apparatus should be able to match with a TOF for energy analysis. Trajectory simulations will be performed using the simulation program SIMION and the effect of space charge accumulation will be taken into account.

## References

[1] A. Scholl, L. Baumgarten, R. Jacquemin, W. Eberhardt, "Ultrafast Spin Dynamics of Ferromagnetic Thin Films Observed by fs Spin-Resolved Two-Photon Photoemission" Phys. Rev. Lett. **79**, 5146-5149 (1997).

[2] V. N. Petrov, M. S. Galaktionov, A. S. Kamochkin, "Comparative tests of conventional and retarding-potential Mott polarimeters" Rev. Sci. Instrum. **72**, 3728-3730 (2001).

# Magnetic-Field-Dependent Fraunhofer Diffraction Pattern by 4f Imaging System in Transparent Magnetooptic Thin Film

**Djati Handoko[a], Je-Ho Shim[a], Dong-Hyun Kim[a*], Tae Kyu Kim[b], and Jaehun Park[c]**

1. Department of Physics, Chungbuk National University, Cheongju 361-763, South Korea
2. Department of Chemistry, Pusan National University, Busan 609-735, South Korea
3. Pohang Accelerator Laboratory, POSTECH, Pohang 790-784, South Korea

* donghyun@chungbuk.ac.kr

**Abstract** Diffraction imaging using X-ray has been attracting huge attention for possible application to investigate ultrafast magnetism. In this work, the sequential changes of Fraunhofer diffraction pattern due to photo-induced nonlinear refractive index changes in a transparent magnetooptic thin film were observed by using the 4f imaging system with phase objects at the entrance aperture. The selection of the aperture shape was observed numerically. We also discussed the power distribution of the temporally observed patterns.

## Introduction

In recent years, the ultrafast optical system has fascinated significant attention for imaging of nanostructured material. In particular, femtosecond visible laser technology combined with magneooptic Kerr or Faraday effect enabled us to explore ultrafast timescale magnetism. On the other hand, diffraction imaging has also been applied to explore ultrafast phenomenon utilizing X-ray light source such as synchrotron and free electron laser. For magnetic materials, where an external applied field is requisite, the Fourier transform holography (FTH) has been known to be useful in observing magnetism on ultrafast timescale[1,2].

Traditionally, FTH is based on 4f optical imaging system. As the easiest system, the 4f coherent imaging, as shown in Fig. 1, is the most basic optical Fourier technique used for image processing. Fourier optics is a powerful tool for the coherent manipulation of optical fields, for the spatial frequency analysis of images, and for the construction of real time correlators. One property that makes optics an exciting tool for frequency analysis is that, in free-space propagation of



J.-Y. Bigot et al. (eds), *Ultrafast Magnetism I*, Springer Proceedings in Physics 159
DOI 10.1007/978-3-319-07743-7_99

light, one can see the effects of lenses and filters on a light beam as the beam propogates through the system to the far field.

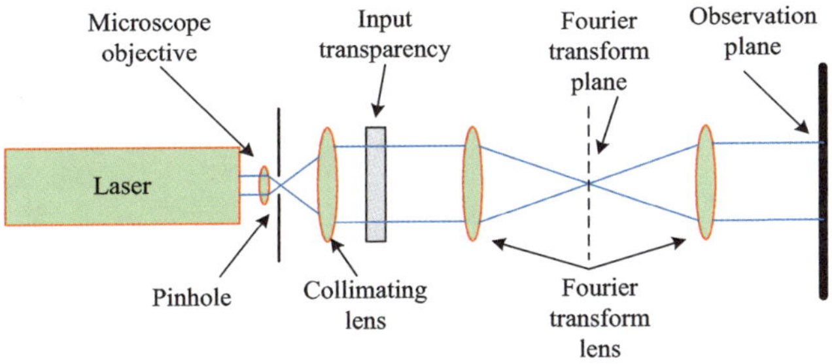


**Fig. 1.** 4f optical imaging system configuration

In this work, we set up a 4f optical imaging system and use it to observe a diffraction pattern from a transparent magneto-optic photonic crystal thin film with nonlinear optical properties. In order to optimize the characterics, we propose a field-dependent diffraction imaging by varying an applied field on the transparent magnetic thin film.

## Theory

The nonlinear image formation inside the 4f system can be described using a simple model based on Fourier optics[3,4]. The object plane is a back focal plane of the first lens and is illuminated by linearly polarized monochromatic plane wave with normally incident angle. Assuming that the input object is placed on the object plane with amplitude transmittance t(x,y), the distribution of amplitude in Fourier plane can be written as

$$S(u,v) = \frac{1}{\lambda f} FT[O(x,y)] \qquad 1$$

where FT denotes the Fourier transform operator, $\lambda$ is the wavelength of incident beam, f is the focal length, $O(x,y) = Et(x,y)$ is the amplitude distribution in object plane and E is the amplitude of the incident beam.

For a magnetic material, if we assume the Faraday effect neglecting nonlinear absoprtion and small phase shifts, we can express the beam intensity at the exit surface of sample as

$$S_{NL}(u,v) = S(u,v) exp(j\Delta\phi_{NL}) \qquad 2$$

where $\Delta\phi_{NL}$ is the nonlinear phase shift. By combining these two equations, one can observe diffraction pattern of the magnetic material thin film.

## Results and discussion

We have simulated the 4f imaging system base on equation (1) and (2) by varying the object plane for rectangular and circular oblique pattern. As shown in Fig 2. the inverse Fourier transform from the diffraction pattern regenerates a similar shape as an original aperture on the object plane.

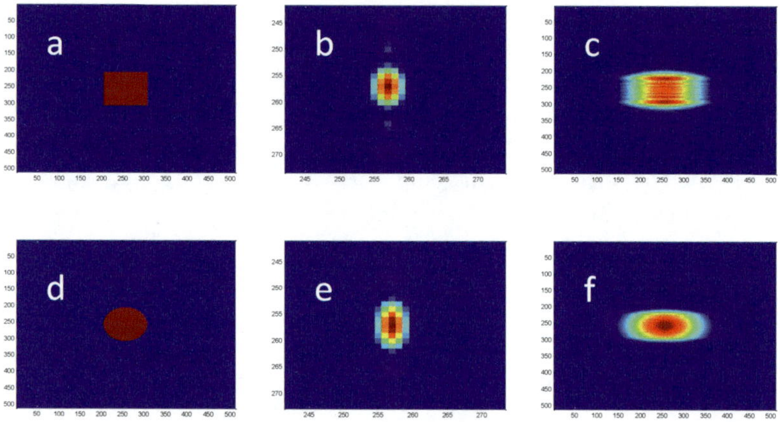


**Fig. 2.** The 4f imaging simulation results for different apertures on object plane.

In conclusion, the 4f optical imaging system based on the Fourier transform has been found to satisfies the consistency condition for variation of apertures and magnetic fields, which might be applied for ultrafast magnetism study experiments using the visible femtosecond laser.

## References

[1] S. Eisebitt , J. Lüning, W. F. Schlotter, M. Lörgen, O. Hellwig, W. Eberhardt and J. Stöhr, “Lensless imaging of magnetic nanostructures by x-ray spectro-holography” Nature **432,** 885-888 (2004).

[2] Thomas A. Duckworth *et al.*, “Holographic imaging of interlayer coupling in Co/Pt/NiFe” New J. Phys. **15**, 023405-023419 (2013).

[3] K. Fedus and G. Boudebs, “Experimental techniques using 4f coherent imaging system for measuring nonlinear refraction” Optics Communications **292,** 140–148 (2013).

[4] Junyi Yang *et al.*, “Measurement of nonlinear refraction using a reversal 4f coherent imaging system with phase objects” Optics Communications **284,** –148 (2013).

# Part IX
# Terahertz Spin Dynamics

# Ultrafast spin precession and transport controlled and probed with terahertz radiation

**T. Kampfrath[1], M. Battiato[2], A. Sell[3], F. Freimuth[4], A. Leitenstorfer[3], M. Wolf[1], R. Huber[5], P.M. Oppeneer[2], M. Münzenberg[6]**

[1]Fritz Haber Institute of the Max Planck Society, Faradayweg 4-6, 14195 Berlin, Germany
[2]Uppsala University, Box 516, 75120 Uppsala, Sweden
[3]University of Konstanz, 78464 Konstanz, Germany
[4]Peter-Grünberg-Institut and Institute for Advanced Simulation, 52425 Jülich, Germany
[5]University of Regensburg, 93053 Regensburg, Germany
[6]Georg-August-Universität Göttingen, Friedrich-Hund-Platz 1, 37077 Göttingen, Germany

kampfrath@fhi-berlin.mpg.de

**Abstract** We present examples of how terahertz (THz) electromagnetic transients can be used to control spin precession in antiferromagnets (through the THz Zeeman torque) and to probe spin transport in magnetic heterostructures (through the THz inverse spin Hall effect), on femtosecond time scales.

## Introduction

THz electromagnetic radiation has a wavelength of the order of 100μm and is located between the realms of electronics and optics. THz transients can be used to probe and even control numerous low-energy excitations including phonons, excitons or Cooper pairs [1]. Here, we show how THz spectroscopy can be applied in the field of spin-based electronics (spintronics) and consider the resonant manipulation of the magnetization of an antiferromagnet [2,3] and the tailored transport of spin angular momentum from a ferromagnetic into a nonmagnetic metal [4].

## THz control over antiferromagnetic magnons

In order to manipulate the orientation of a spin with magnetic moment $\boldsymbol{m}$, the most direct handle is provided through a magnetic field $\boldsymbol{B}$ that exerts the so-called Zeeman torque $\boldsymbol{m}\times\boldsymbol{B}$ onto $\boldsymbol{m}$. To enhance the otherwise weak Zeeman interaction, one can take advantage of resonantly driving a spin wave (magnon) in a magnetically ordered solid. In antiferromagnets, exchange interaction results in high magnon frequencies reaching the THz region [3].

 
J.-Y. Bigot et al. (eds), *Ultrafast Magnetism I*, Springer Proceedings in Physics 159
DOI 10.1007/978-3-319-07743-7_100

Here, we excite a crystal of the textbook antiferromagnet NiO by means of an intense THz pulse with a peak magnetic field of 0.14T (Fig. 1a) and a spectrum fully covering the NiO long-wavelength magnon at 1THz (Fig. 1b). The induced transient magnetization ***M*** is probed with a delayed femtosecond laser pulse whose polarization is rotated by an angle proportional to ***M*** (Faraday effect).

Figure 1c shows the measured Faraday rotation as a function of the pump-probe delay: we observe a long-lived oscillation with a frequency of ~1THz (amplitude spectrum in Fig. 1b), which is a clear hallmark of the 1-THz spin wave of NiO. A more detailed analysis shows that the spin wave is indeed driven by the Zeeman torque of the magnetic field of the THz transient. The long magnon lifetime (Fig. 1c) permits coherent control by using a time-delayed second pump pulse that either enhances or cancels the coherent spin motion [2,3].

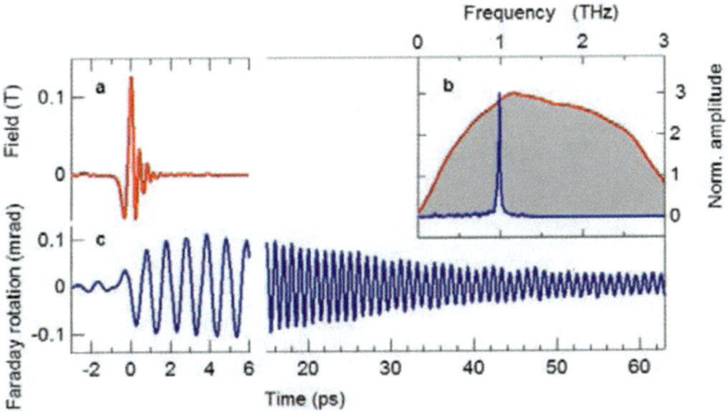


**Fig. 1.** Spin-wave control in NiO by THz magnetic fields. **a** Incident transient magnetic field together with its **b** amplitude spectrum. **c** Measured Faraday rotation of the probe-pulse polarization. Its Fourier spectrum peaks at 1THz (blue curve in **b**), which is a clear signature of the 1-THz long-wavelength magnon of NiO.

## Tailored THz spin currents in magnetic heterostructures

Ultrafast transport of spin angular momentum through a ferromagnetic metal film (such as Fe) can be triggered by a femtosecond laser pulse (Fig. 2a) that excites majority electrons (spin up) from d-type bands with low velocity to sp-type states with high velocity, whereas minority electrons remain slow [5,6].

The resulting spin flow can be modified drastically by covering the ferromagnetic Fe film (thickness 10nm) with a thin (2nm) metallic layer having either low (Ru) or high electron mobility (Au) [4]. Figure 2a shows calculations of the spatiotemporal magnetization dynamics based on a superdiffusive transport theory [5]: as expected from the very different mobility, spins are found to be trapped in the Ru layer whereas only little spin density is accumulated in the Au film.

In the experiment, detection of the spin current is facilitated by the inverse spin Hall effect that deflects spin-up and -down electrons in opposite directions, resulting in the emission of a THz electromagnetic pulse. The measured waveforms (Fig. 2b) are fully reversed upon reversal of the sample magnetization. Note that the emitted THz pulse from the Fe/Au heterostructure is shorter and exhibits components with significantly higher frequencies than the pulse from the Fe/Ru structure. An inversion procedure permits extraction of the total time-dependent charge current, and we find a much shorter current burst for the Fe/Au than for the Fe/Ru sample. The dynamics of the measured charge current is in good qualitative agreement with the spin currents extracted from the calculations of Fig. 2a [4].

Our results may open up a route to the manipulation of ultrashort spin pulses and show that the inverse spin Hall effect is operative also at THz frequencies.

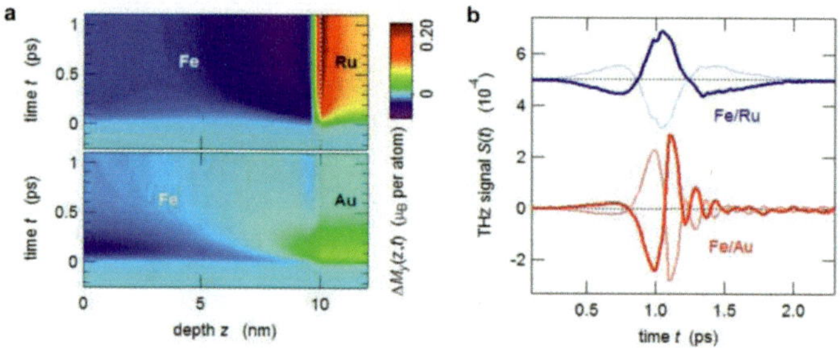


**Fig. 2.** Tailored ultrafast spin transport in magnetic heterostructures. a) Simulated spatiotemporal change in magnetization along the depth of a 10-nm Fe film covered by either 2nm Ru or Au, following excitation with a femtosecond laser pulse (800nm, 1.3mJ $cm^{-2}$). While spins are trapped in the low-mobility Ru, only little spin accumulation is found in high-mobility Au. b) Measured THz transients emitted by the two heterostructures. The signal is inverted when the sample magnetization is reversed. Faster dynamics is observed for Fe/Au, consistent with a).

## References

[1] R. Ulbricht, E. Hendry, J. Shan, T. F. Heinz, and M. Bonn., "Carrier dynamics in semiconductors studied with time-resolved terahertz spectroscopy" Rev. Mod. Phys. 83, 543 (2011).

[2] T. Kampfrath, A. Sell, G. Klatt, A. Pashkin, S. Mährlein, T. Dekorsy, M. Wolf, M. Fiebig, A. Leitenstorfer & R. Huber, "Coherent terahertz control of antiferromagnetic spin waves" Nature Photon. 5, 31 (2011).

[3] A. Pashkin, A. Sell, T. Kampfrath, and R. Huber, "Electric and magnetic terahertz nonlinearities resolved on the sub-cycle scale" New J. Phys. 15, 065003 (2013).

[4] T. Kampfrath et al., "Terahertz spin current pulses controlled by magnetic heterostructures" Nature Nanotech. 8, 256 (2013).

[5] M. Battiato, K. Carva, and P.M. Oppeneer, "Superdiffusive Spin Transport as a Mechanism of Ultrafast Demagnetization" Phys. Rev. Lett. 105, 027203 (2010).

[6] A. Melnikov, I. Razdolski, T. O. Wehling, E. Th. Papaioannou, V. Roddatis, P. Fumagalli, O. Aktsipetrov, A. I. Lichtenstein, and U. Bovensiepen "Ultrafast transport of laser-excited spin polarized carriers in Au/Fe/MgO(001)" Phys. Rev. Lett. 107, 076601 (2011).

# THz spin dynamics: phonon-induced spin order

**K.W. Kim[1,2], M. Porer[1,4], A. Pashkin[1], A. Sell[1], T. Kampfrath[3], A. Leitenstorfer[1], and R. Huber[4]**

1Department of Physics, University of Konstanz, 78464 Konstanz, Germany

2Department of Physics, Chungbuk National University, Cheongju 361-763, Korea

3Fritz-Haber-Institut der Max-Planck-Gesellschaft, Faradayweg 4-6, 14195 Berlin, Germany

4Department of Physics, University of Regensburg, 93053 Regensburg, Germany

rupert.huber@ur.de

**Abstract** Two concepts of ultrafast spin control at THz frequencies are presented: Exploiting a coherent phonon mode we transiently induce and destroy spin order in pnictides at a frequency as high as 5.5 THz. Strong THz magnetic fields control magnons in antiferromagnetic NiO coherently by direct Zeeman coupling.

## Introduction

The terahertz window (1 THz corresponds to an energy of 4 meV) of the electromagnetic spectrum has attracted enormous attention in condensed matter physics, in recent years. This strong interest has been fuelled by the fact that the THz regime contains a rich variety of low-energy dynamics, such as phonons, plasmons, correlation-induced energy gaps or collective magnon modes. While the THz frequency regime had long remained elusive due to a notorious lack of efficient sources and sensitive detectors, the past decade has essentially closed what was once called the "THz gap". Table-top sources of multi-THz pulses have provided phase-locked field transients with multi-octave spanning spectra covering the entire mid- to far-infrared range in a single pulse. Furthermore, peak electric and magnetic field amplitudes beyond 10 GV/m and 33 Tesla, respectively, have been reached in few-cycle multi-THz transients [1,2]. In combination with sensitive electro-optic detection, this toolbox offers an exciting route to the new regime of nonlinear THz optics with sub-cycle temporal resolution [3-6].

Ultrashort THz pulses particularly allow us to study and control spin dynamics on the few-femtosecond scale. First, we harness an optically triggered coherent lattice vibration to induce a transient spin-density wave in BaFe2As2, the parent compound of pnictide superconductors [4]. In a second step, we utilize the magnetic field component of intense THz transients to directly switch on and off co-



J.-Y. Bigot et al. (eds), *Ultrafast Magnetism I*, Springer Proceedings in Physics 159
DOI 10.1007/978-3-319-07743-7_101

herent spin waves in the antiferromagnetic nickel oxide NiO [5]. and coherent magnon control

## Phonon-driven transient spin order in pnictides

The material class of iron-arsenic-based compounds, also called pnictides, has been found to support unconventional superconductivity when properly doped. Yet the microscopic pairing mechanism is largely unknown. In particular, spin and lattice degrees of freedom as well as their interplay have moved into the focus of interest. We employ few-cycle multi-THz pulses to resonantly probe the evolution of a spin density wave gap in the parent compound $BaFe_2As_2$ following excitation with a femtosecond optical pulse [4]. When starting in the low-temperature ground state, optical excitation results in a melting of the spin density order, followed by an ultrafast recovery. In contrast, the spin density wave gap is induced when we excite the normal state above the transition temperature. Very surprisingly, the transient ordering quasi-adiabatically follows a coherent lattice oscillation at a frequency as high as 5.5 THz. Our results attest to a pronounced spin-phonon coupling in pnictides that supports rapid development of a macroscopic order upon small vibrational displacement.

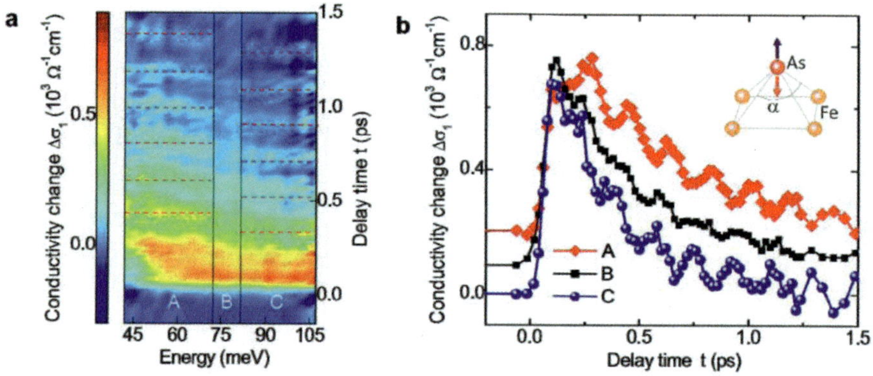


**Fig. 1. a** Two-dimensional map of the photo-induced change of the optical conductivity of $BaFe_2As_2$ at TL = 134 K and **b** its averages in spectral regions A, B and C. The coherent oscillations indicate a periodic transient formation and destruction of antiferromagnetic spin order driven by the $A_{1g}$ vibration at 5.5 THz (inset) which involves the motion of arsenic ions with respect to the Fe square lattice. The Fe-As-Fe angle $\alpha$ is modulated depending on the position of arsenic ions.

## Coherent THz control of antiferromagnetic spin waves in NiO

The most direct and selective excitation of collective spin oscillations is possible by few-cycle THz pulses. THz interaction with matter is routinely mediated by the electric field component of the electromagnetic pulse coupling to the charge degree of freedom. We complement this process by showing that the magnetic component of intense THz transients enables ultrafast control of the spin degree of freedom, by itself [5]: Single-cycle THz pulses switch on and off coherent spin waves in antiferromagnetic NiO, at frequencies as high as 1 THz. An optical probe with a duration of 8 fs follows the THz-induced magnetic dynamics and verifies that the spins are addressed selectively via direct Zeeman interaction. This concept provides a universal ultrafast handle on magnetic excitations in the electronic ground state. In the future, magnon waves may be exploited as local probes while, e.g., orbital degrees of freedom are optically excited. Furthermore, we will discuss the perspective of complete spin reversal in sufficiently strong magnetic fields [6].

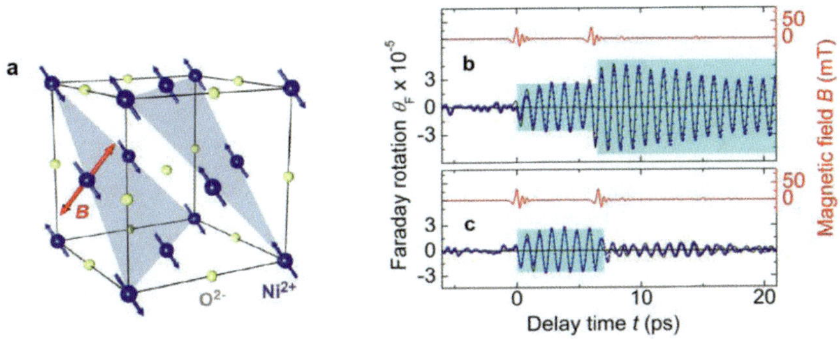


**Fig. 2. a** Crystal lattice of NiO with magnetically ordered spins (blue arrows) and a possible direction of the magnetic THz field B (red double arrow). **b** Faraday rotation (blue dots) induced by a sequence of two THz pulses (red curve) with a mutual delay of $\Delta t = 6$ ps launches a coherent magnon and, subsequently, enhances its amplitude by almost a factor of 2 (shaded background). **c** Double pulse excitation with $\Delta t = 6.5$ ps switches the magnon on and off coherently.

**Acknowledgments** The authors thank the Deutsche Forschungsgemeinschaft (DFG) and the European Research Council for support via Emmy-Noether grant HU 1598/1 and ERC Starting Grant QUANTUMsubCYCLE.

## References

[1] A. Sell et al., "Phase-locked generation and field-resolved detection of widely tunable terahertz pulses with amplitudes exceeding 100 MV/cm" Opt. Lett. **33**, 2767 (2008).

[2] F. Junginger et al., "Single-cycle multi-THz transients with peak fields above 10 MV/cm" Opt. Lett. **35**, 2645-2647 (2010).

[3] F. Junginger et al., "Non-perturbative interband response of a bulk InSb semiconductor driven off-resonantly by terahertz electromagnetic few-cycle pulses" Phys. Rev. Lett. **109**, 147403 (2012).
[4] K. W. Kim et al., "Ultrafast transient generation of spin-density-wave order in the normal state of $BaFe_2As_2$ driven by coherent lattice vibrations" Nature Materials **11**, 497 (2012).
[5] T. Kampfrath. A. Sell et al., "Coherent terahertz control of antiferromagnetic spin waves" Nature Photonics **5**, 31 (2011).
[6] S. Wienholdt, et al., "THz switching of antiferromagnets and ferrimagnets" Phys. Rev. Lett. **108**, 247207 (2012).

# Terahertz Spectroscopy of Femtosecond Spin Dynamics in Orthoferrites

**R.V. Mikhaylovskiy[1], E. Hendry[2], V.V. Kruglyak[2], A. Wu[3], R.V. Pisarev[4], Th. Rasing[1] and A.V. Kimel[1]**

[1]Radboud University Nijmegen, Institute for Molecules and Materials, Heyendaalseweg 135, 6525 AJ Nijmegen, the Netherlands

[2]School of Physics, University of Exeter, Stocker Road, Exeter, EX4 4QL, UK

[3]Shanghai Institute of Ceramics, Chinese Academy of Sciences, Shanghai 200050, China

[4]Ioffe Physical-Technical Institute, Russian Academy of Sciences, 194021 St. Petersburg, Russia

R.Mikhaylovskiy@science.ru.nl

**Abstract** Here we report experiments performed on orthoferrites by means of THz time-domain emission spectroscopy. We show that the femtosecond laser pulses excite both iron and rare-earth sub-systems on sub-picosecond timescale. In particular, the novel route to the optical control of the exchange interaction has been discovered.

## Introduction

Although a remarkable progress in understanding of ultrafast magnetic phenomena has been achieved, the ultimate challenge has not been overcome: the femtosecond control of the magnetic state of matter by employing the exchange interaction which is the strongest, fastest and ubiquitous force in the magnetism, eventually governing the very existence of the magnetic substances. At the same time laser manipulation of the exchange interaction must be feasible in any material in which the magnetic order is governed by indirect exchange via ligand ions (super-exchange). Indeed, since the super-exchange interaction is defined by virtual charge-transfer excitations from ligand to magnetic ion, one can anticipate the feasibility of laser control of magnetism via optical pumping of the electronic transitions mediating the exchange. The effect of the optical modification of the magnetic state has been most conveniently detected by recording the motion of spins triggered by the associated transient torque.

Orthoferrites are an excellent model system for investigation of laser-induced magnetic phenomena, in general, and observing the ultrafast optical control of the exchange interaction, in particular. These materials are a family of canted



J.-Y. Bigot et al. (eds), *Ultrafast Magnetism I*, Springer Proceedings in Physics 159
DOI 10.1007/978-3-319-07743-7_102

antiferromagnets which show an unusual variety of magnetic properties [1]. The spin dynamics of these compounds governed by the exchange interaction belongs to terahertz (THz) range [2] thereby being two orders of magnitude faster than in conventional ferromagnets like permalloy or iron garnets. This fact makes orthoferrites particularly attractive for ultrafast opto-magnetic recording. Indeed, recently laser-induced ultrafast spin reorientation [3] and ultrafast inverse Faraday effect [4] have been successfully demonstrated in orthoferrites.

To date, the ultrafast spin dynamics in orthoferrites has been investigated by all optical pump-probe spectroscopy. In order to understand better the interaction between intense femtosecond pulses and both iron and rare-earth magnetic subsystems of orthoferrites we employed the THz emission spectroscopy. This technique combines sub-picosecond time resolution with high sensitivity with respect to any dipole active modes.

By measuring the electro-magnetic emissions arising from the illumination of the orthoferrites by femtosecond optical pulses of Ti:Sapphire laser we were able to demonstrate that these pulses instantaneously excite many resonances which were not visible with the help of all optical pump-probe spectroscopy at all.

## Results and Discussion

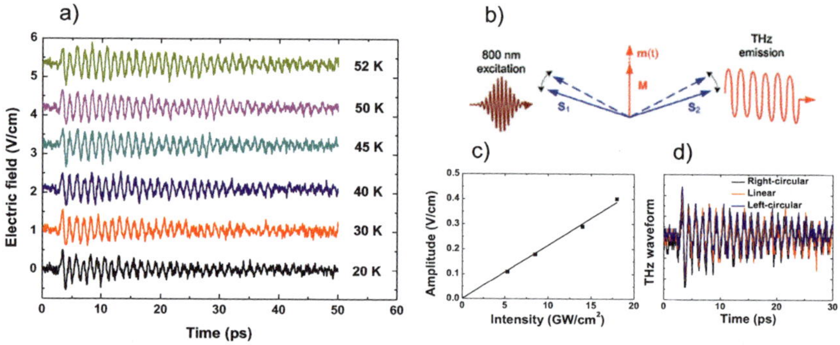


**Fig. 1.** a. The emitted from $TmFeO_3$ signal different temperatures below 55 K. b. Below 55 K the magnetization $\mathbf{M} = \mathbf{S}_1+\mathbf{S}_2$ lies along x-axis in $TmFeO_3$ cut perpendicular to *z*-axis. The optical pump is focused onto the sample along z-axis, while THz emission is collected along this direction at other side of the sample. The THz emission arises from the quasi-antiferromagnetic oscillation $\mathbf{m}(t)$. c. The amplitude of the quasi-antiferromagnetic mode in $TmFeO_3$ at 35 K vs the pump intensity with linear fit. d. The THz waveforms generated in $TmFeO_3$ optical pulses with linear and circular polarization.

By measuring the THz emission arising from optically excited coherent quasi-antiferromagnetic spin oscillations [Fig. 1 (a) and (b)] we demonstrate the evidence for sub-picosecond optical modulation of the energy of the exchange spin-

spin interaction in this compound. Distinct features of such an optical control of the exchange interaction are linear dependence of the effect on the intensity [Fig. 1 (c)] and independence on the direction of propagation and polarization of light [Fig. 1 (d)]. The effect has been observed in several orthoferrites such as $YFeO_3$, $TmFeO_3$ and $ErFeO_3$.

From the strength of the electric field of the emitted THz wave we have been able to estimate that the amplitude of the emitting magnetic dipole is ~ 1 μA cm$^2$ showing that the sub-picosecond laser pulse with the fluence ~ 1 mJ/cm$^2$ was able to change the energy of the exchange interaction over ~ 0.01 % that in turn exerts on spins the effective magnetic field of 0.1 T.

Additionally, we observed the emission arising from the high frequency (1 – 2 THz) dipoles excited between the sub-levels of the $^4I_{15/2}$ ground multiplet of rare-earth ion in $ErFeO_3$. To the best of our knowledge it is the first proven manifestation of the direct optical excitation of the rare-earth sub-system in the orthoferrites.

## Summary

This work demonstrates potential power of the THz techniques in the area of femtomagnetism. Particularly, the THz emission spectroscopy of orthoferrites allowed us to demonstrate the feasibility of sub-picosecond optical control of the exchange interaction that opens a novel means of optical magnetization control.

**Acknowledgments** This work was partially supported by The Netherlands Organization for Scientific Research (NWO), the Foundation for Fundamental Research on Matter (FOM), the European Union's Seventh Framework Program (FP7/2007-2013) Grants No. NMP3-LA-2010-246102 (IFOX), No. 280555 (Go-Fast), No. 214810 (FANTOMAS), the European Research Council under the European Union's Seventh Framework Program (FP7/2007-2013)/ERC Grant Agreement No. 257280 (Femtomagnetism) as well as the program "Invited Scientist" funded by the Russian Ministry of Education and Science (Grant Agreement 14.B37.21.0899).

## References

[1] R. L. White, "Review of Recent Work on the Magnetic and Spectroscopic Properties of the Rare-Earth Orthoferrites" J. Appl. Phys. **40**, 1061 (1969)

[2] G. Srinivasan (Ed.), A. N. Slavin (Ed.) "High frequency processes in magnetic materials" (World Scientific Publishing, 1995), Chap. 2, pp. 56-98.

[3] A.V. Kimel, A. Kirilyuk, A. Tsvetkov, R.V. Pisarev, and Th. Rasing, "Laser-induced ultrafast spin reorientation in the antiferromagnet TmFeO3" Nature **429**, 850 (2004).

[4] A. V. Kimel, A. Kirilyuk, P. A. Usachev, R. V. Pisarev, A. M. Balbashov, and Th. Rasing, "Ultrafast non-thermal control of magnetization by instantaneous photomagnetic pulses" Nature **435**, 655 (2005).

# Contribution of magnetic circular dichroism in all-optical light helicity-dependent magnetic switching

A. Tsukamoto[1], S. Kogure[2], H. Yoshikawa[2], T. Sato[2], A. Itoh[1]

[1]College of Science and Technology, Nihon University, 7-24-1 Narashino-dai, Funabashi, Chiba, 274-8501, Japan

[2]Graduate School of Science and Technology, Nihon University, 7-24-1 Narashino-dai, Funabashi, Chiba, 274-8501, Japan

atsuka@ecs.cst.nihon-u.ac.jp

**Abstract** Compositional dependency of all-optical light helicity-dependent magnetic switching (AO-HDS) in ferromagnetic GdFeCo alloy is presented. It is found that AO-HDS is associated with the collinear sub-lattice magnetization and not with the net magnetization.An explanation of the AO-HDS based on magnetic circular dichroism exactly matches the above features of experiments.

## Introduction

Direct demonstration of all-optical light helicity-dependent magnetic switching (AO-HDS) was observed in ferromagnetic GdFeCo alloys in the absence of an external magnetic field [1], which became subject of intense discussion in modern magnetism. The most obvious explanation via the inverse Faraday effect (IFE) [2,3] could only very qualitatively account for the previously observed features. What is the origin of the helicity dependence in the all-optical switching? We show that all-optical switching with circular polarized (CP) femtosecond laser pulses in ferromagnetic GdFeCo is related to the collinear sub-lattice magnetization and not with the net magnetization. Furthermore, we present an explanation of the AO-HDS based on magnetic circular dichroism (MCD).

## Results and discussion

For our experiments we have chosen GdFeCo, an amorphous ferrimagnetic rear-earth transition metal alloy that is widely used in conventional magneto-optical recording and known for its strong magneto-optical effects. Samples were grown by

 
J.-Y. Bigot et al. (eds), *Ultrafast Magnetism I*, Springer Proceedings in Physics 159
DOI 10.1007/978-3-319-07743-7_103

magnetron sputtering in a multilayer structure: glass/AlTi(10 nm)/SiN(5 nm)/GdFeCo(20 nm)/SiN(60 nm). Net magnetization of GdFeCo is attributed to the difference of magnetization in anti-parallelly coupled rare-earth Gd and transition-metal FeCo. Increasing the amount of Gd, the configuration of sub-lattice (transition metal) magnetization and net magnetization is changed parallel to anti-parallel by across the magnetization compensation composition ratio CM. Using magneto-optical microscopy in combination with femtosecond pulsed light, we measured the compositional dependency (X:22~27) of AO-HDS in $Gd_x(Fe_{87.5}Co_{12.5})_{100-x}$. To excite the material we used regeneratively amplified pulses from a Ti:sapphire laser at a wavelength of 800 nm, pulse width ~90 fs and a repetition rate of 1 kHz.

Figure 1 shows how two different sweeps with the laser beam right-handed circularly polarized (σ+) and left-handed (σ-) circularly polarized affect the initial domain pattern in different ways. The direction of this switching is determined by the helicity of the light pulse. The relation between the direction of reversed net magnetization and helicity of the light changed sign by going across the magnetization compensation composition ratio CM (X:~24.5). We also found that the sign of MCD that might come from transition metal contribution with photon energies in the visible range changed at CM. It was hypothesized that CP light acts as a strong effective magnetic field pulse $H_{OM}$ on the magnetization of the medium through the IFE. The direction of $H_{OM}$ is then defined by the helicity of the light. From basic idea of IFE, effective field should affect to net magnetization. The above results do not match with the description based on the IFE. We already reported that ultrafast heating can act as a sufficient stimulus for magnetization reversal in a ferrimagnet taking into account the multi-sublattice nature[4], and that another type of all-optical switching relies on absorption only via strongly non-equilibrium conditions and transient ferromagnetic like state[5] within a picosecond after excitation. Adequate intensity of multiple-shot laserexposures make possible to obtain helicity-dependent switching, and the intensity window in AO-HDS can be explained quantitatively based on MCD[6].

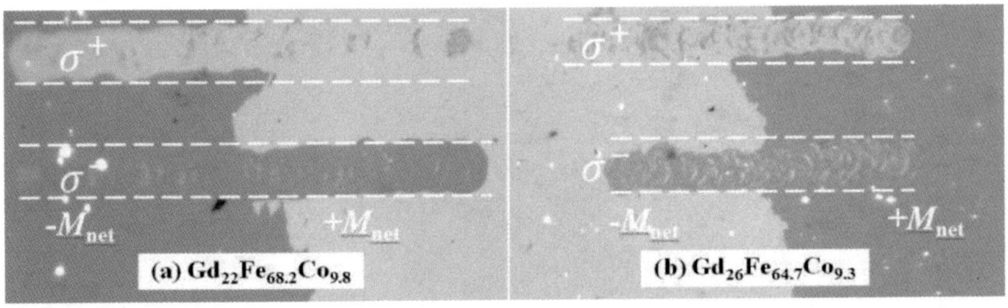


**Fig. 1.**The magneto-optical image of the magnetic thin film after exposure by sweeping of 90 fs circularly polarized laser pulses. Domains with sub-lattice (transition metal) magnetization "up'' and ''down'' could be observed as white and black regions, respectively.Althoughdown-net-magnetization (-$M_{net}$) was switched by right-handed circularly polarized (σ+) in case of (a) $Gd_{22}Fe_{68.2}Co_{9.8}$, up-net-magnetization (+$M_{net}$) was switched in case of (b) $Gd_{26}Fe_{64.7}Co_{9.3}$.

The above results show that the helicity-dependent ultrafast absorption in a multi-sublattice magnetic layer exactly matches the helicity-dependent features in compositional dependent switching experiments.

**Acknowledgments** The authors thank many corroboratorsin references [1][2][5][6] for continuous collaborated work. This work is partially supported by Nihon University Strategic Projects for Academic Research.

## References

[1] C. D. Stanciu, F. Hansteen, A.V. Kimel, A. Kirilyuk, A. Tsukamoto, A. Itoh and Th. Rasing, "All optical femtosecond magnetic recording" Phys. Rev. Lett. **99,** 047601(1-4) (2007).

[2] T.A. Ostler, J. Barker, R.F.L. Evans, R.W. Chantrell, U. Atxitia, O. Chubykalo-Fesenko, S. El Moussaoui, L. Le Guyader, E. Mengotti, L.J. Heyderman, F. Nolting, A. Tsukamoto, A. Itoh, D. Afanasiev, B.A. Ivanov, A.M. Kalashnikova, K. Vahaplar, J. Mentink, A. Kirilyuk, Th. Rasing, and A.V. Kimel, "Ultrafast heating as a sufficient stimulus for magnetization reversal in a ferrimagnet" Nature Communications **3**, 666(1-6) (2012).

[3] L. P. Pitaevskii"Electric Forces in Transparent Dispersive Media"Zh. Eksp.Teor.Fiz.**39**, 1450 (1960) [Sov. Phys. JETP **12**, 1008-1013 (1961)]

[4] J. P. van der Ziel, P. S. Pershan, and L. D. Malmstrom, "Optically-induced magnetization resulting from the inverse Faraday effect" Phys. Rev. Lett. **15**, 190-193 (1965).

[5] I. Radu, K. Vahaplar, C. Stamm, T. Kachel, N. Pontius, H. A. Durr, T. A. Ostler, J. Barker, R. F. L. Evans, R. W. Chantrell, A. Tsukamoto, A. Itoh, A. Kirilyuk, Th. Rasing and A. V. Kimel, "Transient ferromagnetic-like state mediating ultrafast reversal of antiferromagnetically coupled spins"Nature, **472**, 205-208 (2011).

[6] A. R. Khorsand, M. Savoini, A. Kirilyuk, A.V. Kimel, A. Tsukamoto, A. Itoh, and Th. Rasing, "Role of Magnetic Circular Dichroism in All-Optical Magnetic Recording" Phys. Rev. Lett. **108**, 127205(1-5) (2012).